Electronic Properties of Carbon Nanotubes

Electronic Properties of Carbon Nanotubes

Editor

Leroy Sidney

Electronic Properties of Carbon Nanotubes

Edited by **Leroy Sidney**

ISBN: 978-1-68117-240-8
Library of Congress Control Number: 2016934780

© 2017 by
SCITUS Academics LLC,
www.scitusacademics.com
Box No. 4766, 616 Corporate Way,
Suite 2, Valley Cottage,
NY 10989

Notice

Preface

Carbon nanotubes (CNTs) are tubular cylinders of carbon atoms that have extraordinary mechanical, electrical, thermal, optical and chemical properties. CNTs typically have diameters ranging from ‹1 nanometer (nm) up to 50 nm—a nanometer is one thousand millionth of a meter. Typical CNT lengths are several microns—several thousand nanometers long; by contrast, Nanocomp's produced fibers are measured in millimeters—thousands of times longer than all other commercially produced CNTs. They take the form of cylindrical carbon molecules and have novel properties that make them potentially useful in a wide variety of applications in nanotechnology, electronics, optics and other fields of materials science. They exhibit extraordinary strength and unique electrical properties, and are efficient conductors of heat. In the powdery format offered by all CNT producers (but for NTI), applications are limited to the properties possible by this form factor—e.g. additive active ingredients in semiconductors, liquid crystal displays (LCDs), sensors, and other uses in which these powders add some level of functional performance. Due to its fiber length and its form factors, NTI delivers strength and conductivity unlike any other commercial CNT producer, and so can address a much broader array of applications for which its material rivals copper and aluminum in conductivity, and steel, aluminum, carbon fibers and glass composites where strength and lightweight matter. Carbon nanotubes have been a subject of exhaustive research for a wide range of applications. The purpose of this book entitled Properties of Carbon Nanotubes is to give in-depth understanding of the physics and electronic structure of carbon nanotubes. This book discusses fabrication techniques followed by an analysis on the physical properties of carbon nanotubes, including density of states and electronic structures. Eventually, the book follows a significant amount of work in the industry applications of carbon nanotubes.

Table of Contents

CHAPTER 1

Optical-Electronic Properties of Carbon-Nanotubes Based Transparent Conducting Films

Kuan-Ru Chen[1], Hsiu-Feng Yeh[1], Hung-Chih Chen[1], Ta-Jo Liu[1*], Shu-Jiuan Huang[2], Ping-Yao Wu[2], Carlos Tiu[3]

[1]Department of Chemical Engineering, National Tsing Hua University, Hsinchu, Chinese Taipei

[2]Material and Chemical Laboratory, Industrial Technology & Research Institute, Hsinchu, Chinese Taipei

[3]Department of Chemical Engineering, Monash University, HHMelbourneHH, Australia

1. INTRODUCTION

Transparent conductive film or glass is a key component for many optical-electronic devices such as organic light emitting diodes (OLED), organic photovoltaic solar cells (OPV), liquid crystal display (LCD) panels and touch panels, just to name a few. The most critical requirements for transparent conductive film or glass are low sheet resistance and high transparency. Many materials were considered suitable for making transparent conductive film or glass [1-4]. Owing to manufacturing and quality requirements, only indium tin oxide (ITO) film or glass has been commonly used for optical-electronic devices in the current market [5]. Despite its popularity, indium is a rare-earth material, and its price is high [6]. Furthermore, ITO films or glass has to be produced with a vacuum deposition technology. This technology is relatively expensive compared to conventional wet coating processes. Hence, many competitive approaches have been sought to replace ITO film or glass using different materials and coating methods.

Low-cost wet coating processes have been found to be promising in coating nano-scaled conductive media such as silver nanowires [7] or carbon-nanotubes [8-11] on PET film. Silver nanowires can be produced through an efficient chemical approach [12]. However, issues such as how to disperse silver nanowires and reduce their diameter for lower resistance still have to be resolved. Conductive film or glass coated with carbon-nanotubes (CNT) has similar optical-electronic properties, but is also hindered by the dispersion problems. Several effectiveapproaches have been attempted to overcome this issue for multi-walled and single-wall CNT [13-15], but the single-wall CNT (SWCNT) appeared to give better performance [16].

Spin coating is usually applied as a convenient means to prepare samples for laboratory analysis. However, usually over 90% of the coating solution is wasted, and is therefore not suitable for mass production. Recently, several researchers considered different coating methods for CNT solutions. Kim et al. [17] applied spin and spray coating methods for CNT electrode to make organic solar cells. de Andrade et al. [18] compared different technologies for the preparation of CNT networks. They concluded that dip coating and electrophoretic deposition are promising methods for solar cell application. In the present study, three different coating methods were used to make transparent conductive film with a well-dispersed single-wall CNT solution, and the optical and electronic properties of the samples were measured and compared.

The optical requirement for conductive films is that it must be over 85% transparent. The sheet resistance may however vary, depending on special applications. In the present study, it is chosen to be 1000 ohm/sq, which meets the requirement of electrostatic dissipation. The results presented here would be useful for future mass production considerations.

2. EXPERIMENTAL

2.1. Preparation of SWCNT Solution

Single walled carbon nanotubes (SWCNT) were prepared by the floating method in a vertical tube reactor [19] by using alcohol as the carbon source. The alcohol solution with a given composition of ferrocene and thiophene was introduced into the reactor with hydrogen as the carrier gas. The typical reactor temperature was between 1000°C - 1200°C. The SWCNT produced were purified by combining two-step processes of thermal annealing in air and acid treatment [20]. The SWCNT produced had average diameters around 2 to 2.5 nm, with purity >90% based on TGA and G/D ratio (Raman) around 35. The aqueous SWCNT dispersion was prepared by ultrasonication using a tip sonicator with sodium dodecyl benzene sulfonate (SDBS) as surfactant. The concentrations of the SWCNT was 0.1% and the ratio of SWCNT to SDBS was 1:1.5.

2.2. Coating Methods

Three coating devices were selected for making samples. The first device was a laboratory blade coater (Zehntner, ZUA 2000), with a minimum coating gap of 5 μm, as shown in **Figure 1**. The second was a dip coating device shown in **Figure 2**. A machine arm was attached to grab and lift the sample upward from a solution tank. The coating speed could vary between 0.1 - 2 cm/s. The last was a slot die coater as shown in **Figure 3**(a). The slot die was attached to the mount of a patch coater as shown in **Figure 3**(b). The coating solution was delivered by a piston pump (KD scientific, KDS 100) through the slot die, and then coated on the substrate which was fixed on the marble platform of the patch coater.

2.3. Measurements

All the test solutions were coated on the polyethylene terephthalate (PET) films for analysis. A base coat was necessary to prevent the aggregation of SWCNT solution [21]. The PET film was cut into a rectangular shape, 10 cm × 15 cm. All physical properties were measured at fixed positions on the films as marked in **Figure 4**. The coated samples were placed in an oven and heated at 90°C for 5 minutes. Two major properties of the ovendried samples were measured. A four-point probe (MCPT600) was used to detect the sheet resistance of the sam- ples, andan UV-visible spectrophotometer (Varian Cary 50 conc) was used for transparency measurements. The uncoated PET film was used as the reference for comparison. Standard processes were taken to detect the distributions of CNT on the PET films by the scanning electronic microscope (JEOL JSM-5600).

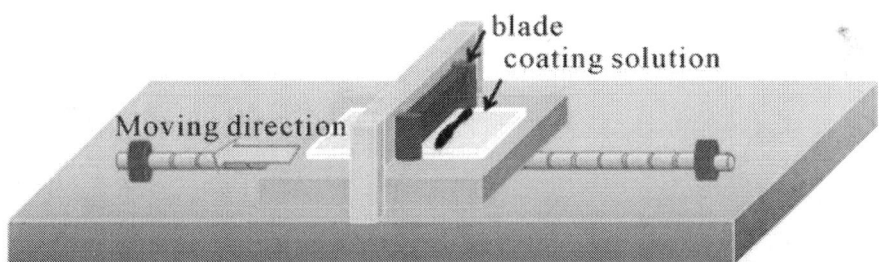

Figure 1. Blade coating operation.

Figure 2. Photo of dip coater.

Figure 3. Photos of (a) the experimental slot die with a shim; (b) the patch coater.

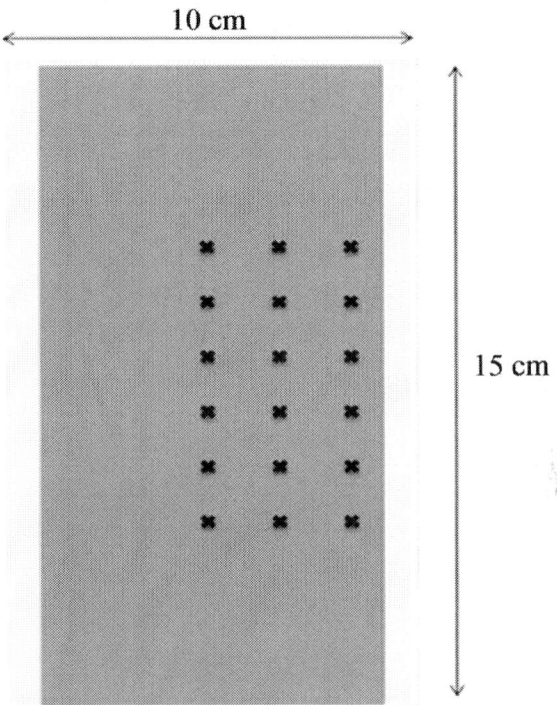

Figure 4. Sample dimensions and marked position for property measurements (10 × 15 cm2).

3. RESULTS AND DISCUSSION

The SWCNT solutions were coated on the PET substrates by the three coating methods. The transparency and sheet resistance of each coated sample were measured and analyzed.

The results obtained on slot die coating are presented first. In order to produce a very thin wet thickness on the slot die coater, the concentration of the SWCNT solutions must be reduced to 0.1%. This yielded a dry film with thickness as low as 5 nm. The dry film thickness t reported here is an average value which depends on flow rate, coating speed and solid content, t can be evaluated with the following formula:

$$t \equiv \frac{Q}{V \cdot W} \cdot S\%$$

(1)

here Q is the volumetric flow rate, V is the coating speed, W is the coated width and S% is the solid content. **Figure 5** presents the results of transparency and sheet resistance as a function of dry film thickness at the coating speed 10 cm/s. The results indicate that both the sheet resistance and transparency decrease as the dry film thickness increases. The increase in dry thickness is due primarily to the increasing amount of CNT accumulated on the PET substrate. Hence, it is expected that both the sheet resistance and transparency will decrease.

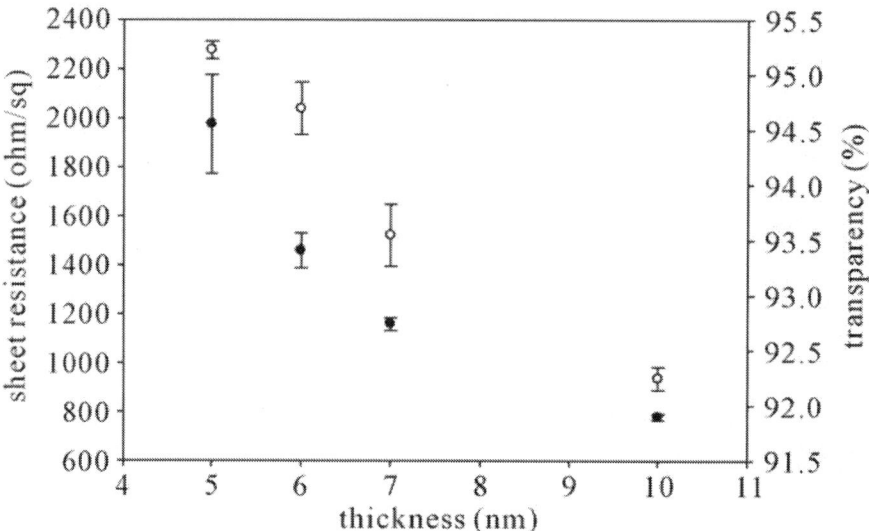

Figure 5. Sheet resistance and transparency as a function of dry film thickness for samples made by slot die coating (●: sheet resistance (ohm/sq); ○: transparency (%)).

The effects of the coating speed on the sheet resistance and transparency for two different dry film thicknesses, 5 nm and 10 nm, are displayed in **Figure 6**. It is seen that both transparency and sheet resistance are independent of the coating speed for the thinner film; whereas these two solid content is known, which is around 7 - 8 nm in present study. It can be seen that both the sheet resistance and transparency decrease markedly as the coating speed increases. The transparency drops from 95% to 91%, and properties decrease slightly as the coating speed increases for the thicker film. The sheet resistance stays around 2000 ohm/sq, and provides a transparency of around 95% for the 5 nm film; where the resistance drops to below 1000 ohm/sq, transparency reduces to about 92% for the 10 nm film when the coating velocity increases from 6 to 10 cm/s. At low coating speed, the coating solution emanating from the slot die exit will expand laterally, but the coating width will contract as the coating speed increases. The lateral movement of coating solution changes the CNT distribution, and affects the two properties. It is noted that the lateral expansion ceases to exist at high coating speed.

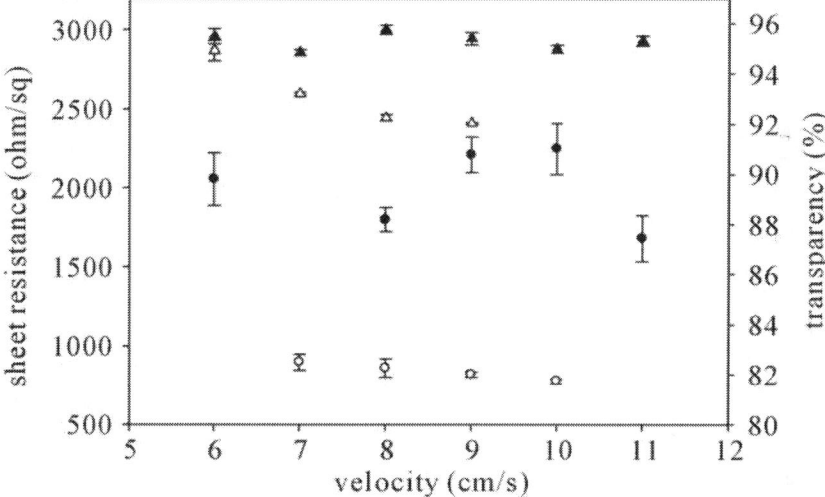

Figure 6. Sheet resistance and transparency as a function of coating speed for slot die coating (●: 5 nm (ohm/sq); ○: 10 nm (ohm/sq); ▲: 5 nm (%); △: 10 nm (%)).

Figure 7 shows the results obtained with a blade coater having a coating gap of 20 μm. The wet film thickness for blade coating is usually less than 50% of the blade gap for dilute Newtonian solutions [22]. The average dry film thickness can be evaluated if the the resistance decreases from 3000 to below 1000 ohm/ sq when the coating speed increases from 1.0 to 6.0 cm/s. This is due to the increase of the coating thickness as the coating speed increases for a blade coater. The present observation is consistent with the conclusion of Yang and Jiang [23]. During the blade coating operation, the coating solution moves in the transverse direction as the blade advances. This lateral movement becomes more significant when the blade speed is faster.

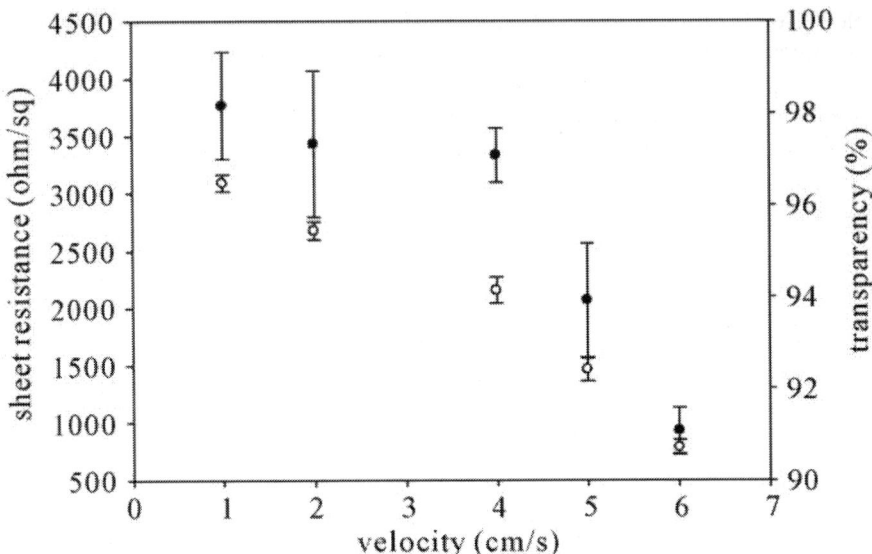

Figure 7. Sheet resistance and transparency as a function of coating speed for blade coating (●: sheet resistance (ohm/sq); ○: transparency (%)).

For a dip coating operation, the wet coated thickness is known to be proportional to the capillary number Ca as follows [24]:

$$T \equiv t_w \left(\frac{\rho g}{\mu v} \right)^{\frac{1}{2}}$$

(2a)

$$T = 0.944 \left(Ca \right)^{\frac{1}{6}}$$

(2b)

here

$$Ca \equiv \frac{\mu v}{\sigma} \quad Ca \ll 1$$

(2c)

Here t_w is the wet film thickness, ρ is fluid density, g is the gravitational factor, μ is the fluid viscosity and σ is the surface tension.

It was estimated the dry film thickness for dip coating is around 5 - 15 nm. Increasing the coating speed will increase Ca, and hence the coated film thickness for a coating solution of constant viscosity and surface tension. Therefore, both transparency and sheet resistance will drop in dip coating operated at higher velocity. The results obtained from dip coating are not presented here in similar plots as in Figures 6 and 7. Instead, they are combined for comparison purposes with the results of slot and blade coatings in the subsequent figures.

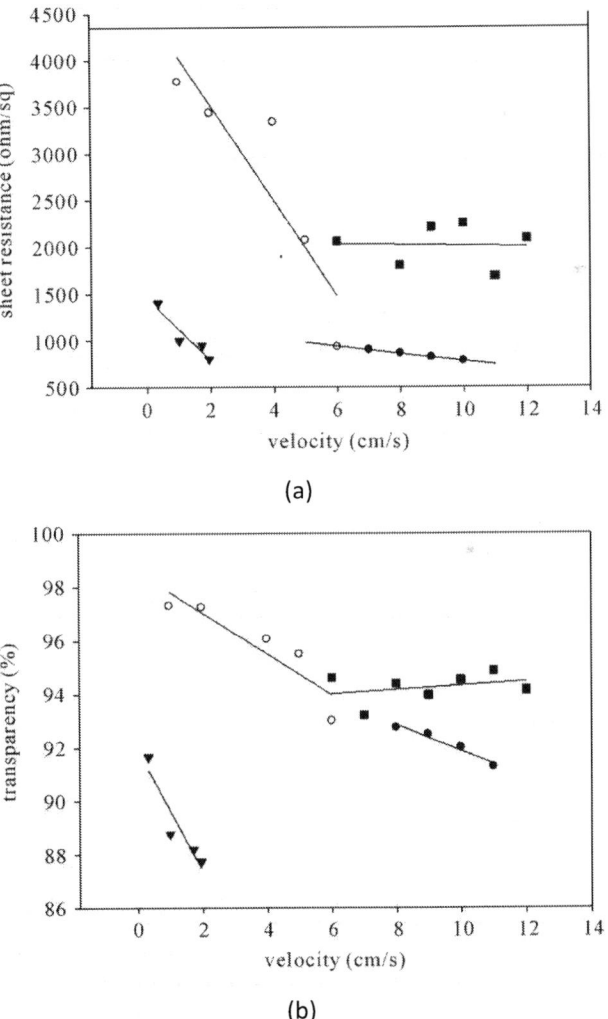

(a)

(b)

Figure 8. (a) Comparison of sheet resistance as a function of coating speed for different coating methods (■: slot 5 nm; ●: slot 10 nm; ○: blade; ▼: dip); (b) Comparison of transparency as a function of coating speed for different coating methods (■: slot 5 nm; ●: slot 10 nm; ○: blade; ▼: dip).

Comparison of the performance of samples based on three different coating methods is displayed in **Figure 8**. For dip and blade coating operations, both sheet resistance and transparency decrease with increasing coating speed. However, in order to meet the requirement of 85% transparency and sheet resistance 1000 ohm/sq, the upper limit of the coating speed for dip coating is only around 2 cm/s, and slightly above 6 cm/s for blade coating. As mentioned earlier, the optic-electronic properties appear to be independent of the coating speed in slot die coating for the 5 nm film and drop slightly for the 10 nm film when operated at a speed as high as 10 cm/s.

An attempt is made to check if a correlation exists between the sheet resistance and the transparency data obtained with the three coating methods, as shown in **Figure 9**. It is noted that the specific requirements of sheet resistance of around 1000 ohm/sq and transparency above 85% are satisfied for all samples obtained using the three coating methods, except for the dip coating operated at a speed of around 2 cm/s. **Figure 9** reveals that there exists a linear correlation between sheet resistance and transparency. Samples with higher transparency exhibit higher sheet resistance.

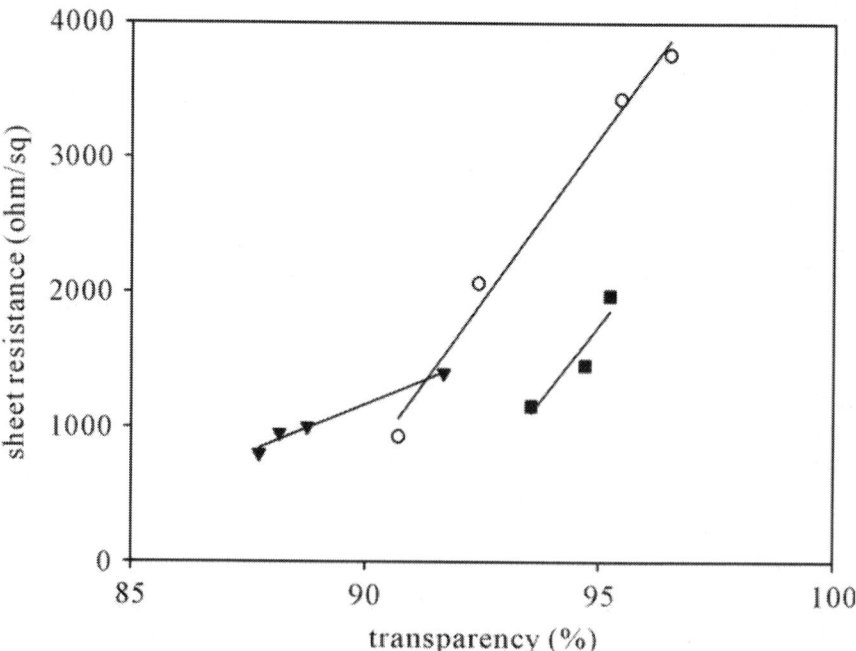

Figure 9. Sheet resistance vs. transparency for the three coating methods (■: slot 5 - 7 nm; ○: blade; ▼: dip).

The data presented in Figures 8 and 9 are based on the average value of sheet resistance and transparency. In addition, the variations of the two properties over the width of the sample are also critical to product quality. The distributions of sheet resistance for three samples are represented by the images shown in **Figure 10**. Figures 10(a) and (b) are two samples of different thickness obtained in slot die coating. For the thinner sample with dry film thickness of around 5 nm, the sheet resistance is quite uniform as shown in **Figure 10**(a). However, once the dry film thickness is increased to around 10 nm, the result in **Figure 10**(b) indicates that the sheet resistance at the two lateral edges increases rapidly. This implies that there are less CNT present in these regions. This is due to the lateral expansion of the coating solution after emanating from the slot die exit, which causes the concentrations of CNT towards the edges to drop sharply. The sheet resistance distribution for blade coating shown in **Figure 10**(c) also indicates that the distribution is not uniform due to the lateral movement of the coating solution, with high CNT concentration appearing in the central region of the sample. Therefore, to obtain a sample with uniform CNT distribution (or sheet resistance), the lateral motion has to be minimized in two-dimensional flow.

The CNT distributions on the PET films can also be observed through SEM images. The samples were taken from the central parts of the PET films. The magnifying power was selected to be 20000×. A bar of 1.0 micrometer length was given in the bottom of each figure for comparison purposes. The surfaces of the coated samples were observed. The CNT concentrations and distributions can be analyzed qualitatively. Figures 11(a) and (b) display the images obtained for two 5 nm thick dry films using the slot die coating at coating speeds of 7 cm/s and 11 cm/s, respectively. It can be seen that CNT are randomly deposited on the PET substrate, and increasing the coating speed does not cause any significant variation of the CNT distribution. The results based on blade coating are presented in Figures 11(c) and (d). Similarly, CNT are randomly distributed on the PET film, but the CNT concentration is increased substantially when the coating speed is increased from 2 cm/s to 6 cm/s as observed in Figures 11(c) and (d). This is due to the increase of the dry film thickness in blade coating at higher speed. The results based on dip coating are shown in Figures 11(e) and (f) respectively for coating speeds of 0.7 cm/s and 2 cm/s. It is seen that CNT are also randomly distributed on the PET film, but the concentrations of CNT are much higher than those obtained from the preceding two methods, and relatively unaffected by the coating speed. It must be noted that the dry films obtained from dip coating are inherently thicker than those obtained from other coating methods, thus producing a higher CNT concentration on the PET surface. The cracks appeared in the SEM images certainly are not acceptable for industrial applications. The present analysis focuses only on the sheet resistance and transparency of the coated CNT films. No binders, adhesives on other materials were added to the CNT solutions.

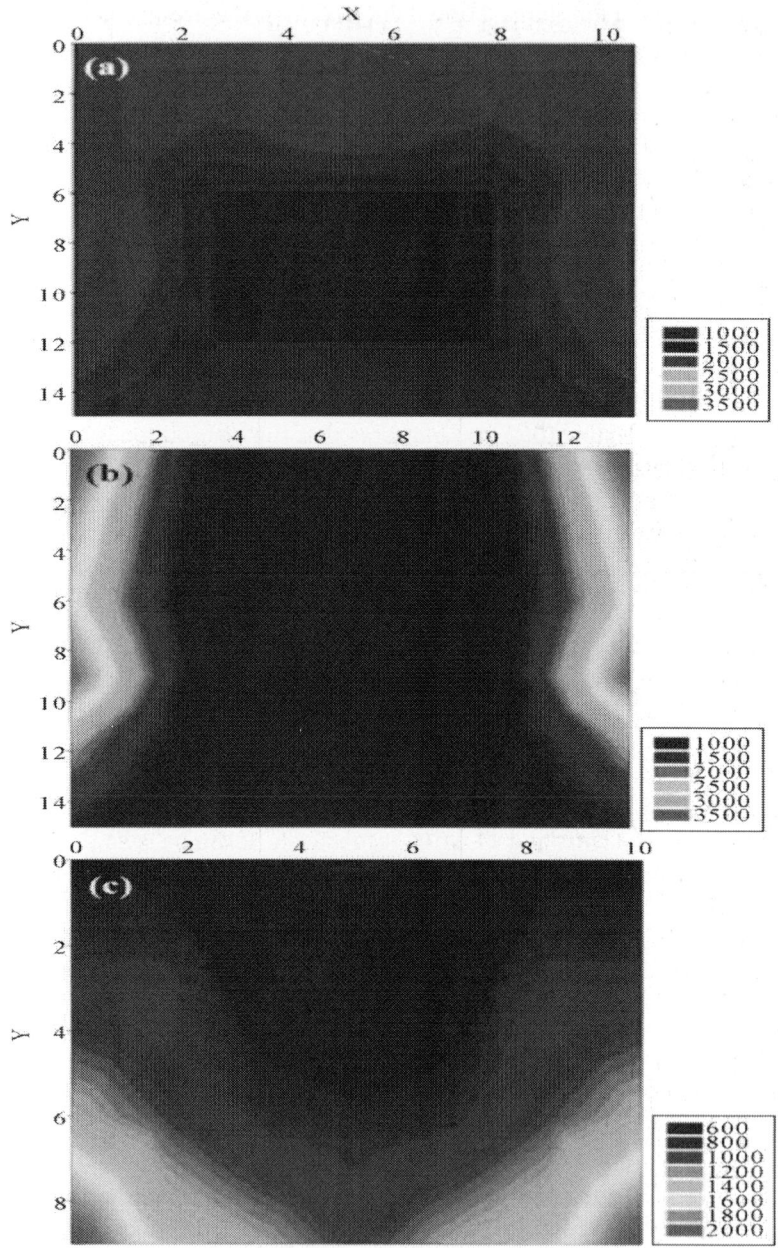

Figure 10. Images that represent sheet resistance. (a) Slot die (5 nm); (b) Slot die (10 nm); (c) Blade coating.

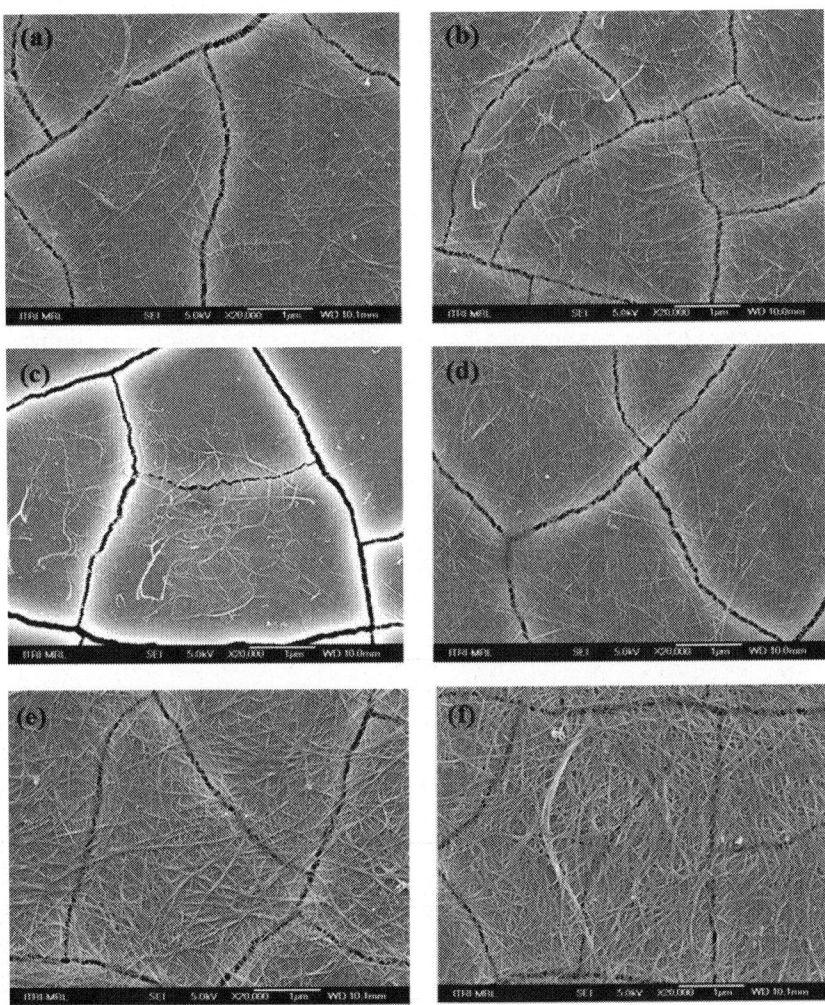

Figure 11. SEM images of samples made by the three coating methods. (a) Slot die V = 7 cm/s; (b) Slot die V = 11 cm/s; (c) Blade coating V = 2 cm/s; (d) Blade coating V = 6 cm/s; (e) Dip coating V = 0.7 cm/s; (f) Dip coating V = 2 cm/s.

4. CONCLUSIONS

The optical-electronicproperties of PET films coated with SWCNT obtained using three different coating methods were measured and compared. The film thickness was found to increase with increasing coating speed for blade and dip coating. The film thickness for slot die coating can be controlled provided the lateral expansion of the coated film can be minimized. Higher film thickness implies high concentration of CNT, which reduces both sheet resistance and transparency of the film.

Samples produced by the three coating methods all meet the criteria of sheet resistance 1000 ohm/sq and 85% transparency at certain coating conditions. However, at coating speed higher than 8 cm/s, only the slot die coating can produce films that meet these two criteria. The SEM images reveal that CNT were randomly distributed on PET films for samples made by the three coating methods. Distributions of sheet resistance were measured for samples obtained by slot die coating and blade coating. Coating solution may expand laterally on the PET film for blade coating and slot die coating at low speeds and high flow rates. The lateral expansion of coating solution will deteriorate the uniformity of sheet resistance. By controlling the lateral movement, slot die coating can produce samples with uniform sheet resistance.

5. ACKNOWLEDGEMENTS

The research work was supported by the National Science Council under Grant No. NSC99-2221-E-007-009. MY3. K.R. Chen was supported by ITRI during this study.

REFERENCES

1. T. Minami, "New n-type Transparent Conducting Oxides," Mrs Bulletin, Vol. 25, No. 8, 2000, pp. 38-44.
2. U. Ozgur, Y. I. Alivov, C. Liu, A. Teke, M. A. Reshchikov, S. Dogan, V. Avrutin, S. J. Cho and H. Morkoc, "A Comprehensive Review of ZnO Materials and Devices," Journal of Applied Physics, Vol. 98, No. 4, 2005, Article ID: 041301.
3. D. H. Zhang and H. L. Ma, "Scattering Mechanisms of Charge Carriers in Transparent Conducting Oxide Films," Applied Physics A—Materials Science & Processing, Vol. 62, No. 5, 1996, pp. 487-492.
4. C. C. Wang, "Deposition of Transparent Conductive Film by Wet Process," Industrial Material Magazine, Vol. 236, 2006, pp. 173-178.
5. S. Ray, R. Banerjee, N. Basu, A. K. Batabyal and A. K. Barua, "Properties of Tin Doped Indium Oxide ThinFilms Prepared by Magnetron Sputtering," Journal of Applied Physics, Vol. 54, No. 6, 1983, pp. 3497-3501.
6. A. Kumar and C. W. Zhou, "The Race to Replace Tin-Doped Indium Oxide: Which Material Will Win?" ACS Nano, Vol. 4, No. 1, 2010, pp. 11-14.

7. L. B. Hu, H. S. Kim, J. Y. Lee, P. Peumans and Y. Cui, "Scalable Coating and Properties of Transparent, Flexible, Silver Nanowire Electrodes," ACS Nano, Vol. 4, No. 5, 2010, pp. 2955-2963.

8. M. Kaempgen, G. S. Duesberg and S. Roth, "Transparent Carbon Nanotube Coatings," Applied Surface Science, Vol. 252, No. 2, 2005, pp. 425-429.

9. O. Hjortstam, P. Isberg, S. Söderholm and H. Dai, "Can We Achieve Ultra-Low Resistivity in Carbon NanotubeBased Metal Composites?" Applied Physics A: Materials Science & Processing, Vol. 78, No. 8, 2004, pp. 1175-1179.

10. L. Hu, D. S. Hecht and G. Gruner, "Percolation in Transparent and Conducting Carbon Nanotube Networks," Nano Letters, Vol. 4, No. 12, 2004, pp. 2513-2517.

11. M. H. A. Ng, L.T. Hartadi, H. Tan and C. H. P. Poa, "Efficient Coating of Transparent and Conductive Carbon Nanotube Thin Films on Plastic Substrates," Nanotechnology, Vol. 19, No. 20, 2008, pp. 205703-205707.

12. S. De, T. M. Higgins, P. E. Lyons, E. M. Doherty, P. N. Nirmalraj, W. J. Blau, J. J. Boland and J. N. Coleman, "Silver Nanowire Networks as Flexible, Transparent, Conducting Films: Extremely High DC to Optical Conductivity Ratios," ACS Nano, Vol. 3, No. 1, 2009, pp. 1767-1774.

13. M. C. Hersam, "Progress towards Monodisperse SingleWalled Carbon Nanotubes," Nature Nanotechnology, Vol. 3, No. 7, 2008, pp. 387-394.

14. M. S. P. Shaffer and K. Koziol, "Polystyrene Grafted Multi-Walled Carbon Nanotubes," Chemical Communications, Vol. 18, 2002, pp. 2074-2075.

15. C. Richard, F. Balavoine, P. Schultz, T. W. Ebbesen and C. Mioskowski, "Supramolecular Self-Assembly of Lipid Derivatives on Carbon Nanotubes," Science, Vol. 300, No. 5620, 2003, pp. 775-778.

16. T. W. Ebbesen, H. J. Lezec, H. Hiura, J. W. Bennett, H. F. Ghaemi and T. Thio, "Electrical Conductivity of Individual Carbon Nanotubes," Nature, Vol. 382, 1996, pp. 54- 56.

17. S. Kim, J. Yim, X. Wang, D. D. C. Bradley, S. Lee and J. C. Demello, "Spinand Spray-Deposited Single-Walled Carbon-Nanotube Electrodes for Organic Solar Cells," Advanced Functional Materials, Vol. 20, No. 14, 2010, pp. 2310-2316.

18. M. J. de Andrade, M. D. Lima, V. Skakalova, C. P. Bergmann and S. Roth, "Electrical Properties of Transparent Carbon Nanotube Networks Prepared through Different Techniques," Physica Status Solidi-Rapid Research Letters, Vol. 1, No. 5, 2007, pp. 178-180.

19. L. J. Ci, Y. H. Li, B. Q. Wei, J. Liang, C. L. Xu and D. H. Wu, "Preparation of Carbon Nanofibers by the Floating Catalyst Method," Carbon, Vol. 38, No. 14, 2000, pp. 1933-1937.

20. J. M. Moon, K. H. An, Y. H. Lee, Y. S. Park, D. J. Bae and G. S. Park, "High-Yield Purification Process of Singlewalled Carbon Nanotubes," The Journal of Physical Chemistry B, Vol. 105, No. 24, 2001, pp. 5677-5681.

21. S. L. Kuo, S. J. Huang and C. M. Hu, "Transparent Conductive Film and Method for Manufacturing the Same," Patent No. US20100040869-A1, 2010.

22. T. M. Sullivan and S. Middleman, "Film Thickness in Blade Coating of Viscous and Viscoelastic Liquids," Journal of Non-Newtonian Fluid Mechanics, Vol. 21, No. 1, 1986, pp. 13-38.

23. H. T. Yang and P. Jiang, "Large-Scale Colloidal SelfAssembly by Doctor Blade Coating," Langmuir, Vol. 26, No. 16, 2010, pp. 13173-13182.

24. L. Landau and B. Levich, "Dragging of a Liquid by a Moving Plate," Acta Physicochimica URSS, Vol. 17, No. 42, 1942, pp. 42-54.

Assembly of Carbon Nanotube Sheets

Mei Zhang[1] and Ray Baughman[2]

[1] Florida State University, The United State of America

[2] University of Texas at Dallas, The United State of America

1. INTRODUCTION

Over the past decades, carbon nanotubes (CNTs) have been actively explored as building blocks for next-generation electronics(Tans et al., 1998; Bachtold et al.,2001; Misewich et al., 2003), optoelectronics, and sensors (Kong et al., 2000; Xia et al., 2003; T. Zhang,2008), including flexible and transparent devices (Ju et al., 2007) as well as stretchable devices (Xu et al., 2011). A critical step in constructing CNT-based devices is assembly of CNTs on a substrate or free-standing for device fabrication, which include alignment, density control, and transfer. Scalable and controlled assembly of CNTs synthesized with diverse methods (*e.g.*, solution fabrication or solid-state fabrication methods) on diverse substrates (*e.g.*, silicon, plastics, rubbers, etc) or free-standing presents a major fabrication challenge that must be overcome if CNTs are to be utilized in practical applications. For assembling CNTs into thin films (or called sheet, buckypaper), there are several different methods or processes in different conditions (*e.g.*, solution or solid-state processes) (Hu et al., 2010). Solution processes start with the CNTs in powder form; the powder is dispersed in an appropriate solvent with or without functionalization. The CNTfilms (buckypapers) are usually made using versions of the ancient art of paper making, by typically long-time filtration of nanotubes dispersed in solvent and peeling the dried nanotubes as a layer from the filter (Rinzler et al.,

1998; Endo et al.,2005). Buckypaper normally has a laminar structure with a random orientation of the bundles of the nanotubes in the plane of the film (Berhan et al., 2004). Interesting variations of the filtration route provide ultra-thin nanotube sheets that are highly transparent and highly conductive(Wu et al., 2004;Hu et al., 2004). While filtration-produced sheets are normally isotropic within the sheet plane, sheets having partial nanotube alignment result from applying high magnetic fields during filtration (Fischer et al.,2003). In other important advances, nanotube films have been fabricated by Langmuir-Blodgett deposition (Y. Kim et al., 2003), casting from oleum (Sreekumar et al., 2003), coating(Ago et al., 1999; Dan et al., 2009), and printing (Zhou et al., 2006;Unidym Inc, 2007).The solid-state processes generally have two approaches. One isto synthesize CNTs by floating catalyst chemical vapor deposition (CVD), either to deposit CNTs on a substrate inside the CVD chamber or to collect the CNT aerogel outside of the chamber on a special substrate and then densify it into a film (Y. Li et al., 2004; Martin, 2010). The catalysts are with the CNTs and the CNTs in the film are usually disoriented. The optimized process control lowers the impurities to less than 5 wt% in the film and a 1.2 meterwide and 10 meter long CNT film has been made (Nanocomp Tech. 2010). The other approach is to synthesize CNTs from the catalysts fixed on substrate to form a array either parallel to the substrate(Kong et al., 1998)or perpendicular to the substrate (also called vertical aligned CNT forest) which are fabricated into the CNT films after the synthesis process by the "domino pushing effect" motion (Ding et al., 2008; Pevzner et al., 2010)or drawing CNTs out of the forest (M. Zhang et al., 2004&2005). The domino pushing of the CNT forest can efficiently ensure that most of the CNTs are aligned tightly in the film. Well aligned CNT sheets are obtained by drawing CNTs from the forest.

For fabricating sheets that have close to single nanotube properties, long nanotubes are needed. Solution fabrication methods work only for short nanotubes since the ability to disperse nanotubes into a liquid and to fabricate oriented nanotube sheets from liquid dispersions decreases with increasing nanofiber length. Solid-state fabrication methods do not disturb the length of the nanotubes and are the methods that benefit from long nanotubes. In this chapter, two processes for assembling well aligned and super-thin CNT sheets are presented. One is to produce a drawable CNT forest, which hasspecial topology,by CVD and then draw CNTs out of forest to form a free-standing CNT sheet. Another process involves synthesis of a patterned CNT array on substrate by CVD and then knocking them down to form an aligned CNT sheet on substrate with the help of the solvent. The CNT growth and the conditions for making drawable CNT forests are discussed.

2. FABRICATION OF CNT SHEET

Since the CNT sheets are fabricated directly from the forests, the synthesis of the CNT forest is an important step. In this chapter, the CNTs are multi-walled CNTs. The CNT forests were synthesized by catalytic CVD using hydrocarbon gas, acetylene or ethylene, as the carbon source (M. Zhang et al., 2004). The following is the basic process and conditions. The catalyst was a ~3 nm thick iron film, which was deposited on a silicon substrate by electron beam evaporation. The substrates were set in the center of a quartz tube furnace. After heating up to 680°C in helium at one atmospheric pressure, 5 molar percent acetylene in helium was introduced at the total flow rate of 580 sccm. Within a few minutes, the dense and vertically aligned CNT forests were grown on substrates. After removing the forest, the substrate is still catalytically active and can be used to grow new forest, indicating a root-growth mode and the presence of the catalysts on the substrate. Based on scanning electron microscopy (SEM), transmission electron microscopy (TEM), and thermo-gravimetric analysis (TGA), the purity of the nanotube forests was very high (more than 99% carbon in the form of CNTs), with less than 1 wt% iron and amorphous carbon, but more importantly, no carbon particles within the CNTswere observed. The CNT sheets are made from the CNT forests by two approaches, knocking down and drawing.

2.1. Knocking Down Approach

The schematic of the knocking down approach is shown in Fig. 1. The line arrays of thin Fe film was made by patterning the resist on Si substrate following the standard lithography process, depositing the catalyst thin film by electron beam evaporation and then the liftingoff the resist mask. The Fe film cracked into nanoparticlesas catalysts for CNT growth during the temperature ramp up in CVD process. The CNTs grew away from the catalyst particles and formed a thin wall array during the CVD process (Figs. 2a to 2c). The CNT wall is

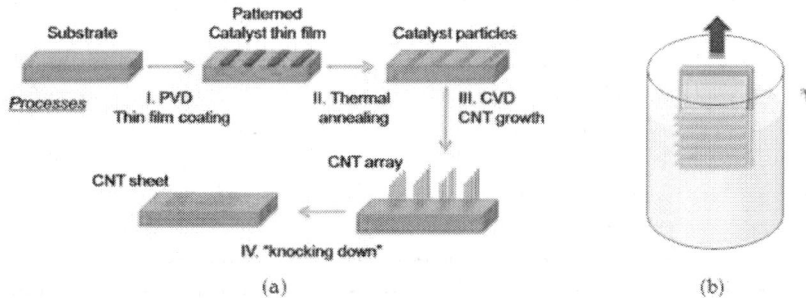

Figure 1. Schematic experimental processes for making an aligned thin CNT film on substrate (a) and knocking down method (b).

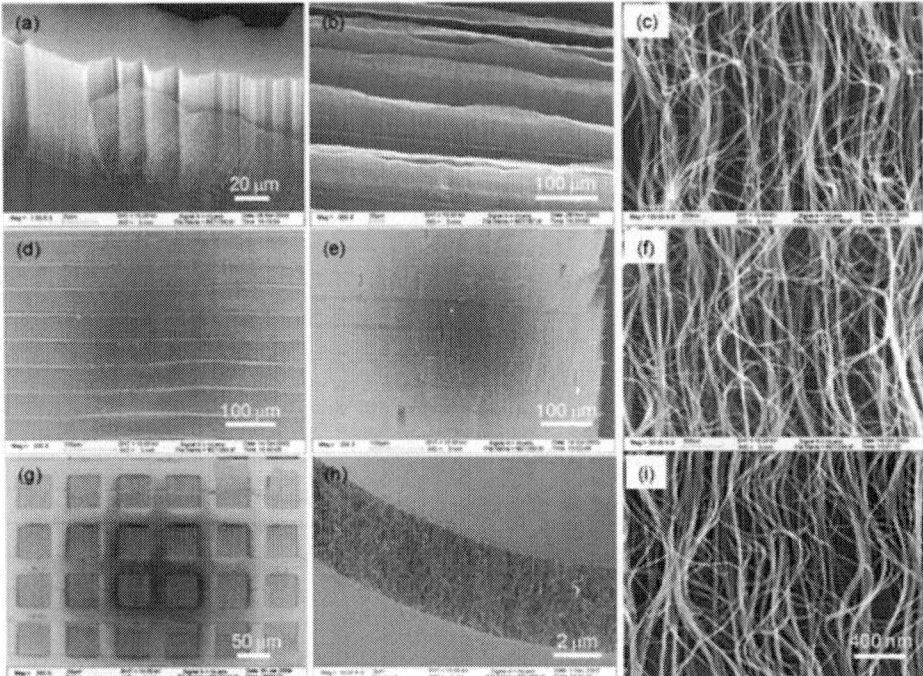

Figure 2. SEM images show CNTs grown perpendicular to a substrate as a forest and a thin wall (a) and a thin wall array (b). (c) SEM image in high magnification shows the structure of the thin wall. The wall is so thin, it is transparent. (d), (e), and (f) show the thin wall array is "knocked" down and CNTs forms an aligned thin film on substrate. (g) and (h) show that the CNT thin films are etched into desired patterns. After removal of the photo resist, the CNTs have clean surface and keep the original structures. (f) and (i) show the detail of (e) and (h), respectively. The dark circle in (d), (e), and (g) is the shadow of the in-lens detector.

very thin, so it is transparent (Fig. 2a). The substrate with the array of CNT walls was drawn through an acetone solution to horizontally redirect the vertical alignment and dried using nitrogen gas (Fig. 1b and Figs. 2d to 2f). The width of the patterned catalyst lines and the growth time of CVD determined the thickness and the height of the CNT walls. The walls do not shrink in height during flattening and drying. Figure 2e shows that a CNT sheet on substrate is formed from the CNT wall array in which the height of the wall is controlled to just 2 times the gap between two catalyst lines so that the CNT sheet on substrate has the thickness of two overlapping walls. The thickness of the CNT sheet could range from a few hundred nanometers to micrometers through the same process by using forest of different thicknesses, namely by controlling the width of the catalyst lines. Substrate-forest interaction and lateral CNT orientation guided densification only in wall thickness direction to essentially flatten the walls to the substrate with strong adhesion. It is because of the surface tension of the liquids and the strong van der Waals interactions that effectively close the CNTs together when liquids were introduced into the thin forest and dried. This self-assembly transformed CNT walls into highly densely packed CNT thin sheets. The CNT sheet was patterned into arbitrarily shaped CNT islands in desired positions by using standard lithographic processes. Figures 2g and 2h are the SEM images showing the patterned CNT sheet by lithography process.The CNT sheets were etched vertically by oxygen/argon reactive ion etching byusing a resist as a mask. Figure 2i is the SEM image of Fig. 2hin high magnification. The surface of the CNTs is clean and the structures of the sheet remain the same, showing that the adhesion with the substrate is sufficient to withstand lithographic processes including heat treatment, immersion into liquids, and drying. The bright and parallel horizontal lines visible in the images are catalyst lines. These lines cannot be totally removed (Fig. 2g). The sheet has to be transferred to another substrate in order for the effects of these lines to be negligible for some applications.

2.2. Drawing Approach

In drawing approach, the catalyst thin film covers all surface of the Si substrate (Fig. 3). The CNTs are synthesized on top of silicon substrate as a vertically aligned forest by CVD andthe transparent nanotube sheets were drawn directly from CNT forests. Draw was initiated using an adhesive strip, like that on a 3M Post-it° Note, or using a blade by cutting into the forest to contact CNTs teased from the forest sidewall. Five-centimeter wide, meter long transparent sheets were made at a meter per minute by hand drawing (M. Zhang et al., 2005).

Despite a measured areal density of only 3 g/cm^2, these 500 cm^2 sheets were self-supporting during drawing. Figure 4shows side-view and top view SEM micrographs of forest and sheet. It indicates that the nanotubes in the forest transition from the highly ordered forest state to a rather disordered intermediate state immediately in front of the forest sidewall, and finally to the highly oriented aerogel state. By taking sequential micrographs through a SEM to form a movie, the process in which the forest nanotubes rotate by about 90° in going from nanotube orientation in the forest to that of the highly oriented stateis captured (Kuznetsov et al., 2011).

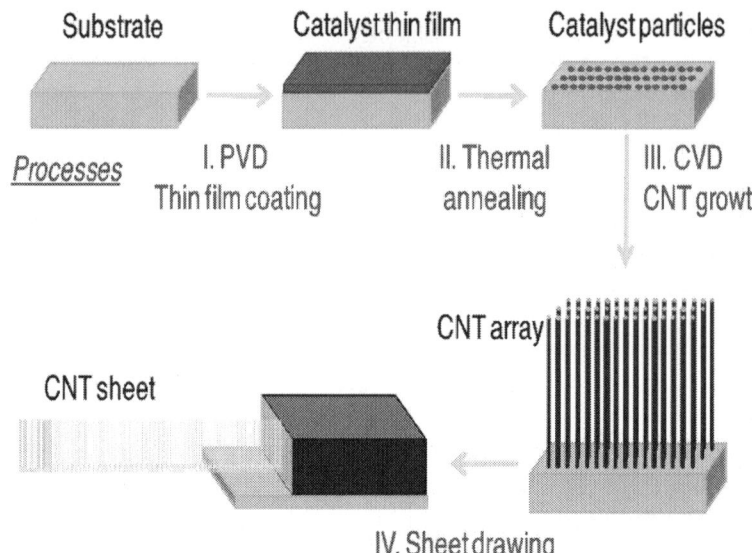

Figure 3. Schematic experimental processes for fabrication of the free-standing CNT sheet.

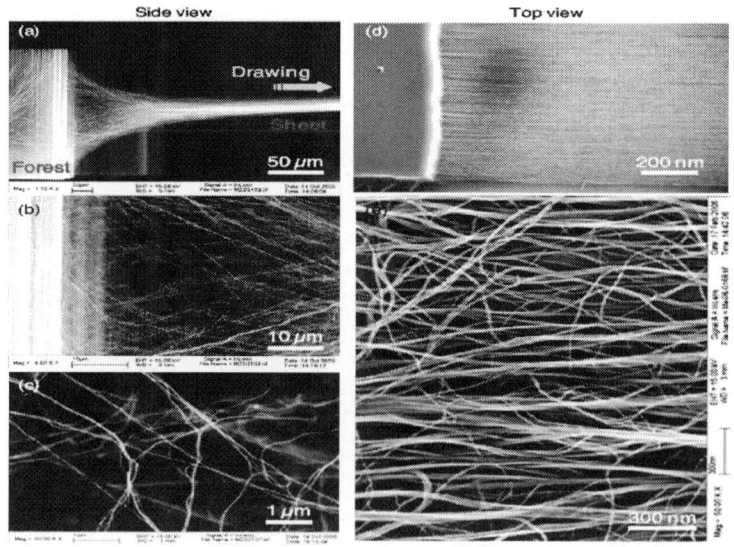

Figure 4. Sem images show that a cnt sheet is drawing from a forest. (a)-(c) side view of the forest and the sheet close to the forest. (b) and (c) show the details in (a) and (b), respectively. (d) and (e) top view of the forest and the cnt sheet.

The forest-drawn CNT sheets can easily be stacked or conveniently assembled into biaxially reinforced sheet arrays. They can be used as conducting layers on non-planar surfaces (Figs. 5a to 5d). These highly anisotropic aerogel sheets can be applied and easily densified into highly oriented sheets having a thickness of 50 nm and a density of 0.5 g/cm^3. We obtain this 360-fold density increase by simply adhering by contact the as-produced sheet to a planar substrate (e.g. glass, many plastics, silicon, gold, copper, aluminum, and steel), vertically immersing the substrate with attached CNT sheet into a liquid (e.g. ethanol) along the nanotube alignment direction, and retracting the substrate from the liquid. Surface tension effects during ethanol evaporation shrink the aerogel sheet thickness to 50 nm. SEM micrographs taken normal to the sheet plane suggest a decrease in nanotube orientation as a result of densification (Fig.4e). This observation is deceptive — the collapse of 20 m sheets to 50 nm sheets without changes in lateral sheet dimensions means that out-of-plane deviations in nanotube orientation become in-plane deviations that are noticeable in the SEM micrographs. The aerogel sheets can be effectively glued to a substrate by contacting selected regions with ethanol, and allowing evaporation to densify the aerogel sheet. Adhesion increases because the collapse of aerogel thickness increases contact area between the nanotubes and the substrate.

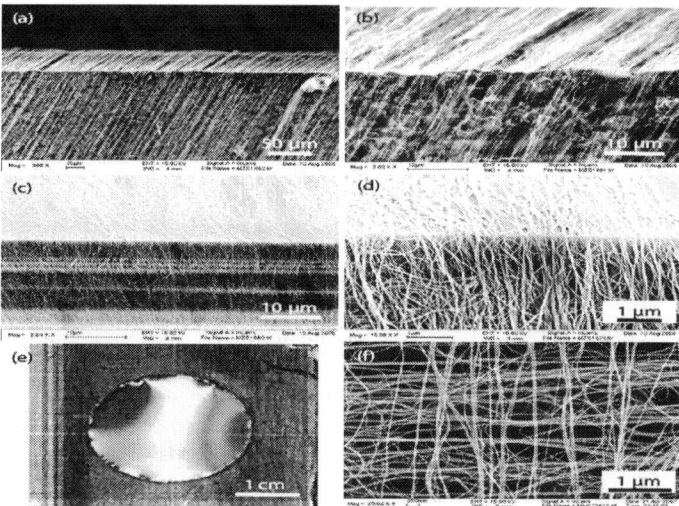

Figure 5. a)-(d) SEM images show a CNT sheet covers a substrate with sharp turns. (b) and (d) are the SEM images in higher magnification of the images in (a) and (c), respectively. (e) A photo image shows cross stacked sheets cover a metal plane with a 2.5 cm-diameter hole after liquid densification. The detail structure is shown in (f).

The aerogel sheets can also be densified into super-thin and free-standing sheet. Figure 5e is a photo image showing that a densified CNT sheet covers a 2.5 cm diameter hole in a metal plane. The super-thin sheet is made by densifing two cross-stacked CNT sheets (Fig. 5f). The nanotube sheets, which combine high transparency with high electronic conductivity, are highly flexible and provide giant gravimetric surface areas. The measured gravimetric strength of orthogonally oriented sheet arrays exceeds that of a high-strength steel sheet (Aliev et al., 2009; M. Zhang et al., 2005). These sheets have been used in laboratory demonstrations for microwave bonding of plastics and for making transparent, highly elastomeric electrodes; planar sources of polarized broadband radiation; conducting appliqués; flexible organic light-emitting diodes; and solar cells (Alive et al., 2009;Ulbricht et al., 2006 & 2007; Williams et al., 2008; M. Zhang et al., 2005).

Many real applications, such as field and thermionic emission electron sources (Kuznetzov et al., 2010; P. Liu et al., 2010; Y. Wei et al., 2008; Xiao et al., 2008; Y. C. Yang et al., 2010), loudspeakers (Alive et al., 2010; Kozlov et al., 2009;Xiao et al., 2008), CNT touch screens (Feng et al., 2010), high strength CNT yarns (Lima et al., 2011; K. Liu et al., 2010; M. Zhang et al., 2004; X. Zhang et al., 2006;Zhong et al., 2010), electrodes for batteries and supercapacitors (H. X. Zhang et al., 2009; R. F.Zhou et al., 2010), CNT/polymer composites (Q. F. Cheng et al., 2010; L. Chen et al., 2009; M.Zhang et al., 2005), and wrappers (Lima et al., 2011) were demonstrated. It is also demonstrated that the CNT sheets can be used as scaffolds for tissue engineering (Galvan-Garcia et al., 2007). It is no doubt that more applications will be developed and practiced.

3. MAKING DRAWABLE CNT FOREST

The CNT draw process does not work for all CNT forests. The experimental results show that the drawability depends strongly on the structural interconnections between CNTs and the network of interconnections between CNT bundles within the forest. The nanotubes in the forest should be intermittently bundled in order to be drawable (Fig. 6). In the forest height direction, this means that a nanotube switches many times from being bundled with a few neighboring nanotubes, to being unbundled, and then to being bundled with a few different neighboring nanotubes. Bundled nanotubes are simultaneously pulled from different elevations in the forest sidewall so that they join with bundled nanotubes that have reached the top and bottom of the forest, thereby minimizing breaks in the resulting fibrils (containing many bundled CNTs) (Figs. 4b and 4c). If there is too little lateral connectivity in the forest, the forest is undrawable because pulling on the forest sidewall just removes a few nanotubes rather than a continuous sheet. If there is too much inter-tube connectivity, only a chunk of forest is extracted before draw terminates.

The interconnections between CNTs and CNT bundles are formed during CNT growth, which are determined by the synthesis process. The CNT synthesis process is a complex process, which is related to the substrate and supporting materials, catalyst materials and their amount, carbon sources and partial pressure (feedstock), carrier gas and gas as an etching agent, total flow rate (gas residual time), process temperature, temperature ramp-up rate and cool-down rate, process pressure, process steps, process time, and many other details, such as history of reaction chamber, contamination, size of chamber and substrate, etc. CNT forests can be produced over a broad range of conditions. However, not every forest is suitable for solid-state fabrication of CNT sheets. This is because the forest needs to meet certain conditions to be a drawable CNT forest as described above. The drawable forest can be fabricated by just using C_2H_2/Ar (without H_2, water or other agents) and Fe thin film on Si substrate (without buffer layers such as SiO_2 and Al_2O_3). The buffer layers and other gases as well as their combinations are necessary to control the size and properties of CNTs. The drawable forests can be fabricated under different synthesis conditions. The CNT area density, height of forest, and purity of the CNTs are considered being the parameters for monitoring and controling the interconnections of CNTs in the forest.

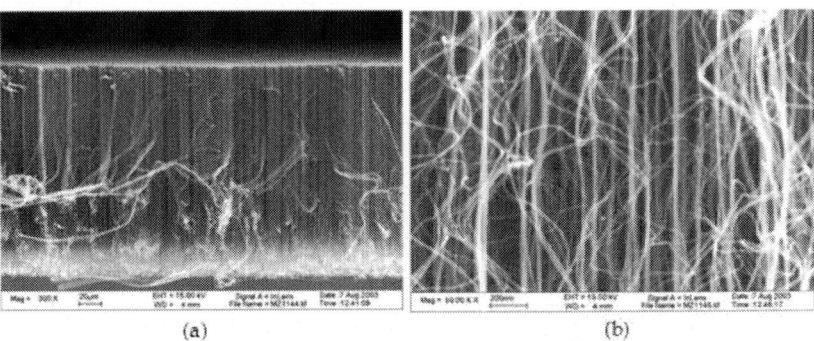

(a)   (b)

Figure 6. SEM images show the side views of a drawable CNT forest (a) and its structure in high magnification (b).

3.1. Structures of Cnt Forest

A single nanotube naturally curves (in bending status) during growth if no external forces exist. The bending stress can come from the nanotube's own weight, interaction with neighbor nanotubes, or limited growing space. As shown in Fig. 7a, a single tube keeps growing straight for a limited length: it falls down to the substrate and turns its growth direction many times during CVD process. A group of CNTs can form a randomly oriented CNT mat or well-aligned CNT arrays, depending on the density of catalyst and their activities under the same synthesis conditions. Figures 7b to 7d show the effects of the number of catalyst particles with similar area density on the formation of a CNT thin sheets array. Figure 7b shows CNT walls that were grown from 0.1μm wide and 40 μm long catalyst lines patterned by e-beam lithography. There were no external forces during CNT growth. The walls bend when their height is over a certain level. The bending directions and angles depend on each wall's morphologies. The nanotubes within each wall confine the nearest neighbors and attract the outermost nanotubes to their neighbors via van der Waals forces, thereby producing oriented growth. However, the CNTs in such thin walls present random curvatures and are tangled (Fig. 7b) because of the weak confinement in thickness direction. As the thickness of the wall increases, the alignment of the CNTs is improved due to the crowding effect. Figure 7c shows ~100 μm high CNT wall array grown from a 0.5 μm wide and 40 μm long catalyst lines in which nanotubes were better aligned (Fig. 7d). In a forest, the CNTs can have different growth rates, which lead to the structure of the forest. Figure 8shows three typical structures. In Fig. 8f, more than 80% of the CNTs are not straight: they periodically bend within fixed intervals throughout their entire length. As a result of this regular bending, a wavy structure resulted. It is believed that the wavy structure is formed because there are roughly two groups of catalysts uniformly distributed on substrate: one is more active and results in higher CNT growth rate than the other. Due to van der Waals forces, which stick nanotubes together whenever they touch, the growth rate of the array is limited by the nanotubes with relatively slow growth rate when catalysts stay on the surface of substrate. The nanotubes with higher growth rate are forced to bend periodically. The period of the wave is related to the ratio of growth rates of these two groups (Fig. 8c). When the distribution of the catalyst activity is relatively narrow and the density is high (Fig. 8a), the forest will have the morphology as shown in Fig. 8d. When the distribution of the catalyst activity results in the distribution of growth rate as shown in Fig. 8b and the density is high, the forest will be formed by the straight CNTs which form bundles while the waved CNTs switch between different bundles as the morphologies show in Figs. 6b and 8e. Such structure is believed to be important for assembling CNT sheets by drawing CNTs directly from the forest.

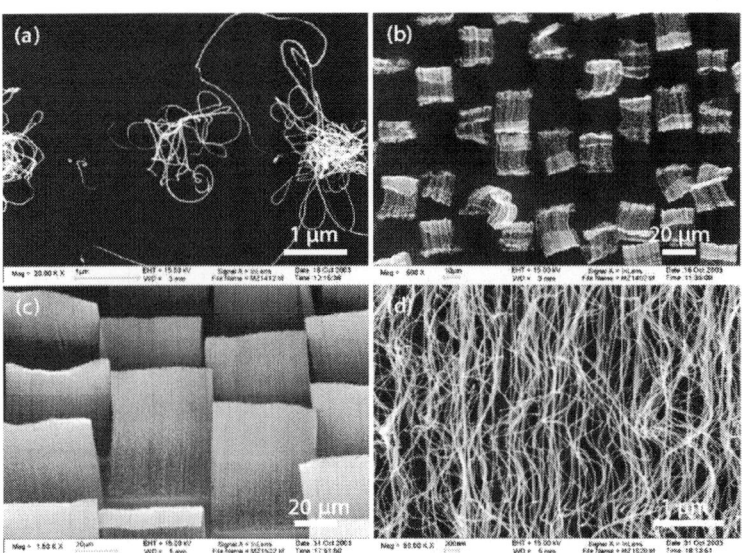

Figure 7. SEM image of CVT walls grow on patterned (a) very thin, (b) 0.1 μm wide and 40 μm long, and (c) 0.5 μm wide and 40 μm long catalyst lines. (d) SEM image of the CNTs in a wall in (c).

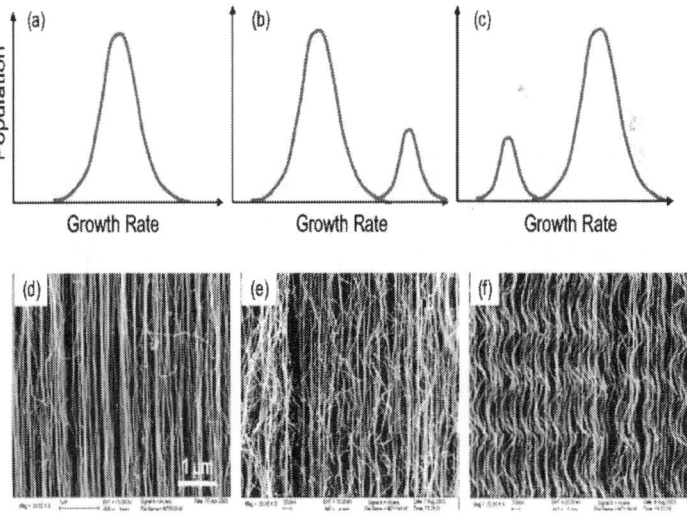

Figure 8. a) to (c) Schematic of the population of the CNTs related with the growth rate and (d) to (f) typical SEM images of the CNT forests. (d), (e), and (f) are in the same scale and they are corresponding to (a), (b), and (c), respectively.

3.2. CNT Area Density

As shown in Fig. 7, CNT area density (number of nanotubes per square centimeter) in the forest is a key factor to establishing the interconnections between CNTs in the forest and the drawability of the forest. The schematic illustration of drawability to the area density is shown in Fig. 9. If the area density is very low, CNTs will lay on the substrate randomly. The CNTs can grow in the out-of-plane direction and form a vertical aligned forest when its area density exceeds a threshold value. If the forest has very high density, the CNTs in the forest will form big bundles and the forest will not be drawable.

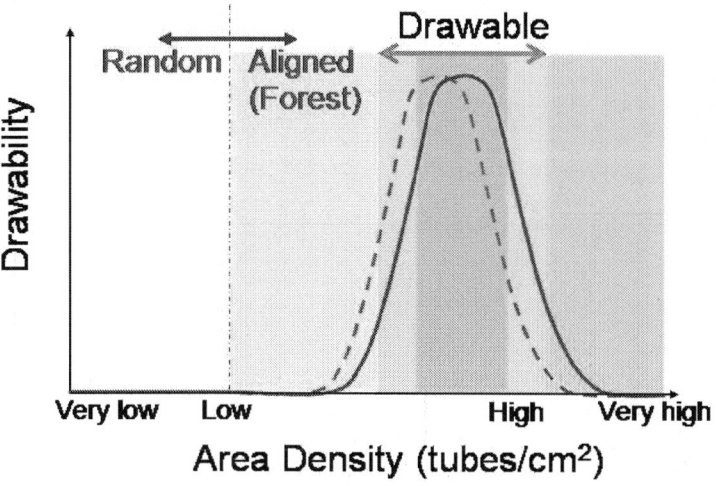

Figure 9. Schematic illustration showing the relationship between the forest drawability and the area density of CNTs. The dash line corresponds to the forest with higher height.

Experimentally, the area density of the forest is calculated from counting the root of the nanotubes on substrate after removal the forest from the substrate. Figure 10 shows the surfaces of the substrates after removing the forest. Each circle dot in the images is the root of a nanotube.

Drawable forests must have a high enough area density for CNTs in the forest to form interconnections. The required area density of the forest is related to the diameter of the nanotubes. When the forest is formed by CNTs with ~10 nm diameters, the well-drawn forest has the area density ~10^{11}/cm^2, and the undrawable forest has an area density of less than 10^{10}/cm^2. The area density needs to be higher for the forest with thinner CNTs.

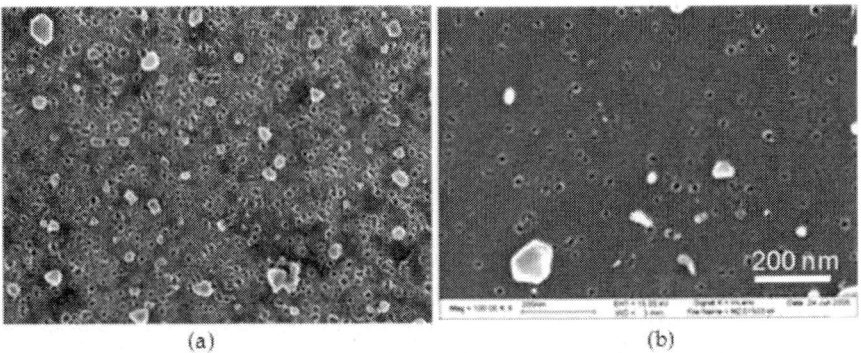

(a)   (b)

Figure 10. SEM images show the surfaces of the substrate after removal CNT forests. (a) and (b) show the typical surfaces of the drawable forest and the un-drawable forest, respectively. Each circle in the image is a root of a CNT. The bright dots are the by-products of the CVD process. Two images are in the same scale.

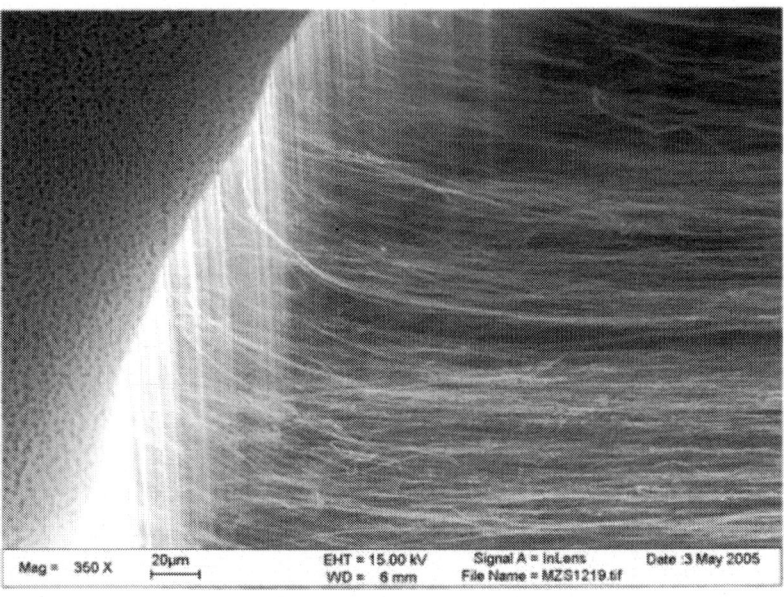

Figure 11. SEM image of a drawable forest. There are holes with ~ 1 μm diameter distributed in the forest.

The effect of area density can be compensated by adjusting the height of the forest (see section 3.4). The area density can also be lowered by controlling the distribution of catalysts through patterning.Figure 11 shows the SEM image of a drawable forest. There are uniformly distributed holes due to the missing catalysts on the substrate. The less CNT interconnectionsin the holes results in the easy draw of the too densely packed CNT forest.

3.3. Purity of the CNTS

Generally, the drawable forests need to be clean. Amorphous carbon (a-C) deposited on CNTs during CVD process might not be avoidable. A proper amount a-C might be helpful in CNT interconnections. However, too much a-C will increase the locking between CNTs and CNT bundles, which will cause the breaking of fibrils during sheet draw. The results of TGA measurement of the drawable forest and TEM observation of the CNT sheet are shown in Fig. 12.

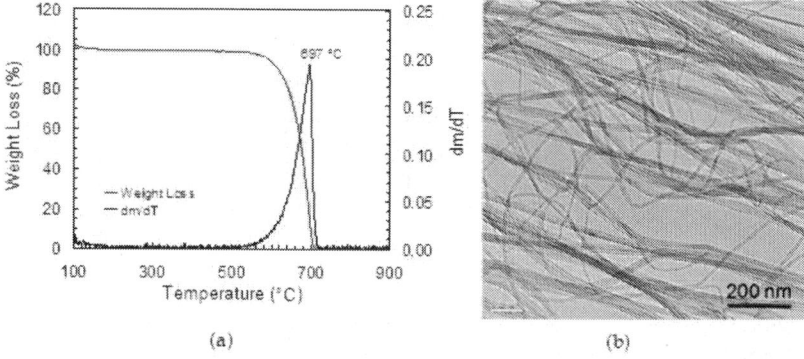

(a)   (b)

Figure 12. a) TGA data shows the thermal stability and the purity of the CNTs and (b) TEM image of a CNT sheet.

3.4. Height of the Forest

As described above, the interconnections of CNTs in the forest play an important role in the drawability of the forest. The CNTs form small bundles, each consisting of a few nanotubes, in the forest with individual nanotubes moving in and out of different bundles. The three-dimensional connectivity caused by intermittently switched bundling is believed to be important for the drawing process. The too-long and too-short CNT forests are not suitable for the solid-state process because the interconnections there either too much or too little for continuously pulling CNTs out of the forest.Figure 13a is the schematic relationship between drawability and length of the forest. CNTs can be drawn from a ~20 μm high forest as shown in Fig. 13b. Experiments demonstrated that the good drawablity has been obtained in up to 500m high forests formed by nanotubes 10 nm in diameter. Two-millimeter high forests formed by tubes 30~50 nm in diameter are demonstratedto be drawableInoue, 2011.

Since the drawability is determined by the interconnections of CNTs in the forest, the very-short and very-high forests could be drawable by adjusting other parameters. For example, very-short forest could be drawable if the area density is high enough and relatively more interconnections occur along their length. For high forests,lowering the area densityand the interconnections between CNTs and their bundles along their length will create the same effect (Fig. 9).

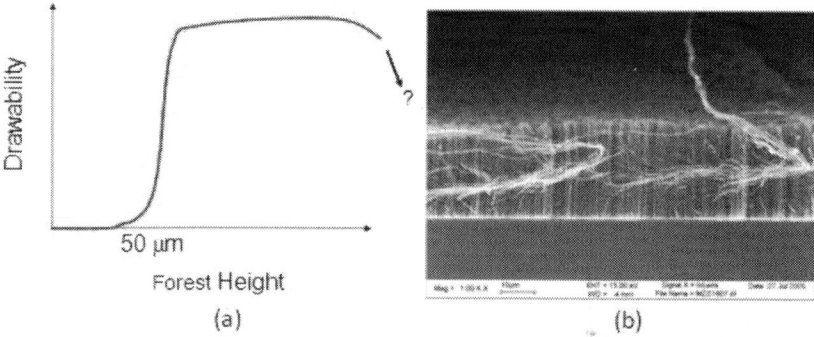

(a)   (b)

Figure 13. a) Schematic of the possible relationship between drawability and forest height. (b) SEM image shows that the CNTs can be drawn continuously from a 20 μm high CNT forest.

The forest-sheet conversion rate and the thickness of as-produced sheet depend on the height and the density of the forest, as well as the bundling level of tubes in the forest. For forests having similar topology, the highest forests were easiest to draw into sheets – most probably because increasing nanotube length increases inter-fibril mechanical coupling within the sheets, and produce a higher forest-sheet conversion rate. A one centimeter length of 245 m high forest converts to about a three-meter-long free-standing CNT sheet. By adjusting the height and density of the forest; converting a one centimeter length forest to over a ten meter long sheet has been achieved. The thickness of the as-produced CNT sheet increased with increasing forest height and was 18 m in SEM images of a sheet drawn from a 245 m high forest.

3.5. Process Temperature

Temperature is another important parameter, which determines other synthesis parameters for CNT growth and the quality of the CNTs. The intensity ratio of the G band (I_G at ~1580 cm^{-1}) and D band (I_D at ~1350 cm^{-1}) in Raman spectra and the initial burning temperature of the CNTs obtained from the TGA are used to evaluate the quality of the CNTs. G band and D band were originated from the Raman active in-plane atomic displacement E_{2g} mode and disorder-included features due to the finite particle size effect or lattice distortion, respectively (Tuinstra & Koenig, 1970). The increase of I_G/I_D indicates that the degree of long-range ordered crystalline perfection of the CNTs increases. The better the crystallization of the graphene layers of the CNTs, the higher the initial burning temperature of the CNTs. Maintaining a spatial homogeneous temperature during the growth process was demonstrated a critical factor for fabricating long CNTs with consistent electrical characteristics (X.Wang et al., 2009). Many researches show that the quality of the CNTs becomes better as the growth temperature increases (Y. T. Lee et al., 2002; C. J. Lee et al., 2001; K. Kim et al., 2005; X. Feng et al., 2009). However, other parameters for CNT growth must be optimized if increasing temperature since the temperature increase has direct influences on the formation and the activity of the catalysts and the feedstock of the carbon atoms. The partial pressure of the hydrocarbon gas usually needs to be lowered at a higher process temperature.

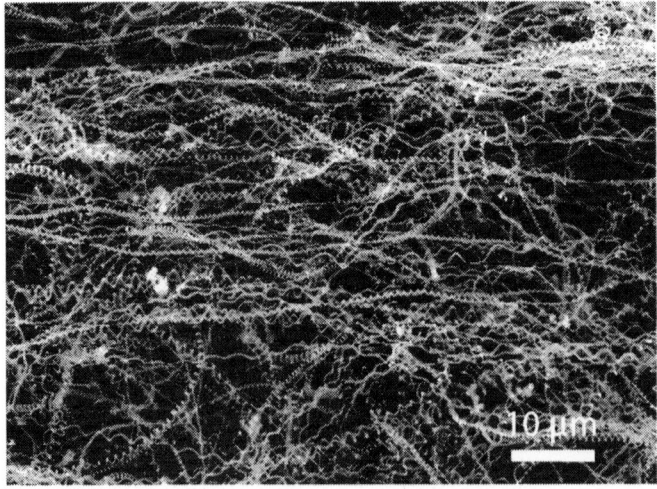

Figure 14. SEM image of large amount of helically coiled carbon nanostructures. Each coil grows with its own diameter and pitch.

4. CONCLUSION

Individual carbon nanotubes are like minute bits of string, and many trillions of these invisible strings must be assembled together to make useful macroscopic articles. This chapter presents the fabrication processes to making CNT sheets. The purity of the nanotubes, the height of the forest, the morphology of the forest, especially the 3D structure by self-assembly during CVD process, and the area density of the nanotubes are the main factors of the drawability of the forest.There are no inherent limitations on either sheet width or length, and no special difficulties arise in maintaining sheet quality during the draw. Currently, the sheets are drawn at up to 2 m/s from special CNT forests (Aliev et al., 2009;Lima, et al., 2011) and the width at up to 20 cm (C. Feng et al., 2010).This solid-state process is scalable for continuous, high-rate production. Extension of the technologies of solid-state sheet fabrication to longer CNTs, as well as to a few walled or single walled nanotubes, are important because longer nanotube lengths will enable properties improvements for active devices by means of enabling closer approach of sheet properties to those of individual nanotubes and the conductivity, transparency, and strength of the sheet could be improved by using thinner nanotubes.

CNTs can also be in the coiled structure (Amelinckxet al., 1994 ; Dunlap, 1992; M. Zhang & J. Li, 2009). Figure 14 shows the helically coiled CNTs grown on iron-coated indium tin oxide substrate by catalytic CVD (M. Zhang et al., 2000).More than 95% of the wires are in helical structures. The coils have various diameters and pitches. They grow out of the substrate and maintain their self-organization well during growth. If the coiled CNT forest is drawable, the sheet will have interesting properties that the straight CNTs could not provide. There is plenty of room to further improve the processes and properties of the CNT sheets.

5. ACKNOWLEDGEMENTS

This work was supported or supported partly by National Science Foundation, the Air Force Research Laboratory, Texas Advanced Technology Program, Robert A. Welch Foundation, and the Strategic Partnership for Research in NanotechnologyConsortium in Texas.

REFERENCES

1. H. Ago, K. Petritsch, M. S. P. Shaffer, A. H. Windle, R. H. Friend, (1999).Composites of carbon nanotubes and conjugated polymers for photovoltaic devices.Adv. Matls. 11 11 12811285.

2. S. Amelinckx, X. B.. Zhang, D. Bernaerts, X. F. Zhang, V. Ivanov, J. B. Nagy, 1994 A Formation Mechanism for CatalyticallyGrownHelix-Shaped Graphite Nanotubes. Science, 29 635639.

3. A. Alive, M. D. Lima, S. Fang, R. H. Baughman, (2010).Underwater sound generation using carbon nanotube projectors.Nano Lett., 10 10 7 23742380 .

4. A. E. Aliev, J. Oh, M. E. Kozlov, A. A. Kuznetsov, S. L. Fang, A. F. Fonseca, R. Ovalle, M. D. Lima, M. H. Haque, Y. N. Gartstein, M. Zhang, Zakhidov, R. H. Baughman, 2009 Giant-Stroke, Superelastic Carbon Nanotube Aerogel Muscles.Science, 323 15751579 .

5. A. Bachtold, P. Hadley, T. Nakanishi, C. Dekker, 2001 Logic Circuits withCarbon Nanotube Transistors. Science, 294 13171320 .

6. L. Berhan, Y. B. Yi, A. M. Sastry, E. Munoz, M. Selvidge, R. Baughman, 2004 Mechanical properties of nanotube sheets: Alterations in joint morphology and achievable moduli in manufacturable materials. J. Appl. Phys. 95 43354346.

7. L. Chen, C. Liu, J. Wang, W. Zhang, C. Hu, S. Fan, 2009 Auxetic Materials with Large Negative Poisson's Ratios Based on Highly OrientiedCarbn Nanotube Structures. Appl. Phys. Lett. 94 253111

8. Q. F. Cheng, J. P. Wang, J. J. Wen, C. H. Liu, K. L. Jiang, Q. Q. Li, S. S. Fan, 2010 Carbon nanotube/epoxy composites fabricated by resin transfer molding. ChengQ. F.WangJ. P.WenJ. J.LiuC. H.JiangK. L.LiQ. Q.FanS. S. (2010). Carbon nanotube/epoxy composites fabricated by resin transfer molding. Carbon, Vol. 48, pp. 260-266., 48 260266.

9. B. . Dan, G. C. Irvin,. Pasquali, M., 2009 Continuous and Scalable Fabrication of Transparent ConductingCarbon Nanotube Films. ACS nano, 3 4 835843.

10. B. I. Dunlap, 1992 Connectingcarbon tubules. Phys. Rev. B, 46 19331936.

11. M. Endo, H. Muramatsu, T. Hayashi, Y. A. Kim, M. Terrones, M. S. Dresselhaus, 2005 Nanotechnology: 'Buckypaper' from coaxial nanotubes. Nature, 433 476

12. X. Feng, K. Liu, X. Xie, R. Zhou, L. Zhang, Q. Li, S. Fan, K. Jiang, 2009 Thermal Analysis Study of the Growth Kinetics of Carbon Nanotubes and Epitaxial Graphene layers on Them. J. Phys. Chem. 113 96239631.

13. C. Feng, K. Liu, J. Wu, L. Liu, J. Cheng, Y. Zhang, Y. Sun, Q. Li, S. Fan, K. Jiang, 2010 Flexible, stretchable, transparent conducting films made from superaligned carbon nanotubes. Adv. Funct. Mater. 20 885891.

14. J. E. Fischer, W. Zhou, J. Vavro, M. C. Llaguno, C. Guthy, R. Haggenmueller, M. J. Casavant, D. E. Walters, R. E. Smalley, 2003 Magnetically aligned single wall carbon nanotube films: Preferred orientation and anisotropic transport properties.Journal of Applied Physics, 93 21572163.

15. P. Galvan-Garcia, E. W. Keefer, F. Yang, M. Zhang, S. L. Fang, A. A. Zakhidov, R. H. Baughman, M. I. Romero, 2007 Robust Cell Migration and Neuronal Growth on Pristine Carbon Nanotube Sheets and Yarns.Journal of Biomaterials Science, 18 12451261 .

16. L. Hu, D. S. Hecht,. Gruener, G. (2010, (2010).Carbon Nanotube Thin Films: Fabrication, Properties, and Applications.Chem. Rev. 110 110 57905844.

17. L. Hu, D. S. Hecht, G. Gruener, (2004).Percolation in transparent and conducting carbon nanotube networks.Nano Letters, 4 4 25132517.

18. Y. Inoue, Y. Suzuki, Y. Minami, J. Muramatsu, Y. Shimamura, K. Suzuki, A. Ghemes, M. Okada, S. Sakakibara, H. Mimura, K. Naito, 2011 Anisotropic Carbon Nanotube Papers Fabricated from Multiwalled Carbon Nanotubes Webs. Carbon, 49 24372443.

19. S. Y. Ju, A. Facchetti, Y. Xuan, J. Liu, F. Ishikawa, P. D. Ye, C. W. Zhou, T. J. Marks, D. B. Janes, 2007 Fabrication of Fully Transparent Nanowire Transistors for Transparent and Flexible Electronics.Nat. Nanotechnol., 2 378384 .

20. K. Kim, K. J. Kim, W. S. Jung, S. Y. Bae, J. Park, J. Choi, . Choo, J., 2005 Investigation on the temperature-dependent growth rate of carbon nanotubes using chemical vapor deposition of ferrocene and acetylene.Chemical Physics Letters, 401 459464.

21. Y. Kim, N. Minami, W. H. Zhu, S. Kazaoui, R. Azumi, M. Matsumoto, (2003). Langmuir-Blodgett films of single-wall carbon nanotubes: layer-by-layer deposition and in-plane orientation of tubes.Jpn. J. Appl. Phys., Part 1 Vol. 42, pp. 7629-.

22. J. . Kong, H. T. Soh, A. M. Cassell, C. F. Quate, . Dai, H., 1998 Synthesis of individual single-walledcarbon nanotubes on patternedsilicon wafers. Nature, 395 878881.

23. J. Kong, N. R. Franklin, C. Zhou, M. G. Chapline, S. Peng, K. Cho, H. Dai, 2000 Nanotube Molecular Wires as Chemical Sensors. Science, 287 622625.

24. M. E. Kozlov, C. S. Haines, J. Oh, M. D. Lima, S. Fang, 2009 Sound of carbon nanotube assemblies. J. Appl. Phys., 106 124311

25. A. A. Kuznetzov, S. B. Lee, M. Zhang, R. H. Baughman, . Zakhidov, A. A., 2010 Electron field emission from transparent multiwalled carbon nanotube sheets for inverted field emission displays. KuznetzovA. A.LeeS. B.ZhangM.BaughmanR. H.Zakhidov.A. A. (2010). Electron field emission from transparent multiwalled carbon nanotube sheets for inverted field emission displays. Carbon, Vol. 48, pp. 41-46., 48 414.

26. A. A. Kuznetsov, A. F. Fonseca, R. H. Baughman, A. A. Zakhidov, (2011). Structural Model for Dry-Drawing of Sheets and Yarns from Carbon Nanotube Forests.ACS Nano, 5 2), 985993.

27. C. J. Lee, J. Park, Y. Huh, J. Y. Lee, (2001).Temperature effect on the growth of carbon nanotubes using thermal chemical vapor deposition.Chemical Physics Letters, 343 343 3338.

28. Y. T. Lee, J. Park, Y. S. Choi, H. Ryu, H. J. Lee, 2002 Temperature-Dependent Growth of Vertically Aligned Carbon Nanotubes in the Range 800–1100 °C.J. Phys. Chem. B, 106 31 76147618.

29. Y. Li, I. A. Kinloch, A. H. Windle, (2004).Direct spinning of carbon nanotube fibers from chemical vapor deposition synthesis.Science, 304 304 276278 .

30. D. Lima, S. Fang, X. Lepró, C. Lewis, R. Ovalle-Robles, J. Carretero-González, E. Castillo-Martínez, M. E. Kozlov, J. Oh, N. Rawat, C. S. Haines, M. H. Haque, V. Aare, S. Stoughton, A. A. Zakhidov, R. H. Baughman, 2011 Biscrolling Nanotube Sheets and Functional Guests into Yarns.Science 331 5155 .

31. K. Liu, Y. H. Sun, R. F. Zhou, H. Y. Zhu, J. P. Wang, L. Liu, S. S. Fan, K. L. Jiang, 2010 Carbon Nanotube Yarns with High Tensile Strength Made by a Twisting and Shrinking Method.Nanotechnology, 21 045708

32. C. Martin, 2010 A Carbon Nano-Wired World.R & Dmagazine, 52 3 June, 2010, 40

33. J. A. Misewich, R. Martel, Ph. Avouris, J. C. Tsang, S. Heinze, J. Tersoff, (2003).Electrically Induced Optical Emission from a Carbon Nanotube FET. Science, 300 300 783786.

34. Nanocomp Technologies Inc. (2011). http://www.nanocomptech.com /html/nanocomp-technology.html

35. A. Pevzner, Y. Engel, R. Elnathan, T. Ducobni, M. Ben-Ishai, K. Reddy, N. Shpaisman, A. Tsukernik, M. Oksman, F. Patolsky, 2010 Knocking Down Highly-Ordered Large-Scale Nanowire Arrays.Nano Lett., 10 12021208 .

36. A. G. Rinzler, J. Liu, H. Dai, P. Nikolaev, C. B. Huffman, F. J. Rodr´ıguez-Mac´ıas, P. J. Boul, A. H. Lu, D. Heymann, D. T. Colbert, R. S. Lee, J. E. Fischer, A. M. Rao, P. C. Eklund, R. E. Smalley, 1998 Large-scale purification of single-wall carbon nanotubes: process, product, and characterization. Applied Physics A, 67 2937.

37. T. V. Sreekumar, T. Liu, S. Kumar, L. M. Ericson, R. H. Hauge, R. E. Smalley, (2003).Single-wall carbon nanotube films.Chemistry of Materials, 15 15 175178 .

38. S. J. Tans, A. R. M. Verschueren, C. Dekker, 1998 Room-temperature transistor based on a single carbon nanotube. Nature, 393 4952.

39. R. Ulbricht, S. B. Lee, K. Inoue, M. Zhang, S. Fang, R. H. Baughman,. Zakhidov, A. A., 2007 Transparent Carbon Nanotube Sheets as 3-D Charge Collectors in Organic Solar Cells. Solar Energy Materials and Solar Cells, 91 416419.

40. R. Ulbricht, X. Jiang, S. B. Lee, K. Inoue, M. Zhang, S. Fang, R. H. Baughman,. Zakhidov, A. A., 2006 Polymetric Solar Cells with Oriented and Strong Transparent Carbon Nanotube Anode.Phys. Stat. Sol. B, 243 35283532.

41. Unidym Inc. 2007 http://www.unidym.com.

42. D. Wang, P. Song, C. Liu, W. Wu, S. Fan, 2008 Highly oriented carbon nanotube papers made of aligned carbon nanotubes. Nanotechnology, 7 075609

43. X. Wang, Q. Li, J. Xie, Zhong. Jin, J. Wang, Y. Li, K. Jiang, S. Fan, 2009 Fabrication of ultralong and electrically uniform single-walled carbon nanotube on clean substrates.Nano Lett. 9 9 31373141.

44. Y. Wei, L. Liu, P. Liu, L. Xiao, K. Jiang, S. Fan, 2008 Scaled fabrication of single-nanotube-tipped ends from carbon nanotube micro-yarns and their field emission applications.Nanotechnology, 19 475707

45. C. D. Williams, R. O. Robles, M. Zhang, S. Li, R. H. Baughman, . Zakhidov, A. A. , 2008 Multiwalled carbon nanotube sheets as transparent electrodes in high brightness organic light-emitting diodes. Applied Physics Letters, 93 183506

46. Z. C. Wu, Z. H. Chen, X. Du, J. M. Logan, J. Sippel, M. Nikolou, K. Kamaras, J. R. Reynolds, D. B. Tanner, A. F. Hebard, . Rinzler, A. G., 2004 Transparent, conductivecarbon nanotube films. Science, 305 12731276.

47. Y. N. Xia, P. D. Yang, Y. G. Sun, Y. Y. Wu, B. Mayers, B. Gates, Y. D. Yin, F. Kim, Y. Q. Yan, 2003 One-Dimensional Nanostructures: Synthesis, Characterization, and Applications. Adv. Mater. 15 353389.

48. L. Xiao, P. Liu, L. Liu, K. Jiang, X. Feng, Y. Wei, L. Qian, S. Fan, T. Zhang, 2008 Barium-Functionalized Multiwalled Carbon Nanotube Yarns as Low-Work-Function Thermionic Cathodes.Appl. Phys. Lett., 92 153108

49. L. Xiao, Z. Chen, C. Feng, L. Liu, Z. Q. Bai, Y. Wang, L. Qian, Y. Y. Zhang, Q. Q. Li, K. L. Jiang, S. S. Fan, 2008 Flexible, Stretchable, Transparent Carbon Nanotube Thin Film Loudspeakers.Nano Lett., 8 45394545 .

50. F. Xu, W. Lu, Y. Zhu, 2011 Controlled 3D Buckling of Silicon Nanowires for Stretchable Electronics.ACS Nano, 5 672678 .

51. M. Zhang, Y. Nakayama, L. Pan, 2000 Synthesis of Carbon Tubule Nanocoils in High YieldUsingIron-Coated Indium Tin Oxide as Catalyst. Jpn. J. Appl. Phys., 39 L1242L1244.

52. M. Zhang, K. R. Atkinson, R. H. Baughman, 2004 Multifunctional carbon nanotube yarns by downsizing an ancient technology. Science, 306 13581361.

53. M. Zhang, S. Fang, A. A. Zakhidov, S. B. Lee, A. E. Aliev, C. D. Williams, K. R. Atkinson, R. H. Baughman, 2005 Strong, Transparent, Multifunctional Carbon Nanotube Sheets.Science, 309 12151219 .

54. M. Zhang, J. Li, (2009).Carbon Nanotube in Different Shapes.Materials Today, 12 12 1218.

55. T. Zhang, S. Mubeen, N. Myung, M. Deshusses, 2008 Recent progress in carbon nanotube-based gas sensors. Nanotechnology, 19 332001332014.

56. H. Zhang, C. Feng, Y. Zhai, K. Jiang, Q. Li, S. Fan, 2009 Cross-Stacked Carbon Nanotube Sheets Uniformly Loaded with SnO2 Nanoparticles: A Novel Binder-Free and High-Capacity Anode Material for Lithium-Ion Batteries. Adv. Mater. 21 22992304.

57. X. Zhang, K. Jiang, C. Feng, P. Liu, L. Zhang, J. Kong, T. Zhang, Q. Li, S. Fan, 2006 Spinning and Processing Continuous Yarns from 4-Inch Wafer Scale Super-Aligned Carbon Nanotube Arrays. Adv. Mater., 18 15051510.

58. G. Zhong, S. Hofmann, F. Yan, H. Telg, J. H. Warner, D. Eder, C. Thomsen, W. I. Milne, J. Robertson, 2009 Acetylene: A Key Growth Precursor for Single-Walled Carbon Nanotube Forests. J. Phys. Chem. C, 113 1732117325.

59. X. H. Zhong, Y. L. Li, Y. K. Liu, X. H. Qiao, Y. Feng, J. Liang, J. Jin, L. Zhu, F. Hou, J. Y. Li, 2010 Continuous Multilayered Carbon Nanotube Yarns.Adv. Mater, 22 692696.

60. Y. Zhou, L. Hu, G. Gruner, A method of printing carbon nanotube thin films. Appl. Phys. Lett. 88 123109

61. R. Zhou, C. Meng, F. Zhu, Q. Li, C. Liu, S. Fan, K. Jiang, 2010 High-performance supercapacitors using a nanoporous current collector made from super-aligned carbon nanotubes. Nanotechnology, 21 345701.

Nitrogen-Doped Carbon Nanotube and Graphene Materials for Oxygen Reduction Reactions

Qiliang Wei [1,2,†], Xin Tong [2,†], Gaixia Zhang [2], Jinli Qiao [3], Qiaojuan Gong [1,*] and Shuhui Sun [2,*]

[1] Department of Applied Chemistry, Yuncheng University, 1155 Fudan West Street, Yun Cheng 04400, China

[2] Institut National de la Recherche Scientifique (INRS), Centre Énergie, Matériaux et Télécommunications, 1650 Boulevard Lionel-Boulet, Varennes, QC J3X 1S2, Canada

[3] College of Environmental Science and Engineering, Donghua University, 2999 Ren'min North Road, Shanghai 201620, China

ABSTRACT

Nitrogen-doped carbon materials, including nitrogen-doped carbon nanotubes (NCNTs) and nitrogen-doped graphene (NG), have attracted increasing attention for oxygen reduction reaction (ORR) in metal-air batteries and fuel cell applications, due to their optimal properties including excellent electronic conductivity, 4e− transfer and superb mechanical properties. Here, the recent progress of NCNTs- and NG-based catalysts for ORR is reviewed. Firstly, the general preparation routes of these two N-doped carbon-allotropes are introduced briefly, and then a special emphasis is placed on the developments of both NCNTs and NG as promising metal-free catalysts and/or catalyst support materials for ORR. All these efficient ORR electrocatalysts feature a low cost, high durability and excellent performance, and are thus the key factors in accelerating the widespread commercialization of metal-air battery and fuel cell technologies.

Keywords: nitrogen-doped carbon nanotubes; nitrogen-doped graphene; metal-free catalysts; ORR

1. INTRODUCTION

Developing highly efficient electrocatalysts to facilitate sluggish cathodic oxygen reduction reaction (ORR) is a key issue in metal-air batteries and fuel cells [1,2,3,4,5]. The ORR mechanism includes two different pathways: (i) a four-electron ($4e^-$) process to produce water directly though the reaction of oxygen, electrons and protons, and (ii) a two-electron ($2e^-$) process to create the intermediate compound (hydrogen peroxide) [6]. The $4e^-$ process is more attractive for cathode catalysts in fuel cells. Although the platinum-based materials are the better choices for the desired $4e^-$ pathway, the use of very expensive and rare platinum is a major impediment to the development and widespread commercialization of fuel cells. Thus, exploring the substitutes for platinum catalysts by employing non-precious metal catalysts is a very promising direction [7]. In this regard, one-dimensional (1D) carbon nanotubes (CNTs) and two-dimensional (2D) graphene (Figure 1) have attracted a great deal of attention for ORR due to their excellent electronic conductivity, huge specific surface area (SSA), as well as excellent thermal and mechanical properties [8]. Interestingly, when the heteroatoms are incorporated in the carbonaceous skeleton, the ORR performance can be greatly enhanced by effectively modulating the chemisorption energy of O_2, catalytic sites, and the reaction mechanism ($2e^-/4e^-$) of catalysts [9]. Among various possible dopants, N-doped carbon materials are attracting much more attention because of their excellent electrocatalytic performance, low cost, excellent stability, and environmental friendliness, thus setting up a new generation of the metal-free catalysts for ORR. Furthermore, when the nitrogen with excessive valence is introduced to the graphitic plane, more π-electrons can be obtained [10]. This feature, together with the significant difference in the electronegativity of N and C, leads to many unique properties to graphitic carbons, including increased n-type carrier concentration, high surface energy, reduced work-function, as well as tunable polarization [11,12,13,14]. As schematically illustrated in Figure 2, three common bonding configurations of N atoms in graphene are demonstrated, including pyrrolic, pyridinic, and graphitic (or quaternary) N [15]. Pyridinic N atoms are located at the edges of graphene planes, and each N atom is bonded to two C atoms and donates one π-electron to the π system. In the case of pyrrolic N atoms, they are incorporated into the heterocyclic rings and each N atom is bonded to two C atoms, contributing two π-electrons to the π system. Graphitic (or quaternary) N refers to the N atoms that replace the carbon atoms in the graphene plane. Such doped N atoms can change the local density state around the Fermi level of N-doped graphitic carbons, which may play a vital role in tailoring the electronic properties and improving their ORR performance [14,16].

On the other hand, metal oxides are also good candidates for ORR catalysts, although they normally suffer from low conductivity, as well as dissolution, sintering, and agglomeration during operation. Consequently, the electrocatalysts show poor electrochemical properties, restricting their applications. NCNTs or NG could effectively buffer the catalyst nanoparticle agglomeration and enhance the electronic conductivity by virtue of their intrinsic excellent conductivity and huge SSA. Therefore, NCNTs and NG can be used as both excellent metal-free electrocatalysts and perfect catalyst support for ORR.

The basic principles and mechanisms behind N doping effectively tailoring the electrical and surface properties of graphitic carbons have been reviewed in some excellent papers [14,17,18]. Here in this review, we place emphasis on the synthesis of NCNTs and NG, and their applications for ORR.

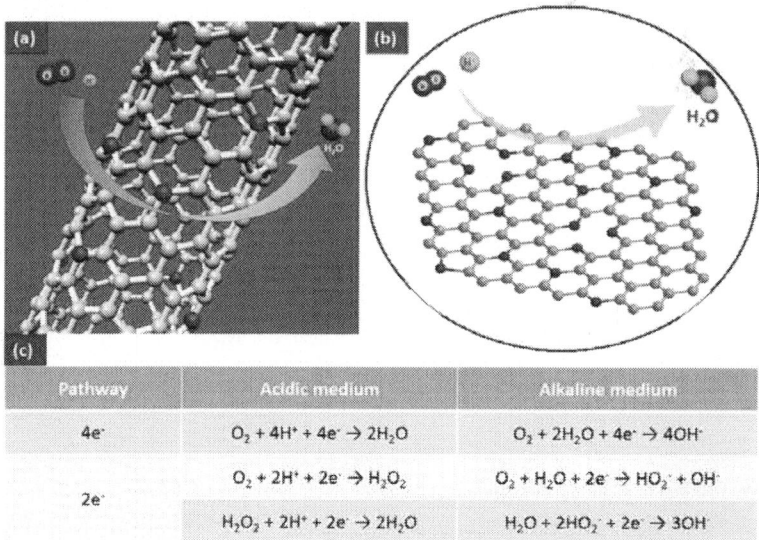

Pathway	Acidic medium	Alkaline medium
4e⁻	$O_2 + 4H^+ + 4e^- \rightarrow 2H_2O$	$O_2 + 2H_2O + 4e^- \rightarrow 4OH^-$
2e⁻	$O_2 + 2H^+ + 2e^- \rightarrow H_2O_2$	$O_2 + H_2O + 2e^- \rightarrow HO_2^- + OH^-$
	$H_2O_2 + 2H^+ + 2e^- \rightarrow 2H_2O$	$H_2O + 2HO_2^- + 2e^- \rightarrow 3OH^-$

Figure 1. Illustration of ORR on **(a)** NCNTs; **(b)** NG and **(c)** ORR pathway in acid and alkaline medium. Reproduced and adapted in part from [19]. Copyright © 2013, Rights Managed by Nature Publishing Group.

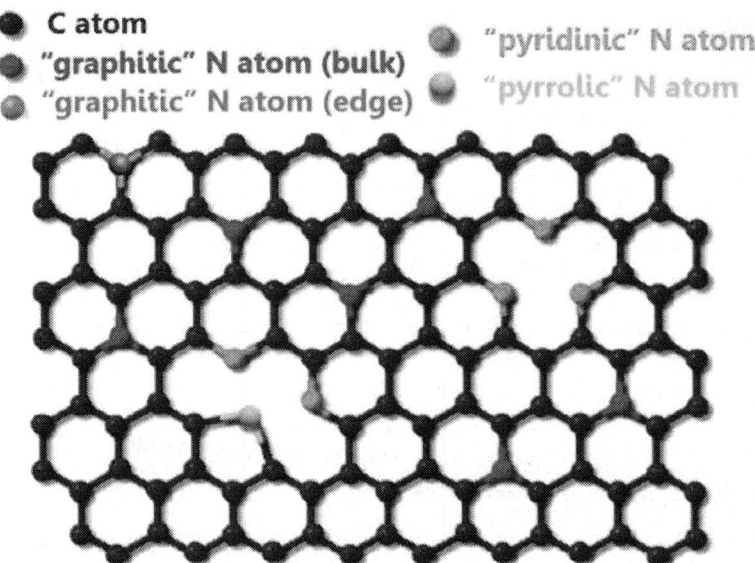

Figure 2. Schematic representation of different types of N atoms (graphitic, pyridinic and pyrrolic N) in NG and NCNTs. Modified with permission from Ref. [20]. Copyright © 2009, American Chemical Society.

2. SYNTHESIS OF NITROGEN-DOPED CARBON

Nitrogen (N) is a neighboring element of carbon in the periodic table, and its electronegativity (3.04) is larger than that of C (2.55). The incorporation of N atom into a graphene lattice plane could modulate the local electronic properties, as it could form strong bonds with carbon atoms because of its comparable atomic size with carbon. Subsequently, it could generate a delocalized conjugated system between the graphene π-system and the lone pair of electrons from N atom. The introduction of N into carbon nanomaterials could improve both reactivity and electrocatalytic performance. As a result, the N-doped carbon materials have been intensively studied among all the available heteroatoms for doping.

2.1. N-Doped Carbon Nanotubes

NCNTs have become a focus as ORR catalysts due to their high activity and excellent stability. In principle, the N-doping methods can be classified to two categories: *in situ* doping and post-treatment doping [17].

2.1.1. In Situ Doping

In situ doping involves the direct incorporation of N heteroatoms into carbon matrix during the preparation process, and it is often used for the preparation of NCNTs. The typical *in situ* doping techniques include high-temperature arc-discharge [21,22], chemical vapor deposition (CVD) [23,24,25,26,27], chemically solvothermal procedures (*ca.* 230–300 °C), [28] and laser ablation methods [29,30]. Thus far, a wide range of N-containing precursors have been used to incorporate N into C matrix with great success. Moreover, the final amount and functionality of N in NCNTs are much more critical for practical applications but could essentially be derived from many different precursors by tuning the synthesis parameters such as temperature of pyrolysis. Among various techniques, CVD is the most promising method to synthesize NCNTs with a different C source (such as methane, acetylene, ethylene, benzene, *etc.*) [31,32,33,34] and N source (such as ethylene diamine, dimethylformamide, imidazole, Fe-Phthalocyanine, benzylamine, *etc.*) [34,35,36,37,38]. For instance, recently, by using a co-pyrolysis route of Fe-Phthalocyanine loaded and PEO_{20}-PPO_{70}-PEO_{20} (P123) retained in mesoporous silica, Wang *et al.* [34] synthesized NCNTs with well-defined morphology and graphitic structure, which exhibited good performance for ORR. Based on CVD, She *et al.* fabricated N-doped 1D macroporous carbonaceous nanotube arrays in anodic alumina oxide (AAO) template, which also showed high performance for ORR [27]. In addition to the precursors and pyrolysis temperatures, for each method, other factors, such as time, gas flow rate, catalysts, also have significant influence on the nitrogen contents and the accurately controlled doping sites [17,28,39,40].

2.1.2. Post-Treatment

NCNTs have also been prepared by various post-treatment methods [41,42]. For instance, Nagaiah *et al.* [41] synthesized NCNTs by post-thermal treating oxidized CNTs with ammonia and used the resultant NCNTs as efficient catalysts for ORR in alkaline medium. However, the post-synthesis treatments [43] normally require high temperature (800–1200 °C) and toxic N precursors (NH_3 or pyridine) which limit their practical application. Moreover, some structural degradation and morphological defects often appear in the materials due to the high temperature treatment [44].

In general, the *in situ* doping tends to form pyrrolic- and/or pyridinic-N atoms, while the graphitic-N in carbon frameworks is normally generated after a high temperature post-treatment [45]. Yet, to obtain an accurate N content and doping sites controllably in these materials is still a challenging problem [17].

2.2. N-Doped Graphene

Compared to doping N into CNTs, the N atom can be more easily introduced into the graphene due to the more open structure in graphene. The N atom could be incorporated into graphene directly during the synthesis of graphene or through post-treatment of graphene oxide (GO) (or graphene). Among numerous methods to produce graphene, CVD, solvothermal fabrication and arc-discharge are normally chosen for *in situ* growth of NG. Compared with the *in situ* synthesis, post-treatment methods which include thermal annealing, plasma or irradiation treatment, or solution treatment are simpler and likely closer to commercialization [46].

2.2.1. In Situ Doping

CVD is one of the important methods to prepare NG [20]. In Liu's group, they used Cu/Si as the catalyst, CH_4 as the C source and NH_3 as the N source to produce few-layers NG under 800 °C for the first time (single-layer graphene can be occasionally detected). On the other hand, by using the sole source that contains both C and N (e.g., acetonitrile [47] and pyridine [48]), N atoms can be simultaneously introduced into the graphene lattice during CVD growth of graphene films. The doping amount of N can be adjusted in the range of 1.2–16 at.% by controlling the gas flow rate and the C source to N source ratio [20,49].

A solvothermal process to obtain NG through the reaction between tetrachloromethane and lithium nitride was also developed by Deng *et al.* [50]. It is a one-pot direct synthesis with just placing the reaction reagent in an autoclave and keeping under N_2 and below 350 °C. It allows scalable synthesis and the nitrogen species can be introduced into the graphene structure with 4.5–16.4 at.% of N.

With the presence of pyridine vapor or NH_3, the arc-discharge technique which is commonly used for preparing carbon-based nanomaterials is also employed to fabricate NG. Rao *et al.* [51,52,53] successfully produced NG with the N content around 0.5–1.5 at.%. However, this process requires complicated purification steps with low yield due to the excessive by-products.

2.2.2. Post-Treatment

Thermal treatment in ammonia atmosphere is an easy and commonly used method to obtain NG by post-modification. Since the N incorporation reactions occur mostly at the defect sites and the edges of graphene, a low N level (e.g., 2.8 at.% in ref.) in graphene is normally obtained in previous reports [54]. In order to get higher N doping, researchers turned their attention to GO which contains a range of reactive oxygen functional groups and more defects to provide more active deposition. In Dai's group [55,56], through thermal annealing of GO under NH_3 atmosphere, the GO nanosheets were reduced and decorated with N simultaneously. At 300 °C, the N-doping process started, while the highest doping level of ~5 at.% N was achieved at 500 °C. The melamine was also used as the N source to synthesize NG and the atomic percentage of N can reach up to 10.1 at.% [57].

Since the chemical defects in graphene play a critical role in the production of NG, some physically based methods such as plasma treatment or ion implantation are used to induce chemical defects [58]. Furthermore, by changing the plasma density or exposure time, the N content can be easily controlled (up to 8.5 at.% N) [59]. For example, Guo *et al.* used N^+-ion irradiation to introduce defects into the plane of the graphene, and then followed by annealing under NH_3 atmosphere to get NG [60]. The level of N doping can also be adjusted by changing the experimental parameters.

In liquid phase environment, the reduction of GO and N doping can be realized simultaneously under the hydrothermal reaction by using N-containing reducing agent such as hydrazine hydrate [61] or urea [62]. At a pH of 10 and temperature of 80 °C, in the presence of hydrazine and ammonia, slightly wrinkled and folded NG sheets (up to 5 at.% N) were obtained. Also, the N-enriched urea could play a key role in the formation of the NG with high N-doping level (10.13 at.%). During the hydrothermal process, NH_3 will release and react with the oxygen-containing groups on GO; meanwhile, the N atoms can dope into a graphene skeleton. Researchers can control the N-doping level through adjusting the experimental parameters, e.g., the mass ratio between GO and the reducing agent, or the reaction temperature.

3. NITROGEN-DOPED CARBON NANOTUBES (NCNTS) FOR OXYGEN REDUCTION REACTION (ORR)

3.1. NCNTs as a Metal-Free Catalyst for ORR

The pioneering work of NCNTs as highly efficient electrocatalysts for ORR in alkaline fuel cells was reported by Gong *et al.* in 2009 [6]. A steady-state output potential of −80 mV and a current density of 4.1 mA/cm^2 at −0.22 V were observed in their study, which is superior to that of −85 mV and 1.1 mA/cm^2 at −0.20 V for a Pt/C electrode. Quantum mechanics calculations show that the carbon atoms adjacent to N dopants have very high positive charge density in order to counterbalance the strong electronic affinity of the N atom. Coupled with aligning the NCNTs, the vertically aligned (VA)-NCNTs show an excellent performance of a 4e$^-$ pathway for ORR. Following this important study, plenty of research has been conducted to fabricate NCNTs [37,41,62,63] and to investigate their electrocatalytic activity from both mechanistic and experimental perspectives [23,38,64,65,66,67,68]. For example, based on B3LYP (a trustworthy calculation for nanomaterials) [69,70,71], Hu et al. [69] investigated the adsorption and activation of triplet O_2 on the surface of NCNTs with different diameters and lengths by density functional theory (DFT). The results showed that N doping sufficiently improved the adsorption ability of O_2 on CNTs [69]. Changing the diameter and length of NCNTs has a large effect on the binding energy between O_2 and NCNT and bond length of O_2, and this result further proves that NCNTs are very promising metal-free catalysts for ORR from a theoretical perspective.

From an experimental perspective, in 2009, Y. Tang et al. [72] synthesized NCNTs via the CVD method using acetonitrile or ethanol as precursors and Ar/H$_2$ as carrier gases. TEM images indicate that the NCNTs are composed of individual nanocups stacked together (Figure 3). Their results indicated that the stacked NCNTs exhibited similar catalytic activity with Pt/CNTs in ORR and they can also be used in the electrochemical detection of H_2O_2 and glucose. Using the CVD method, several other research groups also tried to synthesize NCNTs with different N precursors. Experiments indicate that carbon and N precursors have a significant impact on the morphology and performance of NCNTs. For instance, when ferrocene (catalyst precursor) and imidazole (C and N precursor) were used, the as-synthesized NCNTs had a high N content of 8.54 at.% and a bamboo-like structure [23]; by annealing CNTs and tripyrrolyl[1,3,5]triazine (TPT) mixture in N, the NCNTs annealing at 900 °C exhibited excellent electrochemical performance towards ORR in alkaline medium [73].

In another group, Kundu et al. fabricated NCNTs via the pyrolysis of acetonitrile with cobalt as catalyst at different temperatures in order to control the nitrogen content [63]. The results indicated that NCNTs prepared at lower temperatures had a higher amount of pyridinic groups with more exposed edge planes. Furthermore, they proved that the NCNTs with a higher amount of pyridinic groups possess better catalytic properties for ORR. Later, they synthesized NCNTs using a new approach, i.e., by treating oxidized CNTs with ammonia at 800 °C; the obtained NCNTs exhibited a favorable positive onset potential for ORR, increased reduction current, and excellent stability, demonstrating a very promising cathode catalyst for ORR in alkaline medium [41]. Almost at the same time, Chen and co-workers synthesized NCNTs using various N precursors and/or catalysts [74,75,76,77]. It was concluded from their studies that higher N content and more defects in NCNTs lead to higher ORR performance. Similar conclusions were also drawn by Geng et al. [78]. However, others have found that there is no direct correlation between total N content and the ORR performance; for example, a recent study reported, through post-treatment of few-walled carbon nanotubes (FWCNTs) with polyaniline, a much lower N content (~0.5 at.%). Interestingly, the low N-containing FWCNTs exhibited excellent electrocatalytic activity for ORR as well as higher methanol tolerance properties [79]. Therefore, the exact role of N doping in NCNTs for the ORR activity is still under debate. Until recently, Wågberg et al. [45] investigated how a thermal post-treatment on the N-doped MWCNTs can result in the transformation of pyrrolic and pyridinic N sites into quaternary N sites (N-Q_s), leading to the improvement of ORR performance. They reached the conclusion that the quaternary N valley sites (N-Q_{valley}) are the most active sites in NCNTs for ORR; hence, a 4e$^-$ reduction pathway occurs generally on the N edge defects. Based on this fundamental concept, the chemical functionalization becomes an alternative and effective approach to introducing N into complex carbon nanostructures [80]. Accordingly, Tuci et al. reported a systematic study on the synthesis, characterization, and electrocatalytic property of MWCNTs functionalized with a series of well-defined pyridine groups [81]. They also discussed the role of the electronic charge density distribution at the chemically grafted N heterocycles on the ORR performance. This study therein introduced a deep level of complexity to the understanding of the ultimate role of the pyridine groups on ORR in NCNTs.

All these findings introduced above have significant impacts on catalysis and fuel cell domains. However, most of the CNTs used in these reports were synthesized by the pyrolysis of a nitrogen-containing precursor, and the residual catalyst particles of Fe or Co were removed by the electrochemical method. Though great attention has been paid to the purification process, the effects of metal contaminates in NCNTs on the ORR performance are still controversial, unless NCNTs could be obtained by a metal-free synthetic process. In this regard, by employing water-plasma etching SiO_2/Si wafers, Dai's group reported a simple but effective approach for the growth of densely packed N-doped single-walled CNTs [82]. Figure 4a shows the schematic illustration of the NCNT fabrication process. Typically, the water-plasma was used to etch the SiO_2 coating (30 nm) on the top of the SiO_2/Si wafer to produce uniform SiO_2 nanoparticles, which will act as the catalysts for NCNT growth during the CVD synthesis. As shown in Figure 4b–e, the produced metal-free NCNTs showed superb electrocatalytic activity and excellent durability toward ORR in acidic medium. For the similar purpose of

excluding the possible contribution of metal impurities to ORR catalysis,
Wang *et al.* [64] discovered that, without metal-containing catalysts, N
atoms alone show strong promotion for the self-assembly of NCNTs from
gaseous carbons. Based on this new discovery, pure metal-free CNTs with a
high level of N doping (20 at.%) can be directly synthesized by using
melamine as both the carbon and nitrogen precursor, without any post-
treatment. More importantly, such intact samples can be used to investigate
the intrinsic catalytic activity of NCNTs more clearly; the results indicated
that NCNTs indeed performed very well. Furthermore, Li *et al.* reported that
the concentration of KOH electrolyte also had a large impact on the ORR
performance of the NCNTs [65]. Higher concentration of KOH electrolyte
leads to more negative onset potential and lower current densities. For
example, when the concentration of KOH increased from 0.1 M to 12 M, the
diffusion-limiting current decreased over 100 times. This could be attributed
to the very low oxygen solubility in highly concentrated KOH electrolytes. In
addition, in 3 M and 6 M KOH electrolytes, NCNTs showed competitive
activity with commercial Pt/C catalyst for ORR in alkaline media, and much
better activity than the Ag/C catalyst [65].

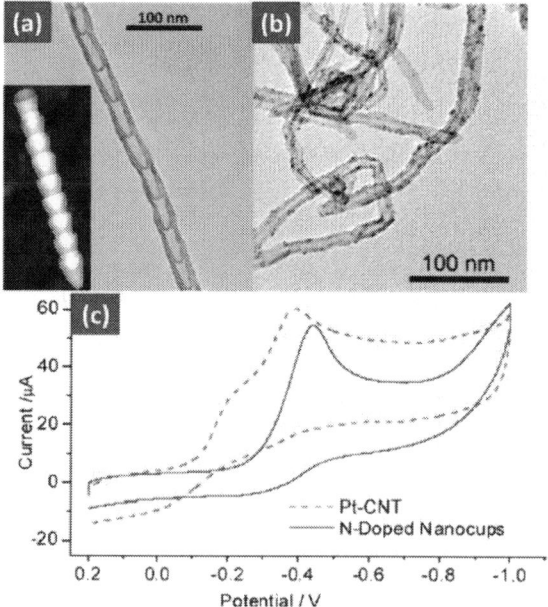

Figure 3. (a,b) TEM image of stacked NCNTs and commercial Pt-CNTs. Inset
in (**a**) is the scheme illustration of the nanocups in stacked NCNTs. (**c**) CV
curves of stacked NCNTs and commercial Pt-CNTs in 0.1 M KOH aqueous

solution saturated with O_2. Reprinted with permission from Ref. [72] Copyright © 2009, American Chemical Society.

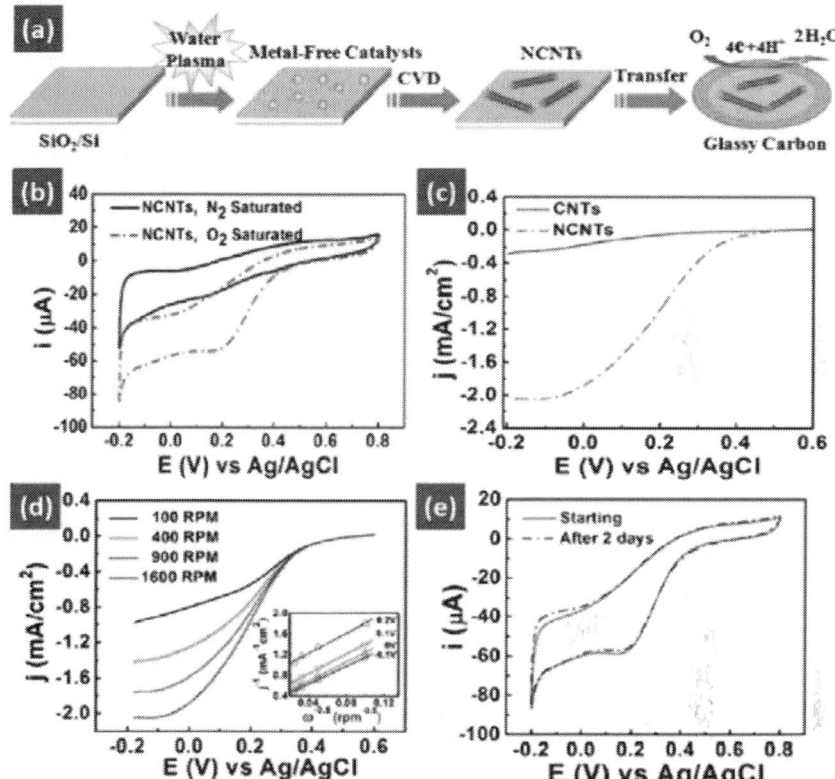

Figure 4. (a) Water-plasma-assisted CVD growth of NCNTs for the ORR; (b) CVs of the NCNTs, 50 mV/s in 0.5 M H_2SO_4 solution saturated with N_2 or O_2; (c) RDE curves of the NCNTs and CNTs in oxygen-saturated 0.5 M H_2SO_4; (d) RDE curves of the NCNT in oxygen-saturated 0.5 M H_2SO_4, inset: Koutecky-Levich plots of the NCNT derived from RDE measurements; (e) The two-day stability measurements of the NCNT by using continuous CV in oxygen-saturated 0.5 M H_2SO_4. Reprinted with permission from [82]. Copyright © 2010, American Chemical Society.

3.2. NCNTs as Catalyst Support Material for ORR

Using CNTs as catalyst supports have attracted significant interest because of their high surface area and excellent electrical conductivity. The N doping creates defects on the surface of pristine CNTs and breaks out its chemical inertness, while preserving its electrical conductivity [83]; moreover, NCNTs contain nitrogenized sites that are electrochemically active. Therefore, NCNTs were also used as excellent supports for catalyst nanoparticles. For instance, Vijayaraghavan *et al.* demonstrated that Pt nanoparticles/NCNTs exhibited enhanced catalytic activity and stability along with N-dopant contents [84]. Later, Sun's group demonstrated that uniform Pt nanoparticles with smaller size and better ORR activity than pure CNTs were obtained from NCNTs [85,86] (Figure 5). The authors also demonstrated that the catalyst stability increased with the increase of N contents in NCNTs [87]. To further take the merits of both carbon and ceramic-based supports for ORR, the Sun group employed the composite nanostructures of NCNTs coated with $TiSi_2O_x$ as Pt catalyst supports, and the results indicated that this composite showed better ORR performance than Pt/NCNT catalysts, thereby illustrating its promise as a catalyst for fuel cells [88]. Chen's group concluded that the NCNTs synthesized from an N-rich precursor solution (ethylenediamine) exhibited superior catalytic activity toward ORR compared with NCNTs grown from a precursor solution with relatively low N content pyridine [89].

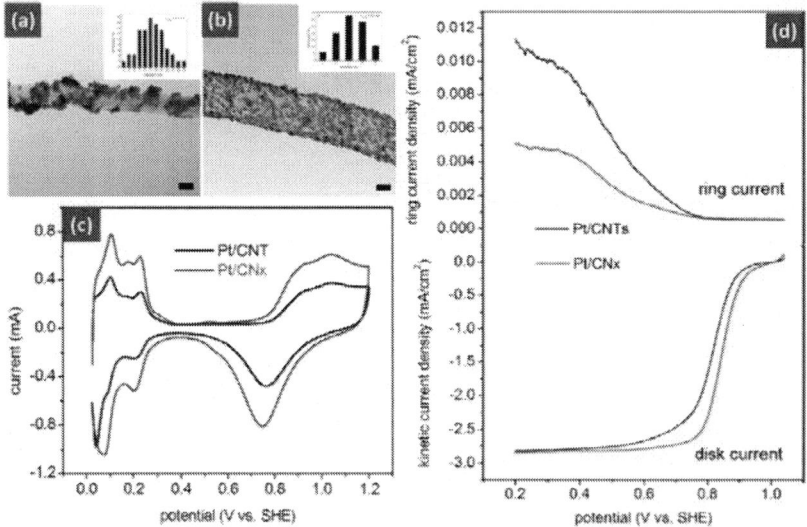

Figure 5. (**a,b**) TEM images and size distribution of Pt/CNTs (**a**) and Pt/CN$_x$(**b**) (scale bars are 20 nm); (**c**) CVs of Pt/CNTs an Pt/CNx 0.5 M H$_2$SO$_4$ with saturated Ar at 50 mV/s; (**d**) RRDE results of Pt/CNTs and Pt/CN$_x$ in 0.5 M H$_2$SO$_4$ saturated with O$_2$ at 5 mV/s at the rotation speed of 1600 rpm at room temperature. Reprinted with permission from [85]. Copyright © 2011, American Chemical Society.

4. NITROGEN-DOPED GRAPHENE (NG) FOR ORR

As discussed above, NCNTs could act as efficient and effective metal-free catalysts for ORR. Carbon atoms adjacent to nitrogen dopants could create a net positive charge density in order to counterbalance the strong electronic affinity of the N atom [6]. Hence the doping of the N atom could readily attract electrons to facilitate the ORR. Similar to NCNTs, coupled with the recent popularity of graphene, NG is also considered an appealing candidate for the applications in ORR where the NCNTs have already been exploited significantly.

4.1. NG as a Metal-Free Catalyst for ORR

Compared with NCNTs, NG has a large surface area and outstanding electrical conductivity; moreover, it also has the unique graphitic basal plane structure that could further facilitate electron transport and supply more active sites.

In 2010, Qu *et al.* first reported the application of NG as catalysts for the ORR [90]. As shown in Figure 6, a free-standing NG film of 4 cm^2 in size consisting of only a few layer sheets was obtained by the CVD method, using gas mixtures of NH_3, CH_4, H_2 and Ar on the Ni catalyst surface. The N content in the as-synthesized NG was *ca.*4 at.%. The RRDE voltammograms measurements were conducted, in alkaline electrolyte, to investigate the catalytic properties of NG, graphene and Pt/C for ORR. From Figure 6b, it can be seen that the graphene electrode showed a 2 e⁻ process for ORR with an onset potential of around −0.45 V. After doping with N, the NG electrode exhibited a one-step, 4 e⁻ pathway for ORR.

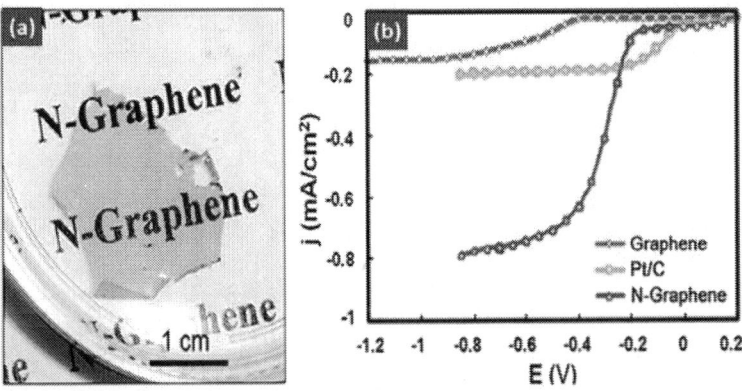

Figure 6. (a) An optical photograph of NG film floating on water; (**b**) LV curves in 0.1 M KOH saturated with air of different samples. Reprinted with permission from [90]. Copyright © 2010, American Chemical Society.

Calculated by the Koutecky-Levich equation, the transferred electron number per O_2 molecule of the NG was 3.6–4. It was found that the steady-state catalytic current density of the NG electrode was three times higher than the commercial Pt/C electrode. Similar to NCNTs, NG has excellent durability and good selectivity for ORR. The accelerated degradation test (ADT), which was carried out by CV in O_2-saturated electrolyte, is used to estimate the stability of the catalyst. In previous work, the graphene showed obviously more stable catalytic performance than Pt/C. Almost no significant loss in the voltammetric charge was observed after even a 100,000-cycle stability test [91]. Another advantage of NG compared to Pt for ORR is that ORR on NG is not greatly affected by methanol [59,90] and CO [90,92]. For instance, a 40% decrease was observed at the Pt/C electrode on the introduction of 2% (w/w) methanol [90], whereas the NG electrode remained unaffected under the identical condition. The high selectivity of NG toward ORR makes it very attractive for implementation in different kinds of fuel cells.

Based on these results, numerous research studies have been conducted on NG for ORR. Some of the typical works are summarized in Table 1. It is notable that the half-wave potential and onset potential for ORR are important criteria for evaluating the activity of an electrocatalyst, and the number of the electron transfer is determined from RRDE measurements to show that whether the electron transfer mechanism is a $2e^-$ dominated process or $4e^-$ dominated process.

In spite of extensive studies, the explanations on the exact catalytic mechanisms of NG (e.g., wherein the N configuration (pyridinic N or graphitic N) is more important for the ORR activity) or even the active sites are still controversial [94,118]. In Sun et al.'s research [55], they found that NG containing 0.3892% quaternary N (the highest N content in three samples) showed the best ORR activity and the relationship between ORR activity and graphitic N contents matched very well. It revealed that graphitic type N plays the vital role for ORR activity. Luo et al. [49] synthesized the graphene layers doped with nearly 100% pyridinic N through the pyrolysis of methane (CH_4) and NH_3 on Cu substrate, and the as-synthesized pyridinic N-doped graphene mainly exhibited a $2e^-$ transfer process for ORR, indicating that pyridinic N may not, as previously expected, effectively promote the $4e^-$ ORR performance of carbon materials.

Table 1. Summary of some typical work dedicated to NG as a metal-free catalyst for ORR.

Synthesis Method and Reactants	N-Content (at.%)	Electrocatalytic Performance	Electron Transfer Number	Ref.
Thermal treatment of glucose and urea	33	NG (25 at.%) shows competitive ORR activities with Pt/C and much better crossover resistance and excellent stability	3.2–3.7	[19]
CVD (C source, ethylene; N source, ammonia, Cu)	up to 16	Higher onset potential as compared to Pt/C	close to 2	[49]
Thermal treatment of GO using melamine	10.1	Much higher ORR activity than graphene	3.4–3.6	[57]
N plasma treatment on graphene	8.5	Higher ORR activity than graphene, and higher durability and selectivity than Pt/C	-	[59]
CVD (C source, methane; N source, ammonia, Cu)	4	Higher activity, better stability and tolerance to crossover than Pt	3.6–4	[90]
Detonation technique with cyanuric chloride and trinitrophenol	12.5	Comparable to that of Pt, more stable and less expensive	3.69	[91]
A resin-based methodology with N-containing resin and metal ions	1.8	The onset potential on the NG electrode is close to that of Pt/C. The current is almost the same for both the Pt/C and NG	2.1–3.9	[92]
Hydrothermal reaction of GO with urea	6.05–7.65	The performance of these NG materials towards ORR is still not as good as that of Pt/C in terms of the half-wave potential and current density	~3	[93]
Covalent functionalize GO using organic molecules and thermal treatment	0.72–4.3	The NG nanosheet exhibited a good electrocatalytic activity through an efficient one-step, 4e⁻ pathway	3.63	[94]
CVD of N-containing aromatic precursor molecules	2.0–2.7	The N dopants in the graphene reduce the ORR overpotential, thereby enhancing the catalytic activity	3.5–4.0	[95]
GO treatment by ammonia hydroxide, heating under ammonia gas, and reaction with melamine	6.0–6.8	Pyridinic N plays a vital role in ORR	3.2–3.7	[96]
Annealing of GO with ammonia and N-containing polymers	2.91–7.56	The higher limiting current density compared to Pt	2.85–3.65	[97]
Thermal reaction between GO and NH₃	2.4–4.6	The onset potential is close to that of Pt/C	~3.8	[98]
Hydrothermal reaction with GO and melamine	26.08	It shows lower ORR activity than Pt/C 40 wt.%	3.2–4.0	[99]
Hydrothermal process using urea and holey GO	8.6	Superb ORR with 4e⁻ pathway and excellent durability	3.85	[100]
Thermally annealing GO with melamine	8.05	The nG-900 exhibits lower activity and onset potential than Pt/C, albeit higher than graphene; excellent stability	3.3–3.7	[101]

On the contrary, in the work of Sheng [57], the NG mainly containing pyridine-like N atoms was obtained by the heat-treatment of GO in the presence of melamine. Since the electrocatalytic activity of the NGs toward

ORR is independent of N-doping level, it may indicate that the pyridine-like N in NGs determines its ORR activity. Pyridinic N, which has a lone electron pair in the plane of the carbon matrix, could donate the electron to the π-bond, attract electrons, and therefore be catalytically active. Some results were shown in many previous works [94,95,96].

In the research of Ruoff's group [97], NG with different N-doping formats was prepared by annealing GO together with different N-containing precursors, such as ammonia and N-containing polymers. It was prone to generate graphitic N and pyridinic N when annealing GO with ammonia, while it tended to form pyridinic and pyrrolic N species when annealing GO with polyaniline or polypyrrole. They found that the total atomic content of N rarely affects the ORR activity under alkaline conditions. Actually, the graphitic N-dominated catalysts exhibit higher catalytic activity and larger limiting current density than that of pyrrolic or pyridinic N-dominated catalysts. However, the pyridinic N could enhance the ORR onset potential and gradually convert the 2e$^-$ dominated pass-way to the 4e$^-$ dominated process. Also, some researchers [119,120,121,122] used the periodic DFT to simulate the ORR at the edge of NG. For example, by taking into account the experimental conditions, i.e., the surface coverage, the water effect, the bias effect and pH, Yu et al. [119] presented a systematic theoretical study on the full reaction path of ORR on NG. They concluded that the rate-determining step is the O(ads) removal from the NG surface. From another perspective, by calculating energy variations during each reaction step using DFT, Zhang and Xia [120] demonstrated that the electrocatalytic activity of NG is related to the atomic charge density distribution and electron spin density The reasons for why NG has catalytic capability (while pristine graphene does not) have also been discussed. From Kim et al.'s results, [121] doping of N in graphene could promote the oxygen adsorption, the first electron transfer, and the selectivity toward the 4e$^-$ reduction pathway. More specifically, they suggested that the outermost graphitic N sites are the main active sites. Meanwhile, they also proposed that the graphitic N site which involves a ring-opening of the cyclic C-N bond at the edge of graphene could result in the pyridinic N, thus, the inter-converts conversion mechanism between pyridinic and graphitic types during the catalytic cycle may reconcile the experimental controversy about what types of N are the ORR active sites for N-doped carbon materials [121].

Besides the doped N species, the morphology of NG also plays a significant role for the ORR properties. During the doping process of graphene, the stacking of graphene sheets is inclined to increase the diffusion resistance of reactants/electrolytes, reduce the specific area, and the exposed active sites. It is thus worth controlling the structure of NG to get more ORR activity. In this regard, there is a great deal of work on the production of N-doped holey graphene [99,100]. For instance, a 3D porous nanostructure which has N-doped holes on individual graphene sheets was synthesized through a hydrothermal process using urea and holey GO by Yu *et al.* [100]. Benefiting from the 3D porous nanostructure, abundant exposed sites, and high-level N doping, the as-prepared material exhibited excellent ORR performance, such as the high limiting current, strong resistance to the methanol crossover, which are competitive with the commercial 20 wt.% Pt/C catalysts.

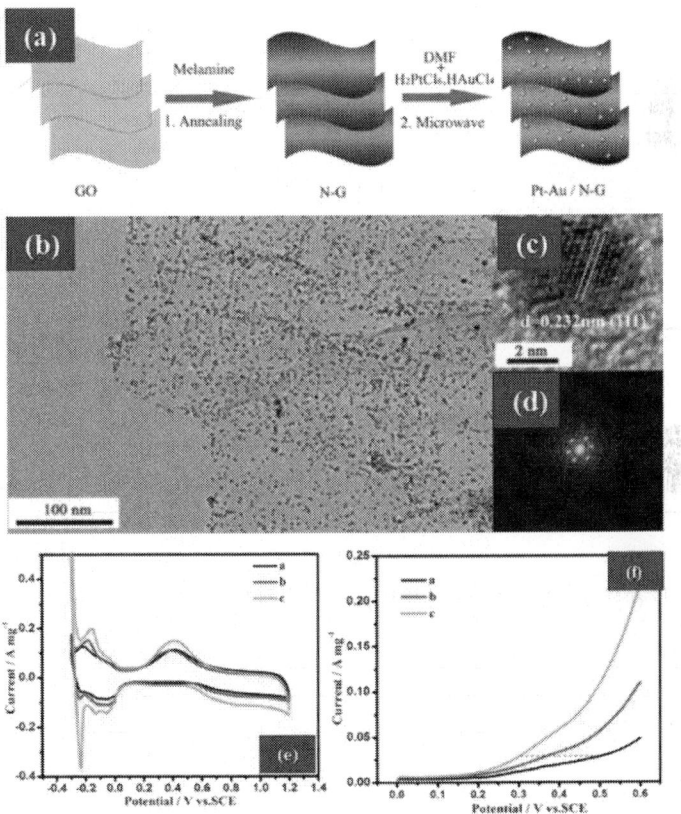

Figure 7. (a) Fabrication of the Pt-Au alloy NPs on the NG sheets; **(b)** TEM of Pt$_3$Au/N-G; **(c)** HRTEM and **(d)** FFTs of a single Pt$_3$Au NP on NG; **(e)** CVs and **(f)** LSV of Pt/C (a, black), Pt$_3$Au/G(b, red) and Pt$_3$Au/N-G catalysts (c, green). Reprinted with permission from [127]. Copyright © 2012, Royal Society of Chemistry.

4.2. NG as Support Material for ORR

The incorporation of N atoms within graphene sheets could contribute more functional groups, higher electron-mobility, and more active sites for catalytic reactions. Also, it is beneficial for facilitating the distribution and uniformity of metal nanoparticles. Moreover, when NG acts as the support, it could enhance the catalytic properties due to the interaction between graphene and metal nanoparticles. Consequently, NG materials have been regarded as one very promising metal catalyst support [123,124,125,126].

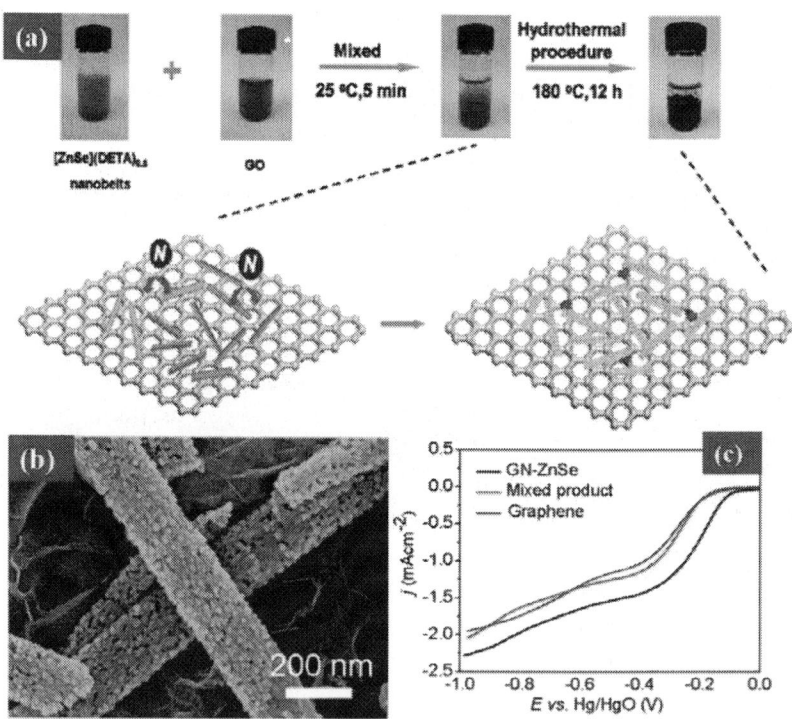

Figure 8. (a) Schematic preparation of NG-ZnSe nanocomposites (blue rods-[ZnSe](DETA)$_{0.5}$ nanobelts; orange rods-ZnSe nanorods; purple balls-N; gray balls-C); **(b)** SEM photograph of ZnSe/NG; **(c)** LV curves in 1.0 M KOH solution with saturated O$_2$ of different electrodes. Reprinted with permission from Ref. [131]. Copyright © 2012, American Chemical Society. Note: in the original paper, the authors refer to "nitrogen-doped graphene" as "GN"; here in this review, for consistency, we named it "NG."

Typically, NG is proposed to be able to stabilize the noble metal nanoparticles, and improve the durability of the catalysts. Moreover, nitrogen doping could introduce active sites for catalytic reactions and also act as anchoring sites for metal nanoparticle deposition. Yang *et al.* fabricated a composite of Pt-Au alloy nanoparticles on NG sheets by a wet-chemistry method [127]. As shown in Figure 7, the NG was synthesized by thermal treatment of GO powder and melamine. Then the solutions of H_2PtCl_6, $HAuCl_4$, NG in DMF and water underwent the microwave irradiation. The as-prepared Pt_3Au-NPs were found to be well dispersed on the NG sheets (Figure 7b) and the HRTEM image in Figure 7c revealed the lattice fringes of the NPs have an interplanar spacing of 0.232 nm. The fast Fourier transforms (FFTs) shown in Figure 7d indicated the single crystallite nature of the Pt_3Au/NG on (111) plane. Figure 7e,f showed that the corresponding potential for Pt_3Au/NG was much lower than the other two samples at a given oxidation current density. Improved electrocatalytict activity was observed due to the small size, uniform dispersion and a high electrochemical active surface area of the nanocomposites. Recently, more studies on NG- or N-rGO-supported Pt electrocatalysts have also been reported; all these results demonstrate the significant function of N doping in producing highly efficient ORR electrocatalysts [128,129,130].

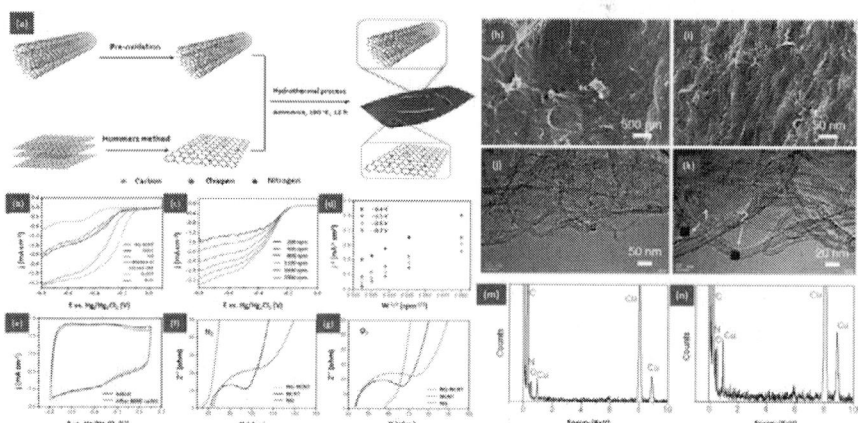

Figure 9. (a) Schematic preparation of the NG-NCNT nanocomposites; (b) LV curves in 0.1 M KOH solution with the rotation speed of 1600 rpm and sweep rate of 20 mV·s^{-1} in oxygen of different samples; (c) LV curves of NG-NCNT with different rotation speeds (sweep rate 20 mV·s^{-1}); (d) K-L plots (i^{-1} *vs.* $\omega^{-1/2}$) at different potentials (*vs.* Hg/Hg$_2$Cl$_2$); (e) CVs of GN-CNT after 8000 cycles with the sweep rate of 150 mV·s^{-1}; (f–g) Impedance data of different samples in 0.1 M KOH solution with saturated N$_2$ and O$_2$, respectively; (h–k) SEM and STEM images of the typical NG-NCNT nanocomposite; (m,n) Elemental analysis image

of the NG and NG-NCNT (the area marked with 1 and 2 in Figure (**k**) respectively. Reprinted with permission from [44]. Copyright © 2013 WILEY-VCH Verlag GmbH & Co. KGaA, Weinheim.

Additionally, it was predicted that non-precious-metal-NG hybrid materials would also lead to enhanced catalytic properties. For instance, Chen *et al.* reported a strategy to synthesize ZnSe/NG nanocomposites (NG-ZnSe) [131]. As shown in Figure 8, [ZnSe](DETA)$_{0.5}$ nanobelts were gradually put into the GO solution, and then the sediments were processed by hydrothermal treatment. As shown in Figure 8b, ZnSe nanorods, which were composed of ZnSe nanoparticles, were grown on a graphene surface. It can be seen from Figure 8c that the NG-ZnSe electrode exhibited higher positive onset potential and larger current for ORR. The improved performance can be attributed to the synergetic effects between NG and alloy nanostructures. There are also a number of similar reports using non-precious metal to produce metal/NG composites, showing potential applications [132,133,134,135,136].

5. THE COMPOSITES OF NCNTS AND NG FOR ORR

As a two-dimensional layer structure of sp^2-hybridized carbon, graphene has strong direction-dependent transport properties and is easily agglomerated and restacked to graphite; therefore, when used as a catalyst, it may result in declined activity. A combination of CNT and graphene may be an effective way to solve this problem [137,138]. Dai's group has demonstrated that CNT-graphene complexes can exhibit excellent activity and stability towards ORR in both acidic and basic electrolytes [139]. Furthermore, based on the STEM-HAADF and EELS mapping results, they speculated that the impurities of nitrogen and iron might be the reason for the excellent ORR properties. While, as illustrated in the previous sections, NCNTs and NG have shown excellent electrocatalytic performance for the ORR compared with pure CNTs or graphene. Therefore, , there have recently been efforts to hybridize these two carbon structures (NCNTs and NG) to obtain a synergy effect to further improve their catalytic performance [138,140]. For example, Ma *et al.* fabricated the 3D NCNTs/graphene composite through the pyrolysis of pyridine over the Ni catalyst supported on graphene sheet [140]. The N content in the NCNTs/NG composite was about 6.6 at.%, compared with the undoped CNTs/G; the doped sample showed higher catalytic activity and selectivity for ORR in the alkaline electrolyte. Another example of highly active N-doped G/CNT composite electrocatalyst for ORR is demonstrated by Ratso and coworkers [141]. N-doped few-layer G/CNT composite was fabricated by the pyrolysis of GO/MWCNT with urea and dicyandiamide. Based on the XPS and RDE results, they concluded that the enhanced electrocatalytic activity is due to a higher content of pyridinic N in the samples, and the higher limiting currents of oxygen reduction can be ascribed by the quaternary N. These results are attractive for alkaline fuel cells. However, these methods require high temperature pyrolysis, during which the morphological defects and structural degradation are probably shown up in the final products [17]. In this regard, Chen *et al.* synthesized NG-NCNT nanocomposite through a hydrothermal process at a much lower temperature (*i.e.*, 180 °C) (Figure 9a) [44]. The diameters of the nanotubes are in the range of 9–15 nm, and the atomic percentages of N content are 3.2 at.% and 1.3 at.% for graphene and CNTs, respectively, which confirm the existence of the N element in both graphene and CNTs. This NG-NCNT displayed a $4e^-$ pathway for ORR with more positive onset potential, large peak current, and good durability (Figure 9b–g). Very recently, however, a hybrid of NCNT and graphene prepared by plasma-enhanced CVD showed inferior ORR activity, [142] which is contradictory to the above-mentioned results. The reason for this discrepancy is still not clear, thus extensive and careful research in this area is still needed.

6. CONCLUSION AND PERSPECTIVES

ORR plays an essential role in energy-related areas, such as metal-air batteries and fuel cells, and traditionally, the Pt-based catalysts are regarded as the best choice for $4e^-$ ORR. Due to the prohibitive price and scarcity of Pt, the development of high performance and inexpensive metal-free and non-noble metal catalysts, to replace Pt, are highly desired, and it plays an important role in promoting the large-scale practical applications of these energy devices. Due to their outstanding properties, such as ultrahigh charge carrier mobility, gigantic thermal conductivity, extremely large surface area, exceptional mechanical strength and flexibility, CNTs and graphene have been extensively explored for ORR. The pristine CNTs and graphene mainly exhibit $2e^-$ pathway for ORR, while N doping has been proved to be a promising way to tailor their properties to promote $4e^-$ ORR which is much more meaningful for energy applications. For N doping in CNTs or graphene, there are mainly two strategies: the first method is the *in situ* doping where nitrogen can be doped into CNTs or graphene nanosheets during the growing process with the addition of proper carbon and nitrogen sources. The second one is the post-treatment process; in this method, CNTs or GO were firstly synthesized, then annealed at high temperatures together with the nitrogen-containing precursors. Despite much progress, it is still not easy to precisely control the N-doping sites and concentration. All of these characteristics affect the ORR properties of NCNTs and NG in the catalytic applications. Therefore, the development of new and more controllable doping methods is still highly desired. Through N doping, various properties, including the surface energy, work function, carrier concentration, and surface polarization, of CNTs and graphene could be tuned, so that NCNTs and NG have become the most promising metal-free catalysts toward $4e^-$ ORR. In general, three common bonding configurations, including graphitic, pyridinic, and pyrrolic N, are normally achieved when doping nitrogen into CNTs and graphene. Different doping strategies would significantly affect the N-doping levels and N types in NCNTs and NG. For example, the *in situ* doping normally generates pyridinic- and/or pyrrolic-N species, while the post-treatment doping is prone to form graphitic-N in carbon frameworks.

In the applications for ORR, from both theoretical and experimental perspectives, researchers have demonstrated that NCNTs and NG show remarkable electrocatalytic performance. In a theoretical context, through DFT simulations, it was shown that in NCNTs and NG, the carbon atoms with higher spin density usually possess more active sites. Through investigating the reaction mechanisms, it was proved that the removal of $O_{(ads)}$ on the surface of nitrogen-doped carbon determines the reaction rate. In the experimental part, the developments of both NCNTs and NG as metal-free ORR catalysts and as the metal catalyst support for ORR are summarized in detail in this review. All the N-doped carbon materials (NCNTs, NG) exhibit higher catalytic performance compared to their pristine counterparts (CNTs, graphene), indicating a great beneficial effect of N doping on the ORR performance. Moreover, the progresses on NCNTs- and NG-based composites for ORR have also been discussed in this review, demonstrating that it is also a very promising research direction for next-generation non-noble metal or metal-free ORR catalysts. Although much progress has been achieved in the area of NCNTs and NG for ORR catalysts, challenges still exist: (i) New and greener methods are required for the large-scale production of NCNTs and NG; (ii) The control of N doping at specific positions in CNTs and graphene is still lacking; (iii) A careful controlling of nitrogen sites, types and concentration is still highly desired; (iv) The deep understanding of oxygen adsorption and reduction on these NCNTs- and NG-based catalysts is still lacking, and therefore, systematic theoretical simulations are also needed, which may boost the developments of N-doping carbon materials for ORR in the future.

ACKNOWLEDGMENTS

We thank the support from Natural Science Foundations of China, the Natural Science Foundation of Shanxi Province, Fonds de Recherche du Québec-Natureet Technologies (FRQNT), the Natural Sciences and

Engineering Research Council of Canada (NSERC), Institut National de la Recherche Scientifique (INRS), and Centre Québécois sur les MateriauxFonctionnels (CQMF) and China Scholarship Council (CSC).

AUTHOR CONTRIBUTIONS

Qiliang Wei was the leading author from the initial draft writing to the finalization of the manuscript. Qiliang Wei and Xin Tong wrote the first draft of the manuscript. All authors contributed as a team to the manuscript plan, revisions, the literature reading, and the proof reading.

REFERENCES

1.   Dai, L.; Xue, Y.; Qu, L.; Choi, H.-J.; Baek, J.-B. Metal-free catalysts for oxygen reduction reaction. *Chem. Rev.***2015**, *115*, 4823–4892.

2.   Cheng, F.; Chen, J. Metal-air batteries: From oxygen reduction electrochemistry to cathode catalysts. *Chem. Soc. Rev.* **2012**, *41*, 2172–2192.

3.   Wang, D.-W.; Su, D. Heterogeneous nanocarbon materials for oxygen reduction reaction. *Energy Environ. Sci.* **2014**, *7*, 576–591.

4.   Feng, L.; Yan, Y.; Chen, Y.; Wang, L. Nitrogen-doped carbon nanotubes as efficient and durable metal-free cathodic catalysts for oxygen reduction in microbial fuel cells. *Energy Environ. Sci.* **2011**, *4*, 1892–1899.

5.   Suntivich, J.; Gasteiger, H.A.; Yabuuchi, N.; Nakanishi, H.; Goodenough, J.B.; Shao-Horn, Y. Design principles for oxygen-reduction activity on perovskite oxide catalysts for fuel cells and metal–air batteries.*Nat. Chem.* **2011**, *3*, 546–550.

6.   Gong, K.; Du, F.; Xia, Z.; Durstock, M.; Dai, L. Nitrogen-doped carbon nanotube arrays with high electrocatalytic activity for oxygen reduction. *Science* **2009**, *323*, 760–764.

7.   Wei, W.; Liang, H.; Parvez, K.; Zhuang, X.; Feng, X.; Müllen, K. Nitrogen-doped carbon nanosheets with size-defined mesopores as highly efficient metal-free catalyst for the oxygen reduction reaction. *Angew. Chem.* **2014**, *126*, 1596–1600.

8. Sun, D.M.; Liu, C.; Ren, W.C.; Cheng, H.M. A review of carbon nanotube-and graphene-based flexible thin-film transistors. *Small* **2013**, *9*, 1188–1205.

9. Yu, D.; Nagelli, E.; Du, F.; Dai, L. Metal-free carbon nanomaterials become more active than metal catalysts and last longer. *J. Phys. Chem. Lett.* **2010**, *1*, 2165–2173.

10. Chun, K.-Y.; Lee, H.S.; Lee, C.J. Nitrogen doping effects on the structure behavior and the field emission performance of double-walled carbon nanotubes. *Carbon* **2009**, *47*, 169–177.

11. Bostwick, A.; Speck, F.; Seyller, T.; Horn, K.; Polini, M.; Asgari, R.; MacDonald, A.H.; Rotenberg, E. Observation of plasmarons in quasi-freestanding doped graphene. *Science* **2010**, *328*, 999–1002.

12. Hwang, J.O.; Park, J.S.; Choi, D.S.; Kim, J.Y.; Lee, S.H.; Lee, K.E.; Kim, Y.-H.; Song, M.H.; Yoo, S.; Kim, S.O. Work function-tunable, N-doped reduced graphene transparent electrodes for high-performance polymer light-emitting diodes. *ACS Nano* **2011**, *6*, 159–167.

13. Czerw, R.; Terrones, M.; Charlier, J.-C.; Blase, X.; Foley, B.; Kamalakaran, R.; Grobert, N.; Terrones, H.; Tekleab, D.; Ajayan, P. Identification of electron donor states in N-doped carbon nanotubes. *Nano Lett.* **2001**, *1*, 457–460.

14. Lee, W.J.; Maiti, U.N.; Lee, J.M.; Lim, J.; Han, T.H.; Kim, S.O. Nitrogen-doped carbon nanotubes and graphene composite structures for energy and catalytic applications. *Chem. Commun.* **2014**, *50*, 6818–6830.

15. Liu, H.; Liu, Y.; Zhu, D. Chemical doping of graphene. *J. Mater. Chem.* **2011**, *21*, 3335–3345.

16. Yang, Z.; Nie, H.; Chen, X.A.; Chen, X.; Huang, S. Recent progress in doped carbon nanomaterials as effective cathode catalysts for fuel cell oxygen reduction reaction. *J. Power Sources* **2013**, *236*, 238–249.

17. Zheng, Y.; Jiao, Y.; Jaroniec, M.; Jin, Y.; Qiao, S.Z. Nanostructured metal-free electrochemical catalysts for highly efficient oxygen reduction. *Small* **2012**, *8*, 3550–3566.

18. Vazquez-Arenas, J.; Higgins, D.; Chen, Z.; Fowler, M.; Chen, Z. Mechanistic analysis of highly active nitrogen-doped carbon nanotubes for the oxygen reduction reaction. *J. Power Sources* **2012**, *205*, 215–221.

19. Zhang, Y.W.; Ge, J.; Wang, L.; Wang, D.H.; Ding, F.; Tao, X.M.; Chen, W. Manageable N-doped graphene for high performance oxygen reduction reaction. *Sci. Rep.* **2013**, *3*.

20. Wei, D.; Liu, Y.; Wang, Y.; Zhang, H.; Huang, L.; Yu, G. Synthesis of N-doped graphene by chemical vapor deposition and its electrical properties. *Nano Lett.* **2009**, *9*, 1752–1758.

21. Jung, S.H.; Kim, M.R.; Jeong, S.H.; Kim, S.U.; Lee, O.J.; Lee, K.H.; Suh, J.H.; Park, C.K. High-yield synthesis of multi-walled carbon nanotubes by arc discharge in liquid nitrogen. *Appl. Phys. A* **2003**, *76*, 285–286.

22. Sun, L.; Wang, C.; Zhou, Y.; Zhang, X.; Cai, B.; Qiu, J. Flowing nitrogen assisted-arc discharge synthesis of nitrogen-doped single-walled carbon nanohorns. *Appl. Surf. Sci.* **2013**, *277*, 88–93.

23. Mo, Z.; Liao, S.; Zheng, Y.; Fu, Z. Preparation of nitrogen-doped carbon nanotube arrays and their catalysis towards cathodic oxygen reduction in acidic and alkaline media. *Carbon* **2012**, *50*, 2620–2627.

24. Sharifi, T.; Nitze, F.; Barzegar, H.R.; Tai, C.-W.; Mazurkiewicz, M.; Malolepszy, A.; Stobinski, L.; Wågberg, T. Nitrogen doped multi walled carbon nanotubes produced by CVD-correlating xps and raman spectroscopy for the study of nitrogen inclusion. *Carbon* **2012**, *50*, 3535–3541.

25. Guo, Q.; Zhao, D.; Liu, S.; Chen, S.; Hanif, M.; Hou, H. Free-standing nitrogen-doped carbon nanotubes at electrospun carbon nanofibers composite as an efficient electrocatalyst for oxygen reduction. *Electrochim. Acta* **2014**, *138*, 318–324.

26. Tao, X.Y.; Zhang, X.B.; Sun, F.Y.; Cheng, J.P.; Liu, F.; Luo, Z.Q. Large-scale CVD synthesis of nitrogen-doped multi-walled carbon nanotubes with controllable nitrogen content on a $Co_xMg_{1-x}MoO_4$ catalyst.*Diamond Relat. Mater.* **2007**, *16*, 425–430.

27. She, X.; Yang, D.; Jing, D.; Yuan, F.; Yang, W.; Guo, L.; Che, Y. Nitrogen-doped one-dimensional (1D) macroporous carbonaceous nanotube arrays and their application in electrocatalytic oxygen reduction reactions. *Nanoscale* **2014**, *6*, 11057–11061.

28. Chen, L.; Xia, K.; Huang, L.; Li, L.; Pei, L.; Fei, S. Facile synthesis and hydrogen storage application of nitrogen-doped carbon nanotubes with bamboo-like structure. *Int. J. Hydrogen Energy* **2013**, *38*, 3297–3303.

29. Shi, W.; Venkatachalam, K.; Gavalas, V.; Qian, D.; Andrews, R.; Bachas, L.G.; Chopra, N. The role of plasma treatment on electrochemical capacitance of undoped and nitrogen doped carbon nanotubes.*Nanomater. Energy* **2013**, *2*, 71–81.

30. Du, Z.; Wang, S.; Kong, C.; Deng, Q.; Wang, G.; Liang, C.; Tang, H. Microwave plasma synthesized nitrogen-doped carbon nanotubes for oxygen reduction. *J. Solid State Electrochem.* **2015**, *19*, 1541–1549.

31. Magrez, A.; Seo, J.W.; Smajda, R.; Mionić, M.; Forró, L. Catalytic CVD synthesis of carbon nanotubes: Towards high yield and low temperature growth. *Materials* **2010**, *3*, 4871–4891.

32. Donato, M.G.; Galvagno, S.; Lanza, M.; Messina, G.; Milone, C.; Piperopoulos, E.; Pistone, A.; Santangelo, S. Influence of carbon source and Fe-catalyst support on the growth of multi-walled carbon nanotubes. *J. Nanosci. Nanotechnol.* **2009**, *9*, 3815–3823.

33. Li, J.; Papadopoulos, C.; Xu, J.M.; Moskovits, M. Highly-ordered carbon nanotube arrays for electronics applications. *Appl. Phys. Lett.* **1999**, *75*, 367–369.

34. Wang, Y.; Cui, X.; Li, Y.; Chen, L.; Chen, H.; Zhang, L.; Shi, J. A co-pyrolysis route to synthesize nitrogen doped multiwall carbon nanotubes for oxygen reduction reaction. *Carbon* **2014**, *68*, 232–239.

35. Ayala, P.; Grüneis, A.; Gemming, T.; Grimm, D.; Kramberger, C.; Rümmeli, M.H.; Freire, F.L.; Kuzmany, H.; Pfeiffer, R.; Barreiro, A. Tailoring N-doped single and double wall carbon nanotubes from a nondiluted carbon/nitrogen feedstock. *J. Phys. Chem. C* **2007**, *111*, 2879–2884.

36. Tang, C.; Golberg, D.; Bando, Y.; Xu, F.; Liu, B. Synthesis and field emission of carbon nanotubular fibers doped with high nitrogen content. *Chem. Commun.* **2003**, 3050–3051.

37. Rao, C.V.; Cabrera, C.R.; Ishikawa, Y. In search of the active site in nitrogen-doped carbon nanotube electrodes for the oxygen reduction reaction. *J. Phys. Chem. Lett.* **2010**, *1*, 2622–2627.

38. Rao, C.V.; Ishikawa, Y. Activity, selectivity, and anion-exchange membrane fuel cell performance of virtually metal-free nitrogen-doped carbon nanotube electrodes for oxygen reduction reaction. *J. Phys. Chem. C* **2012**, *116*, 4340–4346.

39. Guo, Q.; Xie, Y.; Wang, X.; Zhang, S.; Hou, T.; Lv, S. Synthesis of carbon nitride nanotubes with the C_3N_4 stoichiometry via a benzene-thermal process at low temperatures. *Chem. Commun.* **2004**, 26–27.

40. Cao, C.; Huang, F.; Cao, C.; Li, J.; Zhu, H. Synthesis of carbon nitride nanotubes via a catalytic-assembly solvothermal route. *Chem. Mater.* **2004**, *16*, 5213–5215.

41. Nagaiah, T.C.; Kundu, S.; Bron, M.; Muhler, M.; Schuhmann, W. Nitrogen-doped carbon nanotubes as a cathode catalyst for the oxygen reduction reaction in alkaline medium. *Electrochem. Commun.* **2010**, *12*, 338–341.

42. Chan, L.H.; Hong, K.H.; Xiao, D.Q.; Lin, T.C.; Lai, S.H.; Hsieh, W.J.; Shih, H.C. Resolution of the binding configuration in nitrogen-doped carbon nanotubes. *Phys. Rev. B* **2004**, *70*, 125408.

43. Vikkisk, M.; Kruusenberg, I.; Ratso, S.; Joost, U.; Shulga, E.; Kink, I.; Rauwel, P.; Tammeveski, K. Enhanced electrocatalytic activity of nitrogen-doped multi-walled carbon nanotubes towards the oxygen reduction reaction in alkaline media. *RSC Adv.* **2015**, *5*, 59495–59505.

44. Chen, P.; Xiao, T.Y.; Qian, Y.H.; Li, S.S.; Yu, S.H. A nitrogen-doped graphene/carbon nanotube nanocomposite with synergistically enhanced electrochemical activity. *Adv. Mater.* **2013**, *25*, 3192–3196.

45. Sharifi, T.; Hu, G.; Jia, X.; Wågberg, T. Formation of active sites for oxygen reduction reactions by transformation of nitrogen functionalities in nitrogen-doped carbon nanotubes. *ACS Nano* **2012**, *6*, 8904–8912.

46. Wood, K.N.; O'Hayre, R.; Pylypenko, S. Recent progress on nitrogen/carbon structures designed for use in energy and sustainability applications. *Energy Environ. Sci.* **2014**, *7*, 1212–1249.

47. Reddy, A.L.M.; Srivastava, A.; Gowda, S.R.; Gullapalli, H.; Dubey, M.; Ajayan, P.M. Synthesis of nitrogen-doped graphene films for lithium battery application. *ACS Nano* **2010**, *4*, 6337–6342.

48. Jin, Z.; Yao, J.; Kittrell, C.; Tour, J.M. Large-scale growth and characterizations of nitrogen-doped monolayer graphene sheets. *ACS Nano* **2011**, *5*, 4112–4117.

49. Luo, Z.; Lim, S.; Tian, Z.; Shang, J.; Lai, L.; MacDonald, B.; Fu, C.; Shen, Z.; Yu, T.; Lin, J. Pyridinic N doped graphene: Synthesis,

electronic structure, and electrocatalytic property. *J. Mater. Chem.* **2011**, *21*, 8038–8044.

50. Deng, D.; Pan, X.; Yu, L.; Cui, Y.; Jiang, Y.; Qi, J.; Li, W.-X.; Fu, Q.; Ma, X.; Xue, Q.; *et al.* Toward N-doped graphene via solvothermal synthesis. *Chem. Mater.* **2011**, *23*, 1188–1193.

51. Ghosh, A.; Late, D.J.; Panchakarla, L.S.; Govindaraj, A.; Rao, C.N.R. NO_2 and humidity sensing characteristics of few-layer graphenes. *J. Exp. Nanosci.* **2009**, *4*, 313–322.

52. Panchakarla, L.S.; Subrahmanyam, K.S.; Saha, S.K.; Govindaraj, A.; Krishnamurthy, H.R.; Waghmare, U.V.; Rao, C.N.R. Synthesis, structure, and properties of boron-and nitrogen-doped graphene. *Adv. Mater.* **2009**,*21*, 4726–4730.

53. Subrahmanyam, K.S.; Panchakarla, L.S.; Govindaraj, A.; Rao, C.N.R. Simple method of preparing graphene flakes by an arc-discharge method. *J. Phys. Chem. C* **2009**, *113*, 4257–4259.

54. Geng, D.; Chen, Y.; Chen, Y.; Li, Y.; Li, R.; Sun, X.; Ye, S.; Knights, S. High oxygen-reduction activity and durability of nitrogen-doped graphene. *Energy Environ. Sci.* **2011**, *4*, 760–764.

55. Wang, X.; Li, X.; Zhang, L.; Yoon, Y.; Weber, P.K.; Wang, H.; Guo, J.; Dai, H. N-doping of graphene through electrothermal reactions with ammonia. *Science* **2009**, *324*, 768–771.

56. Li, X.; Wang, H.; Robinson, J.T.; Sanchez, H.; Diankov, G.; Dai, H. Simultaneous nitrogen doping and reduction of graphene oxide. *J. Am. Chem. Soc.* **2009**, *131*, 15939–15944.

57. Sheng, Z.-H.; Shao, L.; Chen, J.-J.; Bao, W.-J.; Wang, F.-B.; Xia, X.-H. Catalyst-free synthesis of nitrogen-doped graphene via thermal annealing graphite oxide with melamine and its excellent electrocatalysis.*ACS Nano* **2011**, *5*, 4350–4358.

58. Wang, H.; Maiyalagan, T.; Wang, X. Review on recent progress in nitrogen-doped graphene: Synthesis, characterization, and its potential applications. *ACS Catal.* **2012**, *2*, 781–794.

59. Shao, Y.; Zhang, S.; Engelhard, M.H.; Li, G.; Shao, G.; Wang, Y.; Liu, J.; Aksay, I.A.; Lin, Y. Nitrogen-doped graphene and its electrochemical applications. *J. Mater. Chem.* **2010**, *20*, 7491–7496.

60. Guo, B.; Liu, Q.; Chen, E.; Zhu, H.; Fang, L.; Gong, J.R. Controllable N-doping of graphene. *Nano Lett.* **2010**,*10*, 4975–4980.

61. Long, D.; Li, W.; Ling, L.; Miyawaki, J.; Mochida, I.; Yoon, S.-H. Preparation of nitrogen-doped graphene sheets by a combined chemical and hydrothermal reduction of graphene oxide. *Langmuir* **2010**, *26*, 16096–16102.

62. Sun, L.; Wang, L.; Tian, C.; Tan, T.; Xie, Y.; Shi, K.; Li, M.; Fu, H. Nitrogen-doped graphene with high nitrogen level via a one-step hydrothermal reaction of graphene oxide with urea for superior capacitive energy storage. *RSC Adv.* **2012**, *2*, 4498–4506.

63. Kundu, S.; Nagaiah, T.C.; Xia, W.; Wang, Y.; Dommele, S.V.; Bitter, J.H.; Santa, M.; Grundmeier, G.; Bron, M.; Schuhmann, W. Electrocatalytic activity and stability of nitrogen-containing carbon nanotubes in the oxygen reduction reaction. *J. Phys. Chem. C* **2009**, *113*, 14302–14310.

64. Wang, Z.; Jia, R.; Zheng, J.; Zhao, J.; Li, L.; Song, J.; Zhu, Z. Nitrogen-promoted self-assembly of N-doped carbon nanotubes and their intrinsic catalysis for oxygen reduction in fuel cells. *ACS Nano* **2011**, *5*, 1677–1684.

65. Li, H.; Liu, H.; Jong, Z.; Qu, W.; Geng, D.; Sun, X.; Wang, H. Nitrogen-doped carbon nanotubes with high activity for oxygen reduction in alkaline media. *Int. J. Hydrogen Energy* **2011**, *36*, 2258–2265.]

66. Qiu, Y.; Yin, J.; Hou, H.; Yu, J.; Zuo, X. Preparation of nitrogen-doped carbon submicrotubes by coaxial electrospinning and their electrocatalytic activity for oxygen reduction reaction in acid media. *Electrochim. Acta* **2013**, *96*, 225–229.

67. Wiggins-Camacho, J.D.; Stevenson, K.J. Mechanistic discussion of the oxygen reduction reaction at nitrogen-doped carbon nanotubes. *J. Phys. Chem. C* **2011**, *115*, 20002–20010.

68. Okamoto, Y. First-principles molecular dynamics simulation of O_2 reduction on nitrogen-doped carbon. *Appl. Surf. Sci.* **2009**, *256*, 335–341.

69. Hu, X.; Wu, Y.; Li, H.; Zhang, Z. Adsorption and activation of O_2 on nitrogen-doped carbon nanotubes. *J. Phys. Chem. C* **2010**, *114*, 9603–9607.

70. Barone, V.; Peralta, J.E.; Wert, M.; Heyd, J.; Scuseria, G.E. Density functional theory study of optical transitions in semiconducting single-walled carbon nanotubes. *Nano Lett.* **2005**, *5*, 1621–1624.

71. Nikawa, H.; Yamada, T.; Cao, B.; Mizorogi, N.; Slanina, Z.; Tsuchiya, T.; Akasaka, T.; Yoza, K.; Nagase, S. Missing metallofullerene with C80 cage. *J. Am. Chem. Soc.* **2009**, *131*, 10950–10954.

72. Tang, Y.; Allen, B.L.; Kauffman, D.R.; Star, A. Electrocatalytic activity of nitrogen-doped carbon nanotube cups. *J. Am. Chem. Soc.* **2009**, *131*, 13200–13201.

73. Yang, M.; Yang, D.; Chen, H.; Gao, Y.; Li, H. Nitrogen-doped carbon nanotubes as catalysts for the oxygen reduction reaction in alkaline medium. *J. Power Sources* **2015**, *279*, 28–35.

74. Chen, Z.; Higgins, D.; Tao, H.; Hsu, R.S.; Chen, Z. Highly active nitrogen-doped carbon nanotubes for oxygen reduction reaction in fuel cell applications. *J. Phys. Chem. C* **2009**, *113*, 21008–21013.

75. Chen, Z.; Higgins, D.; Chen, Z. Nitrogen doped carbon nanotubes and their impact on the oxygen reduction reaction in fuel cells. *Carbon* **2010**, *48*, 3057–3065.

76. Higgins, D.; Chen, Z.; Chen, Z. Nitrogen doped carbon nanotubes synthesized from aliphatic diamines for oxygen reduction reaction. *Electrochim. Acta* **2011**, *56*, 1570–1575.

77. Chen, Z.; Higgins, D.; Chen, Z. Electrocatalytic activity of nitrogen doped carbon nanotubes with different morphologies for oxygen reduction reaction. *Electrochim. Acta* **2010**, *55*, 4799–4804.

78. Geng, D.; Liu, H.; Chen, Y.; Li, R.; Sun, X.; Ye, S.; Knights, S. Non-noble metal oxygen reduction electrocatalysts based on carbon nanotubes with controlled nitrogen contents. *J. Power Sources* **2011**, *196*, 1795–1801.

79. Borghei, M.; Kanninen, P.; Lundahl, M.; Susi, T.; Sainio, J.; Anoshkin, I.; Nasibulin, A.; Kallio, T.; Tammeveski, K.; Kauppinen, E. High oxygen reduction activity of few-walled carbon nanotubes with low nitrogen content. *Appl. Catal. B* **2014**, *158*, 233–241.

80. Tuci, G.; Zafferoni, C.; D'Ambrosio, P.; Caporali, S.; Ceppatelli, M.; Rossin, A.; Tsoufis, T.; Innocenti, M.; Giambastiani, G. Tailoring carbon nanotube N-dopants while designing metal-free electrocatalysts for the oxygen reduction reaction in alkaline medium. *ACS Catal.* **2013**, *3*, 2108–2111.

81. Tuci, G.; Zafferoni, C.; Rossin, A.; Milella, A.; Luconi, L.; Innocenti, M.; Truong Phuoc, L.; Duong-Viet, C.; Pham-Huu, C.; Giambastiani, G. Chemically functionalized carbon nanotubes with pyridine groups

as easily tunable N-decorated nanomaterials for the oxygen reduction reaction in alkaline medium. *Chem. Mater.* **2014**, *26*, 3460–3470.

82. Yu, D.; Zhang, Q.; Dai, L. Highly efficient metal-free growth of nitrogen-doped single-walled carbon nanotubes on plasma-etched substrates for oxygen reduction. *J. Am. Chem. Soc.* **2010**, *132*, 15127–15129.

83. Sun, S.; Zhang, G.; Zhong, Y.; Liu, H.; Li, R.; Zhou, X.; Sun, X. Ultrathin single crystal Pt nanowires grown on N-doped carbon nanotubes. *Chem. Commun.* **2009**, 7048–7050.

84. Vijayaraghavan, G.; Stevenson, K.J. Synergistic assembly of dendrimer-templated platinum catalysts on nitrogen-doped carbon nanotube electrodes for oxygen reduction. *Langmuir* **2007**, *23*, 5279–5282.

85. Chen, Y.; Wang, J.; Liu, H.; Banis, M.N.; Li, R.; Sun, X.; Sham, T.-K.; Ye, S.; Knights, S. Nitrogen doping effects on carbon nanotubes and the origin of the enhanced electrocatalytic activity of supported Pt for proton-exchange membrane fuel cells. *J. Phys. Chem. C* **2011**, *115*, 3769–3776.

86. Saha, M.S.; Li, R.; Sun, X.; Ye, S. 3-d composite electrodes for high performance pem fuel cells composed of Pt supported on nitrogen-doped carbon nanotubes grown on carbon paper. *Electrochem. Commun.* **2009**, *11*, 438–441.

87. Chen, Y.; Wang, J.; Liu, H.; Li, R.; Sun, X.; Ye, S.; Knights, S. Enhanced stability of Pt electrocatalysts by nitrogen doping in CNTs for PEM fuel cells. *Electrochem. Commun.* **2009**, *11*, 2071–2076.

88. Banis, M.N.; Sun, S.; Meng, X.; Zhang, Y.; Wang, Z.; Li, R.; Cai, M.; Sham, T.-K.; Sun, X. $TiSi_2O_x$ coated N-doped carbon nanotubes as Pt catalyst support for the oxygen reduction reaction in PEMFCs. *J. Phys. Chem. C* **2013**, *117*, 15457–15467.

89. Higgins, D.C.; Meza, D.; Chen, Z. Nitrogen-doped carbon nanotubes as platinum catalyst supports for oxygen reduction reaction in proton exchange membrane fuel cells. *J. Phys. Chem. C* **2010**, *114*, 21982–21988.

90. Qu, L.; Liu, Y.; Baek, J.-B.; Dai, L. Nitrogen-doped graphene as efficient metal-free electrocatalyst for oxygen reduction in fuel cells. *ACS Nano* **2010**, *4*, 1321–1326.

91. Feng, L.Y.; Chen, Y.G.; Chen, L. Easy-to-operate and low-temperature synthesis of gram-scale nitrogen-doped graphene and its application as cathode catalyst in microbial fuel cells. *ACS Nano* **2011**, *5*, 9611–9618.

92. He, C.Y.; Li, Z.S.; Cai, M.L.; Cai, M.; Wang, J.Q.; Tian, Z.Q.; Zhang, X.; Shen, P.K. A strategy for mass production of self-assembled nitrogen-doped graphene as catalytic materials. *J. Mater. Chem. A* **2013**, *1*, 1401–1406.

93. Wu, J.J.; Zhang, D.; Wang, Y.; Hou, B.R. Electrocatalytic activity of nitrogen-doped graphene synthesized via a one-pot hydrothermal process towards oxygen reduction reaction. *J. Power Sources* **2013**, *227*, 185–190.

94. Park, M.; Lee, T.; Kim, B.-S. Covalent functionalization based heteroatom doped graphene nanosheet as a metal-free electrocatalyst for oxygen reduction reaction. *Nanoscale* **2013**, *5*, 12255–12260.

95. Yasuda, S.; Yu, L.; Kim, J.; Murakoshi, K. Selective nitrogen doping in graphene for oxygen reduction reactions. *Chem. Commun.* **2013**, *49*, 9627–9629.

96. Xing, T.; Zheng, Y.; Li, L.H.; Cowie, B.C.C.; Gunzelmann, D.; Qiao, S.Z.; Huang, S.M.; Chen, Y. Observation of active sites for oxygen reduction reaction on nitrogen-doped multilayer graphene. *ACS Nano* **2014**, *8*, 6856–6862.

97. Lai, L.; Potts, J.R.; Zhan, D.; Wang, L.; Poh, C.K.; Tang, C.; Gong, H.; Shen, Z.; Lin, J.; Ruoff, R.S. Exploration of the active center structure of nitrogen-doped graphene-based catalysts for oxygen reduction reaction. *Energy Environ. Sci.* **2012**, *5*, 7936–7942.

98. Yang, S.; Zhi, L.; Tang, K.; Feng, X.; Maier, J.; Muellen, K. Efficient synthesis of heteroatom (N or S)-doped graphene based on ultrathin graphene oxide-porous silica sheets for oxygen reduction reactions. *Adv. Funct. Mater.* **2012**, *22*, 3634–3640.

99. Jiang, Z.J.; Jiang, Z.Q.; Chen, W.H. The role of holes in improving the performance of nitrogen-doped holey graphene as an active electrode material for supercapacitor and oxygen reduction reaction. *J. Power Sources* **2014**, *251*, 55–65.

100. Yu, D.S.; Wei, L.; Jiang, W.C.; Wang, H.; Sun, B.; Zhang, Q.; Goh, K.L.; Si, R.M.; Chen, Y. Nitrogen doped holey graphene as an efficient

metal-free multifunctional electrochemical catalyst for hydrazine oxidation and oxygen reduction. *Nanoscale* **2013**, *5*, 3457–3464.

101. Lin, Z.Y.; Song, M.K.; Ding, Y.; Liu, Y.; Liu, M.L.; Wong, C.P. Facile preparation of nitrogen-doped graphene as a metal-free catalyst for oxygen reduction reaction. *Phys. Chem. Chem. Phys.* **2012**, *14*, 3381–3387.

102. Lin, Z.Y.; Waller, G.; Liu, Y.; Liu, M.L.; Wong, C.P. Facile synthesis of nitrogen-doped graphene via pyrolysis of graphene oxide and urea, and its electrocatalytic activity toward the oxygen-reduction reaction. *Adv. Energy Mater.* **2012**, *2*, 884–888.

103. Unni, S.M.; Devulapally, S.; Karjule, N.; Kurungot, S. Graphene enriched with pyrrolic coordination of the doped nitrogen as an efficient metal-free electrocatalyst for oxygen reduction. *J. Mater. Chem.* **2012**, *22*, 23506–23513.

104. Zhang, Y.J.; Fugane, K.; Mori, T.; Niu, L.; Ye, J.H. Wet chemical synthesis of nitrogen-doped graphene towards oxygen reduction electrocatalysts without high-temperateure pyrolysis. *J. Mater. Chem.* **2012**, *22*, 6575–6580.

105. Lin, Z.Y.; Waller, G.H.; Liu, Y.; Liu, M.L.; Wong, C.P. Simple preparation of nanoporous few-layer nitrogen-doped graphene for use as an efficient electrocatalyst for oxygen reduction and oxygen evolution reactions. *Carbon* **2013**, *53*, 130–136.

106. Lu, Z.J.; Bao, S.J.; Gou, Y.T.; Cai, C.J.; Ji, C.C.; Xu, M.W.; Song, J.; Wang, R.Y. Nitrogen-doped reduced-graphene oxide as an efficient metal-free electrocatalyst for oxygen reduction in fuel cells. *RSC Adv.* **2013**,*3*, 3990–3995.

107. Pan, F.P.; Jin, J.T.; Fu, X.G.; Liu, Q.; Zhang, J.Y. Advanced oxygen reduction electrocatalyst based on nitrogen-doped graphene derived from edible sugar and urea. *ACS Appl. Mater. Interfaces* **2013**, *5*, 11108–11114.

108. Zheng, B.; Wang, J.; Wang, F.B.; Xia, X.H. Synthesis of nitrogen doped graphene with high electrocatalytic activity toward oxygen reduction reaction. *Electrochem. Commun.* **2013**, *28*, 24–26.

109. Cong, H.P.; Wang, P.; Gong, M.; Yu, S.H. Facile synthesis of mesoporous nitrogen-doped graphene: An efficient methanol-tolerant cathodic catalyst for oxygen reduction reaction. *Nano Energy* **2014**, *3*, 55–63.

110.   Vikkisk, M.; Kruusenberg, I.; Joost, U.; Shulga, E.; Kink, I.; Tammeveski, K. Electrocatalytic oxygen reduction on nitrogen-doped graphene in alkaline media. *Appl. Catal. B* **2014**, *147*, 369–376.

111.   Liu, M.K.; Song, Y.F.; He, S.X.; Tjiu, W.W.; Pan, J.S.; Xia, Y.Y.; Liu, T.X. Nitrogen-doped graphene nanoribbons as efficient metal-free electrocatalysts for oxygen reduction. *ACS Appl. Mater. Interfaces* **2014**,*6*, 4214–4222.

112.   Ouyang, W.; Zeng, D.; Yu, X.; Xie, F.; Zhang, W.; Chen, J.; Yan, J.; Xie, F.; Wang, L.; Meng, H.; *et al*. Exploring the active sites of nitrogen-doped graphene as catalysts for the oxygen reduction reaction. *Int. J. Hydrogen Energy* **2014**, *39*, 15996–16005.

113.   Wang, Z.; Li, B.; Xin, Y.; Liu, J.; Yao, Y.; Zou, Z. Rapid synthesis of nitrogen-doped graphene by microwave heating for oxygen reduction reactions in alkaline electrolyte. *Chin. J. Catal.* **2014**, *35*, 509–513.

114.   Chen, L.; Du, R.; Zhu, J.; Mao, Y.; Xue, C.; Zhang, N.; Hou, Y.; Zhang, J.; Yi, T. Three-dimensional nitrogen-doped graphene nanoribbons aerogel as a highly efficient catalyst for the oxygen reduction reaction. *Small***2015**, *11*, 1423–1429.

115.   Sun, Z.; Masa, J.; Weide, P.; Fairclough, S.M.; Robertson, A.W.; Ebbinghaus, P.; Warner, J.H.; Tsang, S.C.E.; Muhler, M.; Schuhmann, W. High-quality functionalized few-layer graphene: Facile fabrication and doping with nitrogen as a metal-free catalyst for the oxygen reduction reaction. *J. Mater. Chem. A* **2015**, *3*, 15444–15450.

116.   Wu, J.; Ma, L.; Yadav, R.M.; Yang, Y.; Zhang, X.; Vajtai, R.; Lou, J.; Ajayan, P.M. Nitrogen-doped graphene with pyridinic dominance as a highly active and stable electrocatalyst for oxygen reduction. *ACS Appl. Mater. Interfaces* **2015**, *7*, 14763–14769.

117.   Zhou, X.; Bai, Z.; Wu, M.; Qiao, J.; Chen, Z. 3-dimensional porous N-doped graphene foam as a non-precious catalyst for the oxygen reduction reaction. *J. Mater. Chem. A* **2015**, *3*, 3343–3350.

118.   Kong, X.-K.; Chen, C.-L.; Chen, Q.-W. Doped graphene for metal-free catalysis. *Chem. Soc. Rev.* **2014**, *43*, 2841–2857.

119.   Yu, L.; Pan, X.; Cao, X.; Hu, P.; Bao, X. Oxygen reduction reaction mechanism on nitrogen-doped graphene: A density functional theory study. *J. Catal.* **2011**, *282*, 183–190.

120.   Zhang, L.; Xia, Z. Mechanisms of oxygen reduction reaction on nitrogen-doped graphene for fuel cells. *J. Phys. Chem. C* **2011**, *115*, 11170–11176.

121.   Kim, H.; Lee, K.; Woo, S.I.; Jung, Y. On the mechanism of enhanced oxygen reduction reaction in nitrogen-doped graphene nanoribbons. *Phys. Chem. Chem. Phys.* **2011**, *13*, 17505–17510.

122.   Jiao, Y.; Zheng, Y.; Jaroniec, M.; Qiao, S.Z. Origin of the electrocatalytic oxygen reduction activity of graphene-based catalysts: A roadmap to achieve the best performance. *J. Am. Chem. Soc.* **2014**, *136*, 4394–4403.

123.   Sun, M.; Liu, H.; Liu, Y.; Qu, J.; Li, J. Graphene-based transition metal oxide nanocomposites for the oxygen reduction reaction. *Nanoscale* **2015**, *7*, 1250–1269.

124.   Wang, X.; Sun, G.; Routh, P.; Kim, D.-H.; Huang, W.; Chen, P. Heteroatom-doped graphene materials: Syntheses, properties and applications. *Chem. Soc. Rev.* **2014**, *43*, 7067–7098.

125.   Zhou, X.; Qiao, J.; Yang, L.; Zhang, J. A review of graphene-based nanostructural materials for both catalyst supports and metal-free catalysts in pem fuel cell oxygen reduction reactions. *Adv. Energy Mater.* **2014**, *4*.

126.   Li, Q.; Pan, H.; Higgins, D.; Cao, R.; Zhang, G.; Lv, H.; Wu, K.; Cho, J.; Wu, G. Metal-organic framework-derived bamboo-like nitrogen-doped graphene tubes as an active matrix for hybrid oxygen-reduction electrocatalysts. *Small* **2015**, *11*, 1443–1452.

127.   Yang, G.H.; Li, Y.J.; Rana, R.K.; Zhu, J.J. Pt-Au/nitrogen-doped graphene nanocomposites for enhanced electrochemical activities. *J. Mater. Chem. A* **2013**, *1*, 1754–1762.

128.   Imran Jafri, R.; Rajalakshmi, N.; Ramaprabhu, S. Nitrogen doped graphene nanoplatelets as catalyst support for oxygen reduction reaction in proton exchange membrane fuel cell. *J. Mater. Chem.* **2010**, *20*, 7114–7117.

129.   Vinayan, B.P.; Nagar, R.; Rajalakshmi, N.; Ramaprabhu, S. Novel platinum–cobalt alloy nanoparticles dispersed on nitrogen-doped

graphene as a cathode electrocatalyst for PEMFC applications. *Adv. Funct. Mater.* **2012**, *22*, 3519–3526.

130.   Xiong, B.; Zhou, Y.; Zhao, Y.; Wang, J.; Chen, X.; O'Hayre, R.; Shao, Z. The use of nitrogen-doped graphene supporting Pt nanoparticles as a catalyst for methanol electrocatalytic oxidation. *Carbon* **2013**, *52*, 181–192.

131.   Chen, P.; Xiao, T.-Y.; Li, H.-H.; Yang, J.-J.; Wang, Z.; Yao, H.-B.; Yu, S.-H. Nitrogen-doped graphene/ZnSe nanocomposites: Hydrothermal synthesis and their enhanced electrochemical and photocatalytic activities.*ACS Nano* **2012**, *6*, 712–719.

132.   Parvez, K.; Yang, S.B.; Hernandez, Y.; Winter, A.; Turchanin, A.; Feng, X.L.; Mullen, K. Nitrogen-doped graphene and its iron-based composite as efficient electrocatalysts for oxygen reduction reaction. *ACS Nano* **2012**, *6*, 9541–9550.

133.   Bai, J.C.; Zhu, Q.Q.; Lv, Z.X.; Dong, H.Z.; Yu, J.H.; Dong, L.F. Nitrogen-doped graphene as catalysts and catalyst supports for oxygen reduction in both acidic and alkaline solutions. *Int. J. Hydrogen Energy* **2013**,*38*, 1413–1418.

134.   Huang, T.; Mao, S.; Pu, H.; Wen, Z.; Huang, X.; Ci, S.; Chen, J. Nitrogen-doped graphene-vanadium carbide hybrids as a high-performance oxygen reduction reaction electrocatalyst support in alkaline media. *J. Mater. Chem. A* **2013**, *1*, 13404–13410.

135.   Park, H.W.; Lee, D.U.; Nazar, L.F.; Chen, Z.W. Oxygen reduction reaction using MnO_2 nanotubes/nitrogen-doped exfoliated graphene hybrid catalyst for $Li-O_2$ battery applications. *J. Electrochem. Soc.* **2013**, *160*, A344–A350.

136.   Xiao, J.; Bian, X.; Liao, L.; Zhang, S.; Ji, C.; Liu, B. Nitrogen-doped mesoporous graphene as a synergistic electrocatalyst matrix for high-performance oxygen reduction reaction. *ACS Appl. Mater. Interfaces* **2014**, *6*, 17654–17660.

137.   Brownson, D.A.C.; Munro, L.J.; Kampouris, D.K.; Banks, C.E. Electrochemistry of graphene: Not such a beneficial electrode material? *RSC Adv.* **2011**, *1*, 978–988.

138.   Choi, C.H.; Chung, M.W.; Kwon, H.C.; Chung, J.H.; Woo, S.I. Nitrogen-doped graphene/carbon nanotube self-assembly for efficient oxygen reduction reaction in acid media. *Appl. Catal. B* **2014**, *144*, 760–766.

139. Li, Y.; Zhou, W.; Wang, H.; Xie, L.; Liang, Y.; Wei, F.; Idrobo, J.-C.; Pennycook, S.J.; Dai, H. An oxygen reduction electrocatalyst based on carbon nanotube-graphene complexes. *Nat. Nanotechnol.* **2012**, *7*, 394–400.

140. Ma, Y.; Sun, L.; Huang, W.; Zhang, L.; Zhao, J.; Fan, Q.; Huang, W. Three-dimensional nitrogen-doped carbon nanotubes/graphene structure used as a metal-free electrocatalyst for the oxygen reduction reaction. *J. Phys. Chem. C* **2011**, *115*, 24592–24597.

141. Ratso, S.; Kruusenberg, I.; Vikkisk, M.; Joost, U.; Shulga, E.; Kink, I.; Kallio, T.; Tammeveski, K. Highly active nitrogen-doped few-layer graphene/carbon nanotube composite electrocatalyst for oxygen reduction reaction in alkaline media. *Carbon* **2014**, *73*, 361–370.

142. Liu, J.Y.; Wang, Z.; Chen, J.Y.; Wang, X. Nitrogen-doped carbon nanotubes and graphene nanohybrid for oxygen reduction reaction in acidic, alkaline and neutral solutions. *J. Nano Res.* **2015**, *30*, 50–58.

Effect of Tubular Chiralities and Diameters of Single Carbon Nanotubes on Gas Sensing Behavior: A DFT Analysis

A. A. EL-Barbary[1,2*], Kh. M. Eid[1,3], M. A. Kamel[1], H. M. Osman[1], G. H. Ismail[1,2]

[1]Physics Department, Faculty of Education, Ain Shams University, Cairo, Egypt

[2]Physics Department, Faculty of Science, Jazan University, Jazan, KSA

[3]Bukairiayh for Science, Qassim University, Buraydah, KSA

ABSTRACT

Using density functional theory, the adsorption of CO, CO_2, NO and CO_2 gas molecules on different chiralities and diameters of single carbon nanotubes is investigated in terms of energetic, electronic properties and surface reactivity. We found that the adsorption of CO and CO_2 gas molecules is dependent on the chiralities and diameters of CNTs and it is vice versa for NO and NO_2 gas molecules. Also, the electronic character of CNTs is not affected by the adsorption of CO and CO_2 gas molecules while it is strongly affected by NO and NO_2 gas molecules. In addition, it is found that the dipole moments of zig-zag CNTs are always higher than the arm-chair CNTs. Therefore, we conclude that the zig-zag carbon nanotubes are more preferred as gas sensors than the arm-chair carbon nanotubes, especially for detecting NO and NO_2 gas molecules.

Keywords: Carbon Nanotubes, DFT, Gas Sensors

1. INTRODUCTION

Monitoring of combustible gas alarms, gas leak detection, and environmental pollution is of great concern in public security. Advances in nanotechnology give great promise for achieving new sensing materials. Since the discovery of carbon nanotubes in 1991, the single-walled carbon nanotubes (SWCNTs) have been intensively investigated as nanoscale gas sensors because of their great surface areas to bulk ratio and their abilities to modulate electrical properties upon adsorption of various kinds of gas molecules [1] -[17] . The emission of carbon and nitrogen oxides (CO, CO_2, NO and NO_2) results from the combustion of fossil fuels, contributing to both smog and acid precipitation, and affecting both terrestrial and aquatic ecosystems [18]. Although many efforts have been made to use catalysts to reduce the amount of carbon or nitrogen oxides in the air [19] -[25] , an efficient method of sensing and removing carbon and nitrogen oxides is still required.

Because carbon and nitrogen oxides are the most dangerous air pollutants, toxic and global warming gases, our work is concentrated on investigating the effect of tubular chiralities and diameters of single carbon nanotubes on gas sensing behavior for CO, CO_2, NO and NO_2 gas molecules, applying the first principle calculations.

2. COMPUTATIONAL METHODS

All calculations were performed with the density functional theory as implemented within G03W package [26] - [29], using B3LYP exchange-functional and applying basis set 6 - 31g (d,p). Pure carbon nanotubes $(5,0)$ and $(9,0)$, $(5,5)$ and $(6,6)$ are fully optimized with spin average as well as the adsorption of CO, CO_2, NO and NO_2 gas molecules.

The obtained diameters [30] and the adsorption energies of gas molecules on CNTs (E_{ads}) [31] are calculated from the following relations:

$$D = 0.73\left(n^2 + nm + m^2\right)^{1/2}$$

where n and m are integral numbers, the composition of chiral vector.

$$E_{ads} = E_{(nanotube+gas\ molecules)} - E_{nanotube} - E_{gas\ molecules}$$

where $E_{(nanotube+gas\ molecules)}$ is the total energy of nanotube and gas molecules, $E_{nanotube}$ is the energy of the carbon nanotube, and $E_{gas\ molecules}$ is the energy of gas molecules.

3. RESULTS AND DISCUSSION

We will investigate the adsorption of gas molecules, CO, CO_2, NO and NO_2 on four carbon nanotubes with different charilities and diameters $(5,0)$ CNT, $(9,0)$ CNT, $(5,5)$ CNT and $(6,6)$ CNT as shown in Figure 1 and Table 1.

3.1. Adsorption of CO, CO_2, NO and NO_2 Gas Molecules on CNTs

We have adsorbed CO and CO_2 gas molecules vertically on different three positions of $(5,0)$, $(9,0)$, $(5,5)$ and $(6,6)$ CNTs: above a carbon atom (carbon site), above a bond between two carbon atoms (bond site) and above a center of a hexagon ring (vacant site). The calculated adsorption energies of CO and CO_2 gas molecules are listed in Table 2. It is found that the best position and adsorption energy for CO gas molecule is above the bond site on $(9,0)$ CNT with adsorption energy of –0.43 eV, however for CO_2 gas molecule is above the vacant site on $(9,0)$ CNT with adsorption energy of –0.26 eV. Therefore, one can conclude that the best CNT gas sensor for CO and CO_2 gas molecules is the $(9,0)$ CNT.

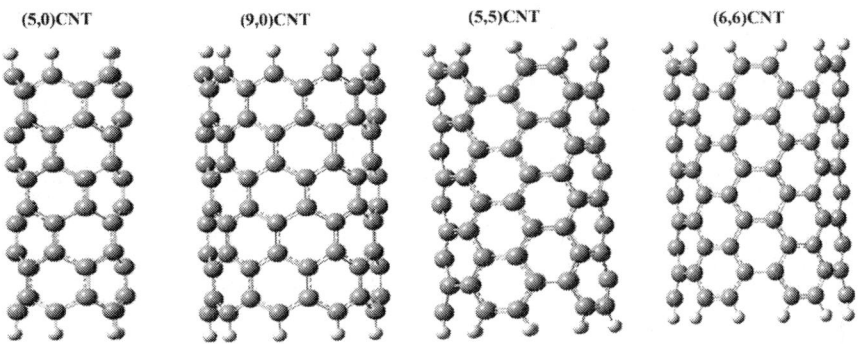

(5,0)CNT (9,0)CNT (5,5)CNT (6,6)CNT

Figure 1. The fully optimized structures of $(5,0)$, $(9,0)$, $(5,5)$ and $(6,6)$ CNTs. Carbon atoms (gray) and hydrogen atoms (white).

Also, we have adsorbed NO and NO_2 gas molecules vertically on different three positions of $(5,0)$, $(9,0)$, $(5,5)$ and $(6,6)$ CNTs: above a carbon site, above a bond site and above a vacant site. The calculated adsorption energies of NO and NO_2 gas molecules are listed in Table 3. It is found that the best adsorption energies of NO gas molecule are on the $(9,0)$ CNT above a bond site, then above a carbon site and after that above a vacant site with adsorption energies of −1.65 eV, −1.55 eV and −1.34 eV, respectively. However, for NO_2 gas molecule is found to be above the bond site on $(9,0)$ CNT with adsorption energy of −1.75 eV. Also, it is noticed that the vacant site is always preferred for NO_2 gas adsorption on all the studied CNTs except for $(9,0)$ CNT. Therefore, one can conclude that all CNTs can be used as gas sensors for NO and NO_2 gas molecules.

From Table 2, Table 3, one can investigate the effect of the chiralities and the diameters on the CNT gas sensors behavior. It is clear that the adsorption of CO and CO_2 gas molecules is dependent on the chiralities and the diameters of CNTs. The adsorption of CO and CO_2 gas molecules is enhanced with increasing the diameter of the zig-zag CNTs. However, the adsorption of NO and NO_2 gas molecules is independent on the chiralities and the diameters of CNTs.

3.2. Energy Gaps of Adsorbed CO, CO_2, NO and NO_2 Gas Molecules on CNTs

From Table 4, it is clear that the adsorption of CO and CO_2 gas molecules on CNTs does not affect the elec tronic character of the CNTs. Also, the band gaps of pristine CNTs and the adsorbed CO and CO_2 gas molecules on CNTs are so close.

Table 1. The configuration structures and diameters of the studied CNTs.

Table 1. The configuration structures and diameters of the studied CNTs.

System	Configuration Structures	Diameters/Å
(5,0) CNT	$C_{60}H_{10}$	3.65
(9,0) CNT	$C_{108}H_{18}$	6.57
(5,5) CNT	$C_{100}H_{20}$	6.32
(6,6) CNT	$C_{120}H_{24}$	7.59

Table 2. The calculated adsorption energies (E_{ads}) of CO and CO_2 above a carbon site, a bond site and a vacant site of pristine $(5,0)$, $(9,0)$, $(5,5)$ and $(6,6)$ CNTs. All energies are given by eV.

System	CO			CO_2		
	Carbon site	Bond site	Vacant site	Carbon site	Bond Site	Vacant Site
(5,0) CNT	0.04	−0.00	−0.04	−0.00	0.00	0.00
(9,0) CNT	0.03	−0.43	−0.24	−0.23	−0.00	−0.26
(5,5) CNT	0.04	0.04	0.04	0.13	0.003	2.19
(6,6) CNT	−0.01	−0.01	−0.03	−0.00	−0.00	−0.16

Table 3. The calculated adsorption energies (E_{ads}) of NO and NO_2 above a carbon site, a bond site and a vacant site of pristine $(5,0)$, $(9,0)$, $(5,5)$ and $(6,6)$ CNTs. All energies are given by eV.

System	NO			NO_2		
	Carbon site	Bond site	Vacant site	Carbon Site	Bond site	Vacant site
(5,0) CNT	−0.35	−0.34	0.29	0.07	−0.36	−1.04
(9,0) CNT	−1.55	−1.65	−1.34	−1.19	−1.75	−1.40
(5,5) CNT	0.06	0.22	0.07	2.26	0.92	−1.32
(6,6) CNT	−0.01	0.25	0.07	0.06	0.35	−1.34

From Table 5, the adsorption of NO and NO_2 gas molecules on CNTs is strongly affected the electronic character of the $(9,0)$ and $(5,0)$ CNTs. However, there is not any change of the electronic character for $(5,5)$ and $(6,6)$ CNTs. The band gap of pristine $(5,0)$ CNT is increased from 0.70 eV to 1.61 eV and to 1.37 eV when NO and NO_2 gas molecules are adsorbed on it, respectively. Also, The band gap of pristine $(9,0)$ CNT is increased from 0.25 eV to 1.34 eV and to 1.25 eV when NO and NO_2 gas molecules are adsorbed on it, respectively. One can conclude that the electronic character of $(5,0)$, $(9,0)$, $(5,5)$ and $(6,6)$ CNTs is not affected by the adsorption of CO and CO_2 gas molecules. The adsorption of NO and NO_2 gas molecules on CNTs is only strongly affected the electronic character of the $(9,0)$ and $(5,0)$ CNTs, however the $(5,5)$ and $(6,6)$ CNTs are not affected at all.

3.3. HOMO-LUMO Orbitals of Adsorbing CO, CO_2, NO and NO_2 Gas Molecules on CNTs

Our calculated band gaps show that the adsorption of CO and CO_2 gas molecules on CNTs is not affected the band gaps of the pristine CNTs, however the adsorption of NO and NO_2 gas molecules is strongly affected the band gaps.

To explain that the molecular orbitals of adsorbing CO, CO_2, NO and NO_2 gas molecules on $(5,0)$, $(9,0)$, $(5,5)$ and $(6,6)$ CNTs are investigated, seeFigure 2, Figure 3. The band gaps of the pristine CNTs are calculated and are listed in Table 4. The HOMO and LUMO energy orbitals for pristine $(5,0)$, $(9,0)$, $(5,5)$ and $(6,6)$ CNTs are found to be (−4.13 eV, −3.42 eV), (−3.93 eV, −3.68 eV), (−4.62 eV, −2.82 eV) and (−4.70 eV, −2.82 eV), respectively. Comparing the HOMO-LUMO energies of the pristine CNTs with ones after the adsorption of CO and CO_2 gas molecules, it is clear that the energy values are so close. Also, it is noticed that there is not any contribution from the gas molecules at the molecular orbitals and the electron density of HOMO and LUMO is distributed over all the carbon atoms of CNTs except for $(9,0)$ CNT is located at the terminals of the tube, see Figure 2. Comparing the HOMO-LUMO energies of the pristine CNTs with ones after the adsorption of NO and NO_2 gas molecules, it is clear that the energy values are so close in case of $(5,5)$ and $(6,6)$ CNTs and are quite far in case of $(5,0)$ and $(9,0)$ CNTs. The HOMO energy levels in case of $(5,0)$ and $(9,0)$ CNTs after adsorbing NO and NO_2 gas molecular are getting deep (lower) in energy however the LUMO energy levels are getting higher in energy. Results in increasing the band gap from 0.70 eV to 1.81 eV in case of $(5,0)$ CNT and from 0.25 eV to 1.34 eV in case of $(9,0)$ CNT. Also, it is noticed that there is representation from the NO gas molecule at LUMO of $(9,0)$ and $(6,6)$ CNTs, seeFigure 3.

Table 4. The calculated energy gaps (E_g) of CO and CO_2 above a carbon site, a bond site and a vacant site of pristine $(5,0)$, $(9,0)$, $(5,5)$ and $(6,6)$ CNTs. All energies are given by eV.

		CO			CO_2		
System	Pristine CNTs	Carbon site	Bond site	Vacant site	Carbon Site	Bond site	Vacant Site
(5,0) CNT	0.70	0.61	0.70	0.70	0.61	0.61	0.60
(9,0) CNT	0.25	0.25	0.25	0.25	0.25	0.24	0.25
(5,5) CNT	1.79	1.80	1.79	1.80	1.80	1.79	1.75
(6,6) CNT	1.90	1.92	1.92	1.92	1.92	1.92	1.92

Table 5. The calculated energy gaps (E_g) of NO and NO_2 above a carbon site, a bond site and a vacant site of pristine $(5,0)$, $(9,0)$, $(5,5)$ and $(6,6)$ CNTs. All energies are given by eV.

	NO			NO_2		
system	Carbon site	Bond Site	Vacant Site	Carbon site	Bond Site	Vacant site
$(5,0)$ CNT	1.61	0.99	0.94	0.61	0.99	1.37
$(9,0)$ CNT	1.34	1.24	1.14	1.25	1.25	1.14
$(5,5)$ CNT	1.79	1.81	1.79	1.72	1.76	1.79
$(6,6)$ CNT	1.92	1.90	1.92	1.94	1.88	1.91

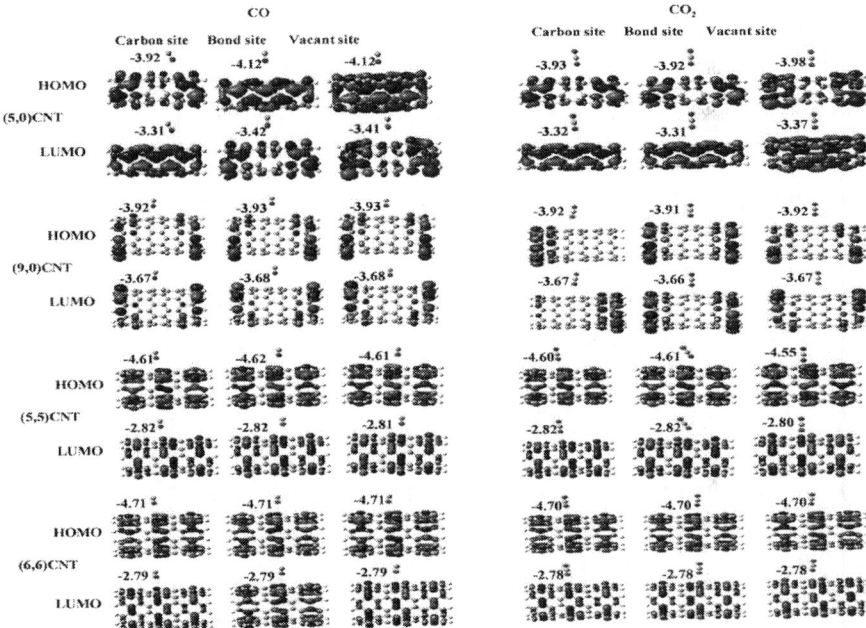

Figure 2. HOMO and LUMO molecular orbitals of adsorbing CO and CO_2 gas molecules on the pristine $(5,0)$, $(9,0)$, $(5,5)$ and $(6,6)$ CNTs. Energies of HOMO and LUMO are listed above the molecular orbitals and are given by eV.

3.4. The Reactivity of CNT Surfaces before and after Adsorbing Gas Molecules

Our calculated band gaps and molecular orbitals show that the adsorption of CO and CO_2 gas molecules on CNTs is not affected neither the band gaps nor the molecular orbitals of the pristine CNTs but the adsorption of NO and NO_2 gas molecules is strongly affected both of the band gaps and the molecular orbitals of $(5,0)$ and $(9,0)$ CNTs. To clear that the reactivity of CNT surfaces before and after adsorbing CO, CO_2, NO and NO_2 gas molecules on $(5,0)$, $(9,0)$, $(5,5)$ and $(6,6)$ CNTs are studied, see Table 6, Table 7. The surface reactivity of the pristine CNTs is calculated and is listed in **Table 6**. The dipole moments of pristine $(5,0)$, $(9,0)$, $(5,5)$ and $(6,6)$ CNTs are found to be 0.54 Debye, 0.20 Debye 0.00 Debye and 0.00 Debye, respectively.

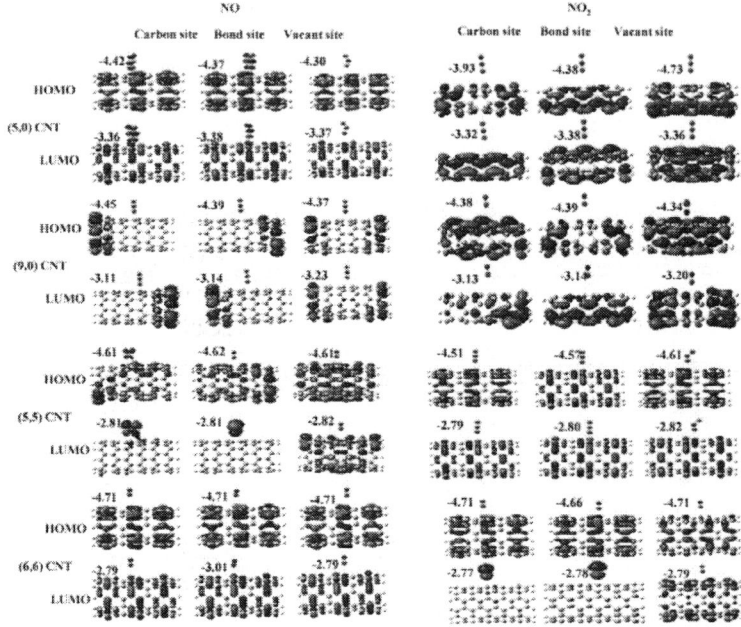

Figure 3. HOMO and LUMO molecular orbitals of adsorbing NO and NO_2 gas molecules on the pristine $(5,0)$, $(9,0)$, $(5,5)$ and $(6,6)$ CNTs. Energies of HOMO and LUMO are listed above their molecular orbitals and are given by eV.

Table 6. The calculated dipole moments of pristine and after adsorbing CO and CO_2 gas molecules above a carbon site, a bond site and a vacant site of $(5,0)$, $(9,0)$, $(5,5)$ and $(6,6)$ CNTs. All dipole moments are given by Debye.

system	Pristine CNTs	CO			CO_2		
		Carbon site	Bond site	Vacant Site	Carbon Site	Bond Site	Vacant Site
$(5,0)$ CNT	0.54	0.54	0.56	0.49	0.46	0.47	0.62
$(9,0)$ CNT	0.20	0.40	0.21	0.10	0.23	0.74	0.30
$(5,5)$ CNT	0.00	0.04	0.12	0.06	0.21	0.25	0.77
$(6,6)$ CNT	0.00	0.06	0.05	0.06	0.23	0.23	0.26

Comparing the dipole moments of the pristine CNTs with ones that are adsorbed the CO and CO_2gas molecules, it is clear that the dipole moment values are so close in case of the adsorption of the CO gas molecule but they are higher in case of the adsorption of the CO_2 gas molecule, seeTable 6. Also, it is noticed that the highest dipole moments after the adsorption of the CO_2 gas molecule are 0.74 Debye (when CO_2 is adsorbed above the bond site of $(9,0)$ CNT) and 0.77 Debye (when CO_2 is adsorbed above the vacant site of $(5,5)$ CNT), respectively. Comparing the dipole moments of the pristine CNTs with ones that are adsorbed the NO and NO_2 gas molecules, it is found that the dipole moments are getting higher. When the NO and NO_2 gas molecules are adsorbed on the vacant sites of CNTs, their dipole moments are either quite close to or are lower than the dipole moments of pristine CNTs, except in case of adsorbing NO_2 on$(5,0)$ CNT, the dipole moment is increased. Also, all the calculated dipole moments of adsorbing NO and NO_2 gas molecules on the

carbon sites of CNTs are increased, except in case of adsorbing NO_2 on $(5,0)$ CNT, the dipole moment is decreased. In case of adsorbing NO and NO_2 gas molecules on the bond sites of CNTs the dipole moments are also increased,

except in case of adsorbing NO_2 on $(9,0)$ CNT is decreased, see Table 7.

Table 7. The calculated dipole moments of pristine and after adsorbing NO and NO_2 gas molecules above a carbon site, a bond site and a vacant site of $(5,0)$, $(9,0)$, $(5,5)$ and $(9'9)$ CNTs. All dipole moments are given by Debye.

	NO			NO_2		
system	Carbon site	Bond Site	Vacant Site	Carbon Site	Bond Site	Vacant site
$(5,0)$ CNT	5.58	1.12	0.36	0.28	0.97	1.47
$(9,0)$ CNT	0.46	0.48	0.11	0.69	0.11	0.07
$(5,5)$ CNT	0.14	0.11	0.11	1.48	0.58	0.13
$(6,6)$ CNT	0.15	0.31	0.17	0.20	0.53	0.17

From Table 6, Table 7, it is clear that the dipole moments of zig-zag $(5,0)$ and $(9,0)$ CNTs are always higher than the arm-chair $(5,5)$ and $(6,6)$ CNTs. Also, it is noticed that the dipole moment of adsorbing NO gas molecule on the bond site of $(5,0)$ CNT is increased by ten times comparing with the dipole moment of pristine $(5,0)$CNT.

4. CONCLUSION

The gas sensing behavior of CNTs, considering a range of different nanotube diameters and chiralities, as well as different adsorption sites is reported. The adsorption of CO, CO_2, NO, and NO_2 gas molecules on the $(5,0)$, $(9,0)$, $(5,5)$ and $(6,6)$ CNTs are studied using B3LYP/6-31 g(d, p). Three different adsorption sites (above a carbon site, a bond site and a vacant site) are applied on CNTs. It is found that the adsorption of CO and CO_2 gas molecules is dependent on the chiralities and the diameters of CNTs and it is enhanced with increasing the diameter of the zig-zag CNTs. However, the adsorption of NO and NO_2 gas molecules is independent on the chiralities and the diameters of CNTs. Also, the electronic character of $(5,0)$, $(9,0)$, $(5,5)$ and $(6,6)$ CNTs is not affected by the adsorption of CO and CO_2 gas molecules. While, the adsorption of NO and NO_2 gas molecules on CNTs is only strongly affected by the electronic character of the $(9,0)$ and $(5,0)$ CNTs but the $(5,5)$ and $(6,6)$ CNTs are not affected at all. It is found that the dipole moments of zig-zag $(5,0)$ and $(9,0)$ CNTs are always higher than the arm-chair $(5,5)$ and $(6,6)$ CNTs. Also, it is noticed that the dipole moment of adsorbing NO gas molecule on the bond site of $(5,0)$ CNT is increased by ten times compared with the dipole moment of pristine $(5,0)$ CNT. Therefore, these findings prove that the zig-zag carbon nanotubes are better than the arm-chair carbon nanotubes as gas sensors, especially for NO and NO_2 gas molecules.

REFERENCES

1. Kong, J., Franklin, N.R., Zhou, C., Chapline, M.G., Peng, S. and Cho, K. (2000) Nanotube Molecular Wires as Chemical Sensors. Science, 287, 622.

2. Snow, E.S., Perkins, F.K., Houser, E.J., Badescu, S.C. and Reinecke, T.L. (2005) Chemical Detection with a Single-Walled Carbon Nanotube Capacitor. Science, 307, 1942-1945.

3. Baei, M.T., Soltani, A.R., Moradi, A.V. and Lemeski, E.T. (2011) Adsorption Properties of NO on (6, 0), (7, 0), and (8, 0) Zigzag Single-Walled Boron Nitride Nanotubes: A Computational Study. Computational and Theoretical Chemistry, 970, 30-35.

4. Breza, M. (2006) Model Studies of $SOCl_2$ Adsorption on Carbon Nanotubes. Journal of Molecular Structure: THEOCHEM, 767, 159-163.

5.   Zhao, J., Buldum, A., Han, J. and Lu, J.P. (2002) Gas Molecule Adsorption in Carbon Nanotubes and Nanotube Bundles. Nanotechnology, 13, 195-200.

6.   Ricca, A. and Bauschlicher Jr., C.W. (2006) The Adsorption of NO on (9, 0) and (10, 0) Carbon Nanotubes. Chemical Physics, 323, 511-518.

7.   Abbas Rafati, A., Majid Hashemianzadeh, S. and Bolboli Nojini, Z. (2008) Electronic Properties of Adsorption Nitrogen Monoxide on Inside and Outside of the Armchair Single Wall Carbon Nanotubes: A Density Functional Theory Calculations. The Journal of Physical Chemistry C, 112, 3597-3604.

8.   Azizi, K., Majid Hashemianzadeh, S. and Bahramifar, Sh. (2011) Density Functional Theory Study of Carbon Monoxide Adsorption on the Inside and Outside of the Armchair Single-Walled Carbon Nanotubes. Current Applied Physics, 11, 776-782.

9.   Ricca, A., Bauschlicher Jr., C.W. (2006) The Physisorption of CH_4 on Graphite and on a (9, 0) Carbon Nanotube. Chemical Physics, 324, 455-458.

10. Santucci, S., Picozzi, S., Di Gregorio, F., Lozzi, L., Cantalini, C., Valentini, L., Kenny, J.M. and Delley, B. (2003) NO_2 and CO Gas Adsorption on Carbon Nanotubes: Experiment and Theory. The Journal of Chemical Physics, 119, 10904-10910.

11. Zanolli, Z. and Charlier, J.C. (2009) Defective Carbon Nanotubes for Single-Molecule Sensing. Physical Review B, 80, 155447.

12. Tang, S. and Cao, Z. (2009) Defect-Induced Chemisorption of Nitrogen Oxides on (10, 0) Single-Walled Carbon Nanotubes: Insights from Density Functional Calculations. The Journal of Chemical Physics, 131, 114706.

13. García-Lastra, J.M., Mowbray, D.J., Thygesen, K.S., Rubio, A. and Jacobsen, K.W. (2010) Modeling Nanoscale Gas Sensors under Realistic Conditions: Computational Screening of Metal-Doped Carbon Nanotubes. Physical Review B, 81, 245429.

14. Denis, P.A. (2008) Methane Adsorption Inside and Outside Pristine and N-Doped Single Wall Carbon Nanotubes. Chemical Physics, 353, 79-86.

15. Yeung, C.S., Liu, L.V. and Wang, Y.A. (2008) Adsorption of Small Gas Molecules onto Pt-Doped Single-Walled Carbon Nanotubes. The Journal of Physical Chemistry C, 112, 7401-7411.

16. Zhao, J.X. and Ding, Y.H. (2008) Theoretical Study of the Interactions of Carbon Monoxide with Rh-Decorated (8, 0) Single-Walled Carbon Nanotubes, Materials Chemistry and Physics, 110, 411-416.

17. An, W. and Turner, C.H. (2009) Electronic Structure Calculations of Gas Adsorption on Boron Doped Carbon Nanotubes Sensitized with Tungsten. Chemical Physics, 482, 274-280.

18. Sayago, I., Santos, H., Horrillo, M.C., Aleixandre, M., Fernández, M.J., Terrado, E., Tacchini, I., Aroz, R., Maser, W.K., Benito, A.M., Martínez, M.T., Gutiérrez, J. and Munoz, E. (2008) Carbon Nanotube Networks as Gas Sensors for NO_2 Detection. Talanta, 77, 758-764.

19. Li, X.M., Tian, W.Q., Dong, Q., Huang, X.R., Sun, C.C. and Jiang, L. (2011) Substitutional Doping of BN Nanotube by Transition Metal: A Density Functional Theory Simulation. Computational and Theoretical Chemistry, 964, 199-206.

20. Chen, H.L., Wu, S.Y., Chen, H.T., Chang, J.G., Ju, S.P., Tsai, C. and Hsu, L.C. (2010) Theoretical Study on Adsorption and Dissociation of NO_2 Molecule on Fe(1 1 1) Surface. Langmuir, 26, 7157-7164.

21. Wickham, D.T., Banse, B.A. and Koel, B.E. (1991) Adsorption of Nitrogen Dioxide and Nitric Oxide on Pd(1 1 1). Surface Science, 243, 83-95.

22. Jirsak, T., Kuhn, M. and Rodriguez, J.A. (2000) Chemistry of NO_2 on Mo(1 1 0): Decomposition Reactions and Formation of MoO_2. Surface Science, 457, 254-266.

23. Huang, W., Jiang, Z., Jiao, J., Tan, D., Zhai, R. and Bao, X. (2002) Decomposition of NO on Pt(1 1 0): Formation of a New Oxygen Adsorption State. Surface Science, 506, 287-292.

24. Hellman, A., Panas, I. and Grönbeck, H. (2008) NO_2 Dissociation on Ag(1 1 1) Revisited by Theory. Journal of Chemical Physics, 128, 104704-104709.

25. Yen, M.Y. and Ho, J.J. (2010) Density-Functional Study for the NO_x (x = 1, 2) Dissociation Mechanism on the Cu(1 1 1) Surface. Chemical Physics, 373, 300-306.

26. Frisch, M.J., Trucks, G.W., Schlegel, H.B., Scuseria, G.E., Robb, M.A., Cheeseman, J.R., Zakrzewski, V.G., Montgomery, J.A., Stratmann, R.E., Burant, J.C., Dapprich, S., Millam, J.M., Daniels, A.D., Kudin, K.N., Strain, M.C., Farkas, O., Tomasi, J., Barone, V., Cossi, M., Cammi, R., Mennucci, B., Pomelli, C., Adamo, C., Clifford, S., Ochterski, J.,

Petersson, G.A., Ayala, P.Y., Cui, Q., Morokuma, K., Malick, D.K., Rabuck, A.D., Raghavachari, K., Foresman, J.B., Cioslowski, J., Ortiz, J.V., Stefanov, B.B., Liu, G., Liashenko, A., Piskorz, P., Komaromi, I., Gomperts, R., Martin, R.L., Fox, D.J., Keith, T., Al-Lamham, M.A., Peng, C.Y., Nanayakkara, A., Gonzalez, C., Challacombe, M., Gill, P.M.W., Johnson, B.G., Chen, W., Wong, M.W., Andres, J.L., Head-Gordon, M., Replogle, E.S. and Pople, J.A. (2004) Gaussian. Wallingford CT, Inc., Wallingford.

27.   EL-Barbary, A.A., Lebda, H.I. and Kamel, M.A. (2009) The High Conductivity of Defect Fullerene C_{40} Cage. Computational Materials Science, 46, 128.

28.   El-Barbary, A.A., Eid, K.M., Kamel, M.A. and Hassan, M.M. (2013) Band Gap Engineering in Short Heteronanotube Segments via Monovacancy Defects. Computational Materials Science, 69, 87-94.

29.   EL-Barbary, A.A., Ismail, G.H. and Babeer, A.M. (2013) Effect of Monovacancy Defects on Adsorbing of CO, CO_2, NO and NO_2 on Carbon Nanotubes: First Principle Calculations. Journal of Surface Engineered Materials and Advanced Technology, 3, 287-294.

30.   Nalwa, H. (2002) Nanostructured Materials and Nanotechnology. Academic Press, San Diego.

31.   Chang, H., Lee, J.D., Lee, S.M. and Lee, Y.H. (2001) Adsorption of NH3 and NO2 Molecules on Carbon Nanotubes. Applied Physics Letters, 79, 3863.

Sensing Properties of Multiwalled Carbon Nanotubes Grown in MW Plasma Torch: Electronic and Electrochemical Behavior, Gas Sensing, Field Emission, IR Absorption

Petra Majzlíková [1,2], Jiří Sedláček [1,2], Jan Prášek [1,2], Jan Pekárek [1,2], Vojtěch Svatoš [1], Alexander G. Bannov [3], Ondřej Jašek [3,4], Petr Synek [3], Marek Eliáš [3,4], Lenka Zajíčková [3,4] and Jaromír Hubálek [1,2,*]

[1] Central European Institute of Technology, Brno University of Technology, Technická 3058/10, CZ-61600 Brno, Czech Republic

[2] Centre of Sensors, Information and Communication Systems, Faculty of Electrical Engineering and Communication, Technická 3058/10, CZ-61600 Brno, Czech Republic

[3] Central European Institute of Technology, Masaryk University, Kamenice 5, CZ-62500 Brno, Czech Republic

[4] Department of Physical Electronics, Faculty of Science, Masaryk University, Kotlářská 2, CZ-61137 Brno, Czech Republic

ABSTRACT

Vertically aligned multi-walled carbon nanotubes (VA-MWCNTs) with an average diameter below 80 nm and a thickness of the uniform VA-MWCNT layer of about 16 μm were grown in microwave plasma torch and tested for selected functional properties. IR absorption important for a construction of bolometers was studied by Fourier transform infrared spectroscopy. Basic electrochemical characterization was performed by cyclic voltammetry. Comparing the obtained results with the standard or MWCNT-modified screen-printed electrodes, the prepared VA-MWCNT electrodes indicated their high potential for the construction of electrochemical sensors. Resistive CNT gas sensor revealed a good sensitivity to ammonia taking into account room temperature operation. Field emission detected from CNTs was suitable for the pressure sensing application based on the measurement of emission current in the diode structure with bending diaphragm. The advantages of microwave plasma torch growth of CNTs, **i.e.**, fast processing and versatility of the process, can be therefore fully exploited for the integration of surface-bound grown CNTs into various sensing structures.

Keywords: carbon nanotubes; microwave torch; plasma enhanced chemical vapor deposition; electronic properties; electrochemical sensor; gas sensor; field emission; IR absorption

1. INTRODUCTION

Carbon nanotubes (CNTs), a synthetic carbon allotrope, are made of sp^2 hybridized carbon atoms. Single-walled CNT (SWCNT) is a graphene sheet rolled-up into a seamless cylinder and multi-walled CNT (MWCNT) is composed of several such cylinders, nested concentrically [1,2]. CNTs are often synthesized by a chemical vapor deposition (CVD) in the presence of a catalyst, nanoparticles of transition metals. In thermal CVD, a carbon-containing gas mixture is heated typically to 550–1100 °C by a conventional heat source. Plasma enhanced CVD (PECVD) activates the gas mixture by ignition of plasma discharge but a separate heating of the substrate might be required too [3]. General arguments for the PECVD include low-temperature, easily achieved vertical alignment and large area processing [4].The CVD methods are used for a CNT volume-synthesis and a surface-bound growth of vertically aligned and micropatterned CNT arrays [5]. The vertically-aligned CNTs (VA-CNTs) are highly desirable for the integration into functional devices. The aligned growth can produce CNTs free from amorphous carbon with a very narrow range of tube lengths and diameters, which is an additional advantage for many applications [6].

Unique physical and electrical properties of CNTs, *i.e.*, high electrical conductivity, remarkable mechanical strength, and thermal and chemical stability, predestinate them for many nanotechnology-based applications in electronic, optical and biomedical devices, sensors and composites [7,8]. The application potential of CNTs depends on their properties that are given by their structure and the form in which they are applied. Separated MWCNTs exhibited a non-linearity in I-V characteristics that were relatively large in case of the contacts made by underlying gold electrodes but almost diminished if the contacts were placed on the top of partially etched nanotubes [9]. Theoretically predicted impedance of SWCNT bundles at high frequencies is quite complex, employing resistance R, inductance L and capacitance C, and their validation proved to be difficult [10]. The experimentally proposed circuit model of SWCNTs contains a parallel RC element resulting from the contacts in series with R and L representing the SWCNT intrinsic behavior [10,11]. Similarly, an equivalent circuit model consisting of RC networks is constructed to simulate the electrical responses of MWCNT/polymer composites [12,13].

The optical properties of CNTs help to understand their structure [14], evaluate their purity [15] and open new applications for optical sensing [16,17]. MWCNT-based infrared detectors have received much attention due to MWCNT band gap of 0.4–6.0 eV and high absorption efficiency in IR [18]. A bolometer based on CNT/polymer composites was constructed for the detection of infrared radiation from the range 0.2–20 µm [19] and it was shown that the sensitivity and response time of the CNT-based bolometers can be substantially improved by an appropriate functionalization and selection of organic matrix [19,20].

CNTs exhibit also a great potential for electrochemical sensing due to their unique electrical properties, high surface area, fast heterogeneous electron transfer, and electrochemical stability [21–25]. The CNTs implemented as a VA-CNT film provide other advantages such as a controlled growth in defined areas and an easy modification of their surface demanded for particular sensing or biosensing applications [6]. The as-prepared VA-CNT-based sensors have been successfully applied to detect rutin [26] and salbutamol [27], a prohibited drug in sports. The VA-CNT thin films have also been tested as candidate platforms for DNA immobilization and detection of DNA hybridization [28]. The VA-CNT electrode modified by gold nanoparticles has been used for non-enzymatic detection of uric acid [29] and a platinum nanoparticle-modified VA-CNT electrode for detection of L-cysteine [30].

The CNTs belong to a group of new materials that have been extensively tested for gas sensing during the last 10 years and much effort has been put into the development of gas sensors working at room temperature. In spite of many papers devoted to SWCNT-based gas sensors [31–33], the MWCNTs can be also successfully used [34–36] and are preferable because of their low costs. The construction of room temperature ammonia (NH$_3$) sensors is an important task because NH$_3$ is a dangerous gas having a negative influence on the environment and humans. Cui *et al.* created a sensor with MWCNTs decorated by Ag nanocrystals that exhibited an enhanced response of 9% to 10,000 ppm of NH_3 as well as fast recovery [37]. A high sensitivity—6.2% to 4000 ppm of NH_3—was proven for $Co_{1-x}Ni_xFe_2O_4$/MWCNT composites [38]. Varghese *et al.* investigated sensitivity of resistive and capacitive MWCNT sensors for different gases such as water vapors, NH_3, CO_2 and CO [39].

CNTs have been also considered as promising field emitters due to their low turn-on filed, long emitter lifetime and good emission stability [40]. The field-emitter configuration should have the highest aspect ratio and low work function at its surface. The first extensive study of field emission from CNTs has been published by Bonard*et al.* in 1998 [41]. Later on, CNTs were used as field-emitters in flat-panel field-emission displays [42] or electron sources in electron microscopes [43].

In the present work, a promising application potential of VA-MWCNTs grown by PECVD in microwave (MW) plasma torch [44,45] is explored in detail. Previous studies of the CNTs deposited by the MW plasma torch revealed possible improvements of the process [46] and provided a basic structural characterization of the CNTs using scanning and transmission electron microscopies and Raman spectroscopy [47]. The technology based on the MW torch is a high speed process that takes only 60–120 s including catalyst activation (restructuralization of a catalytic thin film into nanoparticles) and does not require any external heating source of the substrate. Besides starting the CNT growth with a thin catalytic film deposited on the substrate in a separate technological step, it offers the possibility to prepare catalytic nanoparticles during the same process, *i.e.*, using the MW plasma torch [48]. Another advantage is a successful preparation of vertically aligned MWCNTs directly on Si without using a barrier SiO_2 layer [3]. A direct contact between the VA-CNTs and the Si substrate can be very important for some applications and it enables the construction of a field-emission based pressure sensor [49]. Besides the field emission, other functional properties (electrical and electrochemical, gas sensing, IR absorption) of the CNTs from the MW plasma torch are tested. Thus, the present work investigates and summarizes the CNT properties directly related to particular sensing applications and in some cases (field-emission pressure sensor, electrochemical sensor, gas sensor) describes the sensor structure and its preparation.

2. EXPERIMENTAL SECTION

2.1. CNT Deposition and Structural Characterization

The VA-MWCNT layers were deposited from $Ar/CH_4/H_2$ mixtures using the atmospheric pressure MW plasma torch operated at the power of 210 W. The flow rates of Ar, CH_4 and H_2 were Q_{Ar} = 700 sccm, Q_{CH4} = 19–38 sccm and Q_{H2} = 250 sccm, respectively. The substrate for the CNT growth was heated by the interaction with plasma, and its temperature (950–1050 K) was regulated by the distance from the plasma torch nozzle. The CNT growth time, that included also the catalyst activation phase, varied from 60 to120 s.

The aim of the present work was to investigate functional properties of the CNTs deposited by MW torch and, in some cases, integrate them into sensing devices. Most of the measurement structures were based on Si due to its compatibility with microelectronic chips and microelectromechanical systems (MEMS). CNTs were grown either on a polished single crystal silicon (c-Si) substrate, the c-Si coated with an adhesive metallic interlayer or the c-Si coated with in a thermal silicon oxide (SiO_2) film. The latter is used if the application requires a dielectric or thermal separation of the CNTs. The thickness of SiO_2 film did not play a significant role and was chosen arbitrarily. The substrate details are given in the following sections describing each particular measurement structure.

A vacuum evaporated Fe film, 5 nm in thickness, or Fe nanoparticles (NPs) deposited on the substrate by MW torch from iron pentacarbonyl ($Fe(CO)_5$) vapors were used as catalysts. The details of the NPs deposition are described in previous papers [50,51]. The Ar flow through the central torch nozzle and through the blower with liquid $Fe(CO)_5$ were 700 and 28 sccm, respectively. The Ar flow of 28 sccm through the blower with the liquid corresponds to 0.1 sccm of $Fe(CO)_5$. The deposition time was 10 s.

Surfaces and cross sections of the prepared CNT samples were checked by the Tescan MIRA II LMU scanning electron microscope (SEM, TESCAN, Brno, Czech Republic) using 15 kV acceleration voltage.

2.2. Characterization of Electrical Behavior of CNTs

The CNTs for electrical characterization, such as I-V characteristics and sheet resistance measurements, were grown in the MW plasma torch on p-type, boron-doped, c-Si substrates (8 mm × 8 mm, thickness 525 μm, resistivity <6 $\Omega \cdot cm^{-1}$) coated by the SiO_2 film, 300 nm in thickness, and the top Fe catalytic layer. The square chips fully covered with CNT films were finished by vacuum evaporation of gold pads (0.5 mm × 0.5 mm) in each sample corner.

The chips were investigated at the probe station Cascade M150 connected to the Keithley SCS-4200 semiconductor analyzer (Keithley Instruments, Inc., Cleveland, OH, USA). The I-V characteristics of the films were measured between the opposite corners in the voltage range from −5 V to +5 V. The specific electrical resistance was determined by Van der Pauw measurement which was carried out automatically by the Keithley SCS-4200 analyzer.

For the impedance spectroscopy, the CNTs were grown on a SITAL glass-ceramic substrate (10 mm × 15 mm, thickness 525 µm) using Fe catalytic layer. This substrate was chosen to suppress the effect of Si substrate properties during the impedance spectroscopy measurement. The VA-MWCNTs were deposited on a central circular area, 6 mm in diameter. Two circular gold contacts of 3 mm in diameter were prepared by vacuum evaporation. The impedance was measured using the Agilent E4980A Precision LCR meter (Agilent Technologies, Santa Clara, CA, USA) and data acquisition by LabView software (National Instruments, Austin, TX, USA). The measurements were performed in the frequency range from 20 Hz to 2 MHz with the voltage level of 0.5 V.

2.3. IR Absorption

The FTIR measurement procedure is a simple way to compare samples with or without grown CNTs in order to consider CNTs as possible IR detector. The CNTs for infrared absorption measurements were grown on the same substrates as described in the previous Section 2.2, *i.e.*, c-Si substrate with the SiO_2/Fe double layer. In the case of the bolometer, thin SiO_2 layer is needed for construction of low thickness MEMS diaphragm to suppress thermal loses to the substrate mass [52].

The IR absorption of VA-CNTs on the Si substrate covered by 300 nm thick SiO_2 film was investigated with the Fourier transform infrared (FTIR) spectrometer Nicolet iS50 in the attenuated total reflection (ATR) mode. The ATR crystal was pressed against the sample and the measurement was performed in the wavelength range 2.5–22.5 µm. The absorbance of the VA-CNT sample was compared to the absorbance of Si substrate covered by 300 nm thick SiO_2 film and 5 nm thick Fe film used as catalyst of the CNT growth.

2.4. Electrochemical Characterization

The preparation of samples for the electrochemical measurements started with n-type, highly antimony-doped, c-Si substrates (5 mm × 30 mm, thickness 525 µm, resistivity <0.02 $\Omega \cdot cm^{-1}$) coated by the SiO_2 film, 300 nm in thickness. The working electrode had dimensions 4.5 mm × 4.5 mm and consisted of the CNTs deposited in the MW plasma torch (Section 2.1). Before the CNT growth, the SiO_2 insulating film was removed from the working electrode area by a wet chemical etching in buffered HF and the area of the working electrode was covered by magnetron sputtered Ti (10 nm)/Ta (250 nm) double layer and vacuum evaporated top Fe film, the catalyst for the CNT growth. The Ti/Ta coating was necessary to ensure a good adhesion of the CNTs that otherwise peeled off the substrate when immersed in an electrolyte. The highly doped Si substrate was needed for a good electrical connection between the CNT layer and the contact pads which were situated on the opposite end of the substrate with the same dimensions as the electrode.

Electrochemical response of the $[Fe(CN)_6]^{4-/3-}$ redox couple mediated by the CNT electrode was investigated by the cyclic voltammetry (CV) with AUTOLAB PGSTAT 204 potentiostat/galvanostat controlled by Nova 1.10 software (Metrohm Autolab B.V., Utrecht, The Netherlands). A standard three-electrode voltammetric cell employing an Ag/AgCl reference electrode (type 6.0729.100, Metrohm, Herisau, Switzerland) and a platinum auxiliary electrode (type 6.0343.000, Metrohm) was used for all the experiments. The electrolyte was an equimolar solution of 2.5 mM potassium ferrocyanide and potassium ferricyanide ($[Fe(CN)_6]^{4-/3-}$) in 0.1 M KCl. The cyclic voltammograms were recorded in the potential range from -1 V to $+1$ V with scan rates (υ) from 5 to 500 $mV \cdot s^{-1}$.

2.5. Gas Sensing Properties

The CNTs for testing the gas sensing properties were prepared on the p-type, boron-doped, c-Si substrates (8 mm × 8 mm, thickness 525 μm, resistivity <6 $\Omega \cdot cm^{-1}$) coated with the 92 nm thick SiO_2 film. The catalytic Fe nanoparticles were deposited by the MW plasma torch (see Section 2.1) in the central area, 4 mm × 4 mm, of the substrate. This form of the Fe catalyst was chosen for the growth of a less dense CNT mesh because the gas sensing application requires a large surface area and dense CNTs mask each other. The CNTs were grown from Fe NPs in the MW plasma torch (Ar = 700 sccm, H_2 = 250 sccm, CH_4 = 38 sccm, deposition time 60 s, deposition temperature 973 K) as described in Section 2.1. The measurement chip was finished by vacuum evaporated gold contacts (15 nm NiCr adhesion layer with 350 nm Au layer on the top) with a size of 2 mm × 6 mm centered symmetrically to the middle of the sensor.

The gas sensing properties were determined as a change of the sample resistance during its exposure to a gaseous analyte, either ammonia (NH_3) or isobutane (iC_4H_{10}). The measurements were performed in a custom-built gas station equipped with two gas channels and one chamber for two sensors' characterization at once. One gas channel is used for the synthetic air as a carrier gas. The second gas channel supplies diluted analytes, NH_3 or iC_4H_{10} in nitrogen. The response of the sensors was determined at different analyte concentrations, namely 100 ppm, 250 ppm and 500 ppm. Before each measurement, a sample conditioning was carried out for 30 min at 200 °C in the air flow of 1000 sccm. The sensor response was defined as

$$\Delta R/R_0 = \left((R - R_0)/R_0\right) \times 100\% \tag{1}$$

where R is the resistance of the sensor exposed to the analyte and R_0 is the sensor resistance in pure air. The sensitivity tests were performed at two temperatures, room temperature and 200 °C. The resistance R_0 was determined from 60 min measurement in air flow of 500 sccm. The sensor response to analyte was measured as three 10 min cycles (for 100 ppm, 250 ppm and 500 ppm of analyte) alternated with three 10 min cycles in air flow. The total gas flow rate was kept constant at 500 sccm.

2.6. Field Emission Properties for Pressure Sensing

The MEMS field emission pressure sensor was designed as a diode structure (see Figure 1). It consisted of two n-type, highly antimony-doped, c-Si electrodes (10 mm × 15 mm, thickness 525 μm, resistivity <0.02 $\Omega \cdot cm^{-1}$). One of them was anisotropically etched to a bending diaphragm. The other was coated by an emissive material, the VA-MWCNTs deposited in the MW plasma torch (see Section 2.1). The highly doped Si substrate was needed for a good electrical connection between the CNT layer and the contact pads. The Fe catalytic film, required for the growth of CNTs, was deposited in the center of the substrate on the area of 4 mm × 4 mm. A native oxide film on Si was removed by HF prior to the deposition of the Fe film, thus ensuring the electrical contact between CNTs and Si. The field-emission pressure sensor is proposed to be constructed by the separation of the electrodes with a dielectric layer creating an integrated evacuated volume. The dielectric layer can be made of Pyrex or Simax glass using anodic bonding technology or made of glass frit using a screen printing process.

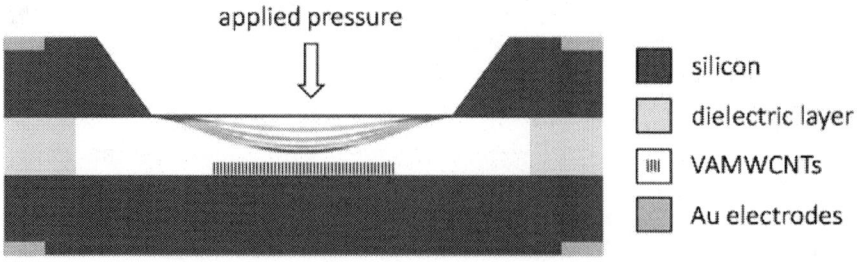

Figure 1. Schematic view of the MEMS pressure sensor with carbon nanotubes emitters.

The measurements of the field emission properties were carried out in a vacuum chamber at pressure lower than 10^{-4} Pa. A diaphragm bending was simulated by the linear nano-motion drive SmarAct enabling precise changes of the distance between the two electrodes inside the vacuum chamber with the step from 50 nm to1000 nm. The initial distance of 120 μm was established using a solid dielectric foil that was then removed and the emitter-to-anode distance was set up with the SmarAct drive from 84 μm to120 μm. The measurement voltage from 0 to150 V was automatically applied using software communicating via GPIB with the voltage supply.

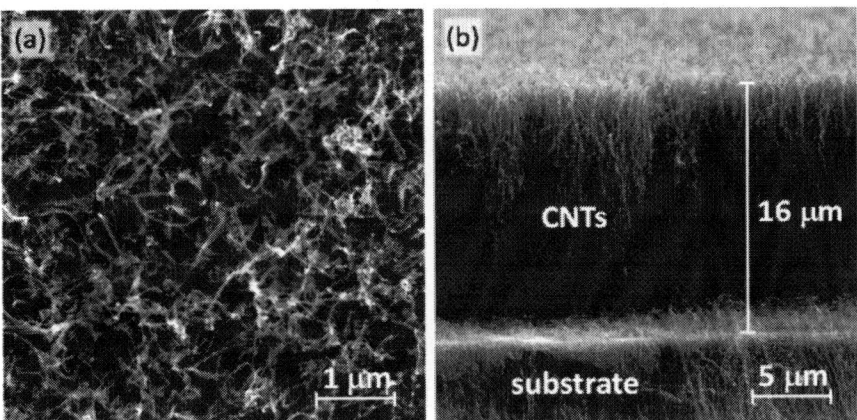

Figure 2. Typical SEM micrographs of the silicon substrate covered with VA-MWCNTs prepared in MW plasma torch (Ar = 700 sccm, H_2 = 250 sccm, CH_4= 25 sccm, deposition time 60 s, deposition temperature 973 K): (**a**) top view of VA-MWCNTs and (**b**) cross-sectional view of the VA-MWCNT film.

3. RESULTS AND DISCUSSION

3.1. SEM of CNTs

A fast growth of VA-MWCNT films on the c-Si and c-Si/SiO_2 substrates has been achieved in MW plasma torch operated at atmospheric pressure without an external heating source [44,45]. The characterization of the VA-MWCNT films by SEM, transmission electron microscopy (TEM), Raman spectroscopy and the influence of process parameters on the CNT growth were discussed in detail in our previous publications [3,46,47]. Typical SEM images of prepared VA-MWCNT film are shown in Figure 2. Although the cross-sectional view in Figure 2bconfirms the vertical alignment of the CNT film, the top view (Figure 2a) reveals that the alignment at the end of nanotubes is not perfect. The CNTs having a high aspect ratio, less than 80 nm in the diameter and a length of about 16 μm, are curled at the top end due to different heights. Therefore, top view micrographs cannot provide sufficient information about the structure of all the CNT film.

3.2. Electrical Properties of CNTs

The measured I-V characteristics of the VA-MWCNT samples were nearly linear as documented in Figure 3afor the sample shown in Figure 2. The resistance was in the range of 1–1.3 kΩ. The specific electrical resistance was about 0.5 Ω·cm as calculated from Van der Pauw measurement and the film thickness of 16 μm determined by SEM. A small nonlinearity was revealed when the difference of the sample resistance and its linear approximation was plotted (Figure 3b). The deviation from linear behavior was governed by an exponential growth with a small exponent. The nonlinearity of the CNT resistance did not exceed 6 Ω which corresponded to 0.5% of the film resistance.

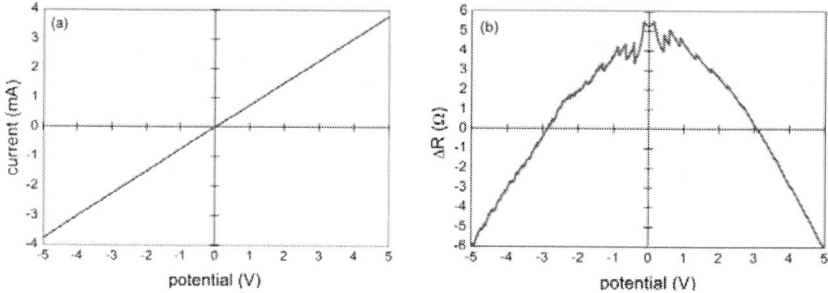

Figure 3. (a) I-V characteristic and (b) deviation of measured resistance from linear regression.

Fitting of the impedance measurements confirmed a simple RC model of two resistances and capacitances in parallel (Figure 4). According to Plombon *et al.* [10], the contact pads added a significant part of the RC circuit element. The R_C/C_C circuit represents the contact impedance, and R_{CNT}/C_{CNT} stands for the impedance of the vertically aligned CNT film. The resistances of the contact and the CNT films were 500 and 700 Ω, respectively. The capacitance of the contacts was much higher, about 15 nF, than the capacitance of the CNT film, 3.5 nF. It means that the surface of nanotubes is not pure enough to create a good contact. Adsorbed molecules such as water, CO_2 and O_2 can create a dielectric film that contributes to its high capacitance.

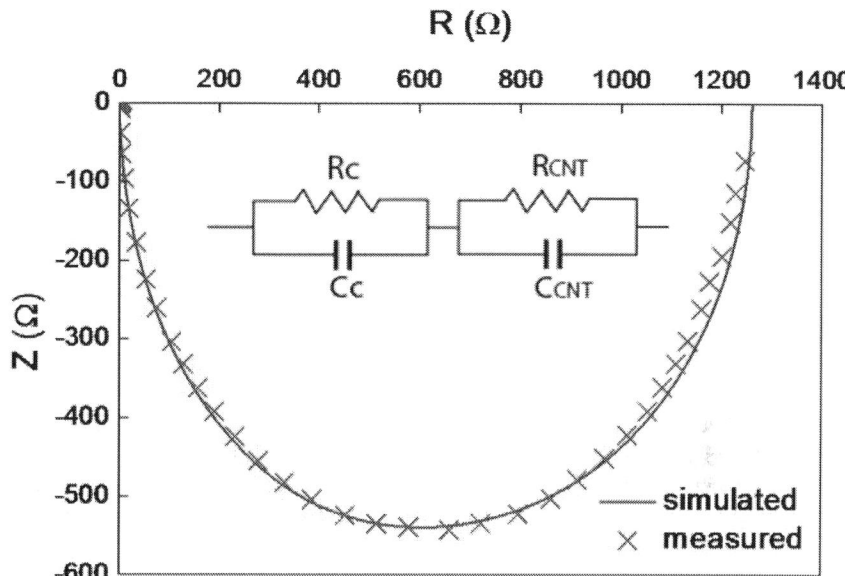

Figure 4. Impedance characteristics, crossed markers are measured data, dashed line is simulated according the inset equivalent circuit.

3.3. IR Absorption

The results of the ATR-FTIR study are shown in Figure 5. They indicate that the VA-MWCNTs can be affectionately applied as a possible absorption layer for IR detection. The mean absorbance value of nearly 80% was obtained. In Figure 5, the interval from 2.5 μm to 7.5 μm represents the absorbance of substrate. The local low points of the curve at approximately 9.0 μm show the typical progression for the atmospheric humidity. The absorbance of the CNTs is mostly seen in the interval from 8 μm to 22 μm which therefore includes the atmospheric window for IR detection.

From the physical principle of the material absorption, it has been well known the particle dipole moment is necessary. No modification is required for CNTs to create the dipole moment according to this measurement. The absorption in the IR region causes changes of vibrational and rotational status of the molecules. The absorption intensity depends on the IR photon energy which can be transferred to the molecule and this depends on the change of the dipole moment that occurs as a result of molecular vibration. As a consequence, a particle will absorb the IR light only if the absorption causes a change in the dipole moment. The absorption frequency is dependent on the vibrational frequency of the molecule.

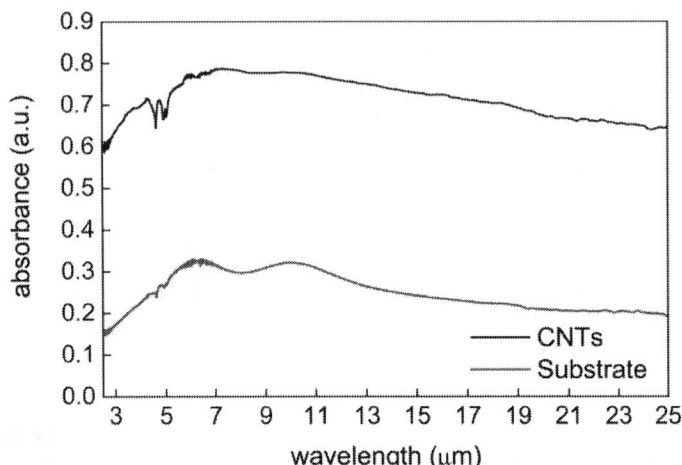

Figure 5. ATR-FTIR spectra obtained for the substrate with the catalytic layer (red line), and the CNT structure (black line).

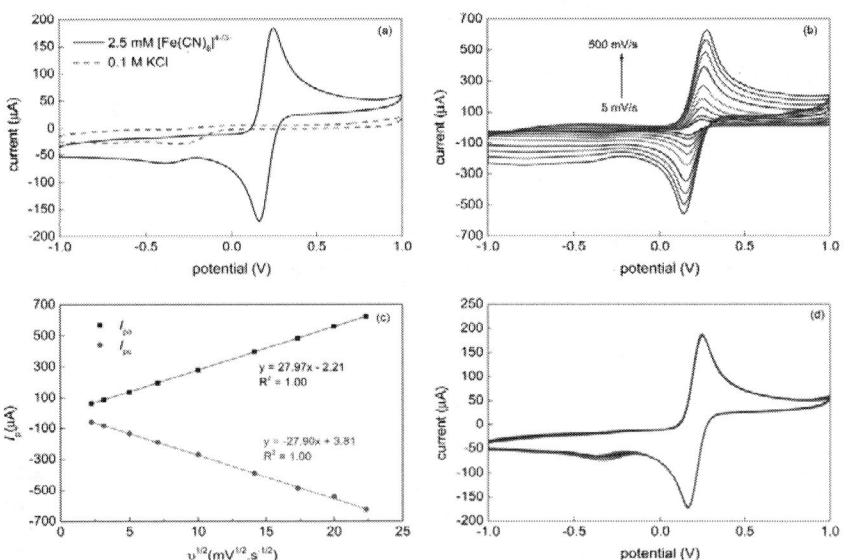

Figure 6. (a) Cyclic voltammograms obtained for 0.1 M KCl and 2.5 mM $[Fe(CN)6]^{4-/3-}$ in 0.1 M KCl at υ = 50 mV·s^{-1} using VAMWCNT-based working electrode; **(b)** Cyclic voltammograms obtained for 2.5 mM $[Fe(CN)6]^{4-/3-}$ in 0.1 M KCl at various scan rates (υ = 5, 10, 25, 50, 100, 200, 300, 400 and 500 mV·s^{-1}) using VAMWCNT-based working electrode; **(c)**I_{pa}vs. $\upsilon^{1/2}$ and I_{pc}vs. $\upsilon^{1/2}$ curves; **(d)** 10 cycles of CV obtained in 2.5 mM $[Fe(CN)6]^{4-/3-}$ in 0.1 M KCl at υ = 50 mV·s^{-1} using VAMWCNT-based working electrode.

3.4. Electrochemical Properties

A representative cyclic voltammogram recorded at 50 mV·s^{-1} with the VA-MWCNT electrode and 2.5 mM [Fe(CN)$_6$]$^{4-/3-}$ (1:1) solution in 0.1 M KCl is shown in Figure 6a. Two well-defined symmetric redox peaks separated by $\Delta E_p = (E_{pa} - E_{pc}) = 83$ mV were observed. The ratio of the anodic and cathodic peak currents reached unity ($I_{pa}/I_{pc} = 1.01$). The results indicated that the VA-MWCNT electrode promote electron transfer quite well.

The effect of varying scan rates was studied for the scan rate range 5–500 mV·s^{-1}. The corresponding cyclic voltammograms are given in Figure 6b. The anodic (I_{pa}) and cathodic (I_{pc}) peak currents varied linearly with the square root of the scan rate ($u^{1/2}$), as shown in Figure 6c. It demonstrates that the electrode process is controlled by a diffusion. The stability of the VA-MWCNT electrode was studied for the [Fe(CN)$_6$]$^{4-/3-}$ in 0.1 M KCl at 50 mV·s^{-1} using 10 cycles of CV. The results, depicted in Figure 6d, revealed that both, the oxidative and reductive, peak currents of the studied redox couple remained practically constant throughout all 10 potential cycles and, therefore, the CNT-based working electrodes with the Ti/Ta adhesive interlayers are suitable for repeated measurements.

Comparing the obtained results with the standard or MWCNT-modified screen-printed electrodes [53], the prepared VA-MWCNT electrodes indicate their high potential for construction of electrochemical sensors or biosensors detecting substances in aqueous solutions.

3.5. Gas Sensing Properties

The results of the CNT sensor response to NH$_3$ at room temperatures and 200 °C are shown in Figure 7a. Since the sensor resistance increased during ammonia exposure, it is concluded that free carriers—the holes—in MWCNTs were neutralized by electrons coming from adsorbed NH$_3$ and, therefore, MWCNTs exhibited p-type nature. This phenomenon has been already described in previous studies. Hoa et al. supposed that MWCNT is a p-type semiconductor and the adsorption of electron donor compound such as NH$_3$ decreases the charge carrier concentration, thus inducing an increase of the resistance [54]. The same mechanism of interaction of adsorbed electron-donor compound with p-type CNTs was also described in [34,55–57]. The main interaction mechanism of ammonia with CNTs is a reversible adsorption. Vikramaditya et al. assume that NH$_3$ is physisorbed on non-doped CNTs [58]. Thus, the sensing mechanism of ammonia with CNTs consists of two main steps, (i) physical adsorption of gas molecules and (ii) a charge transfer between adsorbed molecules and CNTs [59,60]. Testing of the sensor sensitivity to isobutane was carried out at room temperature (Figure 8) and revealed that the sensor is insensitive to a non-polar molecule such as iC$_4$H$_{10}$.

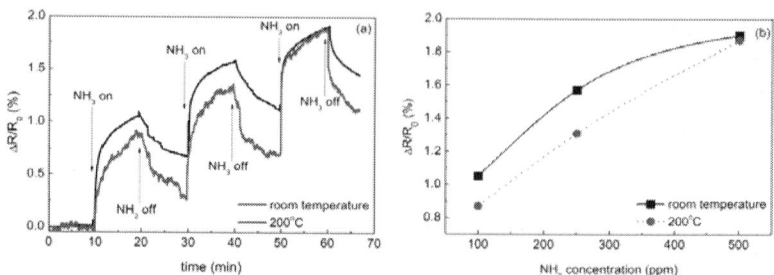

Figure 7. (a) Time dependent response of the CNT sensor to NH$_3$ at room temperature and at 200 °C and **(b)** the response of the sensor determined after 10 min of the exposure as a function of NH$_3$ concentration.

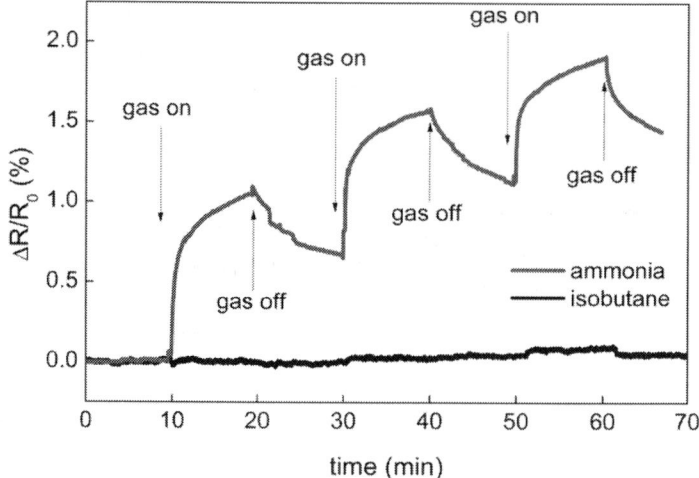

Figure 8. Time dependent response of the CNT sensor to NH$_3$ and iC$_4$H$_{10}$ at room temperature.

According to the data presented in Figure 7b, the sensor exhibited higher response to NH$_3$ at the room temperature than at 200 °C. Since the adsorption is an exothermal process, it is enhanced with the decreasing temperature. The response $\Delta R/R_0$ ranged from 0.87% to 1.9%. The dependence of the sensor response $\Delta R/R_0$ determined after 10 min of the exposure on the concentration of NH$_3$ was almost linear at 200 °C whereas the relation became strongly nonlinear at room temperature. It can be linked to an incomplete desorption of NH$_3$ at room temperature.

The obtained response of the sensors to NH$_3$ was quite high taking into account that MWCNTs were not treated or modified for sensing. The highest response, 1.9%, was achieved at room temperature for 500 ppm of NH_3. For comparison, Hoa *et al.* [54] achieved 8% response for 6% of NH_3 in N_2, *i.e.*, for 60,000 ppm. Here, the reported sensor was not tested for such a high concentration of NH_3 but its response to a 120× lower concentration was only four times lower. A response of 9%, but to a lower concentration of 1% (10,000 ppm), was reported by Cui *et al.* for Ag-decorated MWCNTs [37].

In conclusion, the described sensor preparation technique is quite simple and allows achieving a sufficiently high response. The direct nanotube growth on the substrate allows managing the response by synthesis conditions that is critical for a scale-up production. The present CNT sensor can be further improved by a suitable functionalization and/or a decoration of the CNTs with metallic nanoparticles using the same set-up or a different procedure.

3.6. Field Emission Properties for Pressure Sensing

The measurements of field emission current were carried out multiple times at 10 emitter-to-anode distances from 84 µm to120 µm. The current density in dependence on the electric field intensity calculated from measured results is shown in Figure 9a and the corresponding Fowler-Nordheim (F-N) plot is in Figure 9b. Straight lines in the F-N plot indicates the quantum mechanical tunneling characteristic of the electron field emission and are based on the F-N equation

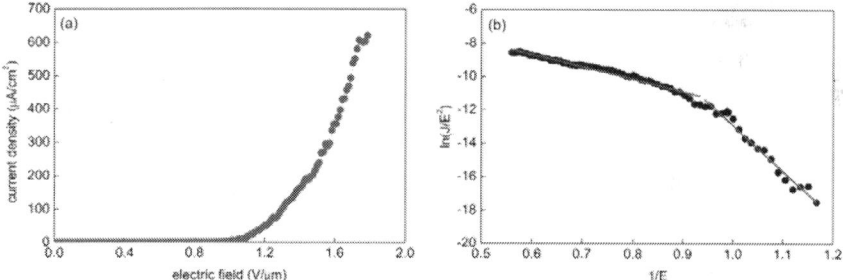

Figure 9. (a) Current density in dependence on field intensity for the CNT array field emitters; **(b)** The relevant Fowler-Nordheim curve.

$$I = (A\alpha\beta^2 V^2 / x^2\phi)\exp[-B\phi^{3/2}x/(\beta V)] \qquad (2)$$

where I is the emission current, A = 1.56 × 10^{-6} A·V^{-2}·eV, B = 6.83 × 10^9 eV$^{-3/2}$·V·m^{-1}, α is the emission area, β is the field enhancement factor, ϕ is the work function, x is the distance between the anode and the emitter, and V is the applied voltage. Emission of many electrons at a low applied voltage requires a low work function (ϕ) and a high field enhancement factor (β). Using the F-N plot, $ln(J/E^2)$ vs. $1/E$, the field enhancement factor can be determined from the slope of the straight line. The measured data can be divided into two sections, below and above 0.9 $\mu m \cdot V^{-1}$, that have different enhancement factors. Such behavior of the plot is not unusual and can be caused by various reasons, such as resistance, gas absorption, localized states or interaction between emitters [61,62].

The turn-on field, defined as the field intensity enabling the emission current density of 10 $\mu A \cdot cm^{-2}$, was determined to be below 1 $V \cdot \mu m^{-1}$. The current density of hundreds of μA was achieved already at 1.8 $V \cdot \mu m^{-1}$. The measured data follow the Fowler-Nordheim law in Equation (2) concerning the dependence of the emission current on the electrode distance when the applied voltage is fixed. Similarly, for the fixed distance between the electrodes, the emission current increases with increasing voltage. This confirms the expected behavior of our field emission electrode as proposed in the pressure sensor design. From the measured characteristics, one could also conclude that it is of advantage to operate at a higher electric field and shorter distances. In these conditions, a bigger change of the emission current with changing diaphragm deformation (pressure) leads to the higher sensitivity of the sensor, i.e., higher $\Delta I/\Delta p$ with the identical Δd. This effect can be strengthened with the proper choice of diaphragm [49].

4. CONCLUSIONS

Selected functional properties of VA-MWCNT films were investigated in different types of measurement devices that integrated CNTs prepared in a microwave plasma torch. The average diameter of prepared MWCNTs was lower than 80 nm and the thickness of uniform VA-MWCNT layers was about 16 μm. Electrical measurements of CNTs showed their resistive behavior with a low resistance of 0.5 $\Omega \cdot cm$, and non-linearity of about 0.5%. A parallel capacitance proposed in previously published papers was confirmed by dynamic measurements. The FTIR analysis demonstrated a high absorption in the IR range of 8–22 μm allowing VA-MWCNT application in IR thermometers such as bolometers usually used in thermo-vision. Cyclic voltammetry using redox couple of $[Fe(CN)_6]^{4-/3-}$ were used to investigate the electrochemical response of the working electrodes coated by plasma grown VA-MWCNTs. The obtained results indicated good electrochemical properties (I_{pa}/I_{pc} = 1.01, ΔE_p = 83 mV) and confirmed that studied VA-MWCNT working electrodes can be successfully used in the construction of electrochemical sensors.

The constructed resistive sensor based on VA-MWCNTs exhibited increasing resistance of the VA-MWCNT layer when exposed to NH_3 that must be explained by the p-type semiconducting behavior of CNTs when adsorbed ammonia gas molecules donate electrons to CNTs. The field emission properties of the electrode with VA-MWCNTs grown directly on silicon were investigated with the aim to use the electrode in pressure sensing applications. The measured dependencies showed that emission current from the CNTs is stable, and a relatively low noise can be achieved for smaller electrode distances and higher voltage between the electrodes. The characteristics were reversible, with low turn-on field (1 $V \cdot \mu m^{-1}$), and current of hundreds of μA for the electric field was around 1.8 $V \cdot \mu m^{-1}$. It indicated that plasma-grown VA-MWCNTs are well suited for pressure sensing.

ACKNOWLEDGMENTS

This work has been performed in laboratories supported by the operational program Research and Development for Innovation, by the SIX project CZ.1.05/2.1.00/03.0072 and supported by Grant Agency of the Czech Republic under the contracts GACR P205/10/1374 and GA13-19947S. Petr Synek would like to acknowledge the project "Employment of Newly Graduated Doctors of Science for Scientific Excellence" (CZ.1.07/2.3.00/30.0009) co-financed from European Social Fund and the state budget of the Czech Republic.

AUTHOR CONTRIBUTIONS

P.M. and J.P. (Jan Prášek) designed, performed, evaluated and wrote the experiments related to the electrochemical and surface characterization and finalized the paper; J.S. designed and performed the experiments related to the standard electrical characterization which was analyzed, evaluated and written by J.H.; J.P. (Jan Pekárek) designed, performed, evaluated and wrote the experiments related to the field emission properties; V.S. designed, performed, evaluated and wrote the experiments related to the optical characterization; A.G.B. designed, performed, evaluated and wrote the experiments related to the gas sensing properties; O.J., P.S. and M.E. fabricated electrodes and described the fabrication process; L.Z. described the fabrication process and made the final proofs.

REFERENCES

1. Meyyappan, M. *Carbon Nanotubes: Science and Applications*, 1st ed.; CRC Press: Boca Raton, FL, USA, 2005; p. 304.

2. Backes, C. *Noncovalent Functionalization of Carbon Nanotubes: Fundamental Aspects of Dispersion and Separation in Water*; Springer: Berlin, Germany, 2012; p. 203.

3. Zajíčková, L.; Jašek, O.; Eliáš, M.; Synek, P.; Lazar, L.; Schneeweiss, O.; Hanzlíková, R. Synthesis of carbon nanotubes by plasma-enhanced chemical vapor deposition in an atmospheric-pressure microwave torch.*Pure Appl. Chem.* **2010**, *82*, 1259–1272.

4. Yen, J.H.; Leu, I.C.; Wu, M.T.; Lin, C.C.; Hon, M.H. Density control for carbon nanotube arrays synthesized by ICP-CVD using AAO/Si as a nanotemplate. *Electrochem. Solid State Lett.* **2004**, *7*, H29–H31.

5. Prášek, J.; Drbohlavová, J.; Chomoucká, J.; Hubálek, J.; Jašek, O.; Adam, V.; Kizek, R. Methods for carbon nanotubes synthesis-review. *J. Mater. Chem.* **2011**, *21*, 15872–15884.

6. Chen, H.; Roy, A.; Baek, J.B.; Zhu, L.; Qu, J.; Dai, L.M. Controlled growth and modification of vertically-aligned carbon nanotubes for multifunctional applications. *Mater. Sci. Eng. R-Rep.* **2010**, *70*, 63–91.

7. Tsierkezos, N.G.; Szroeder, P.; Ritter, U. Multi-walled carbon nanotubes as electrode materials for electrochemical studies of organometallic compounds in organic solvent media. *Monatshefte Chem.-Chem. Mon.* **2011**, *142*, 233–242.

8. Tsierkezos, N.G.; Szroeder, P.; Ritter, U. Application of Films Consisting of Carbon Nanoparticles for Electrochemical Detection of Redox Systems in Organic Solvent Media. *Fuller. Nanotub. Carbon Nanostruct.***2011**, *19*, 505–516.

9. Ahlskog, M.; Hakonen, P.; Paalanen, M.; Roschier, L.; Tarkiainen, R. Multiwalled carbon nanotubes as building blocks in nanoelectronics. *J. Low Temp. Phys.* **2001**, *124*, 335–352.

10. Plombon, J.J.; O'Brien, K.P.; Gstrein, F.; Dubin, V.M.; Jiao, Y. High-frequency electrical properties of individual and bundled carbon nanotubes. *Appl. Phys. Lett.* **2007**, *90*.

11. Ksenevich, V.K.; Gorbachuk, N.I.; Poklonski, N.A.; Samuilov, V.A.; Kozlov, M.E.; Wieck, A.D. Impedance of Single-Walled Carbon Nanotube Fibers. *Fuller. Nanotub. Carbon Nanostruct.* **2012**, *20*, 434–438.

12. Geng, S.N.; Wang, P.; Ding, T.H. Impedance characteristics and electrical modelling of multi-walled carbon nanotube/silicone rubber composites. *Compos. Sci. Technol.* **2011**, *72*, 36–40.

13. Allaoui, A.; Hoa, S.V.; Pugh, M.D. The electronic transport properties and microstructure of carbon nanofiber/epoxy composites. *Compos. Sci. Technol.* **2008**, *68*, 410–416.

14. Popov, V.N. Carbon nanotubes: Properties and application. *Mater. Sci. Eng. R Rep.* **2004**, *43*, 61–102.

15. Goak, J.C.; Lee, H.S.; Han, J.H.; Park, J.-Y.; Seo, Y.; Kim, K.B.; Lee, N. New metric for evaluating the purity of single-walled carbon nanotubes using ultraviolet–visible-near infrared absorption spectroscopy. *Carbon* **2014**, *75*, 68–80.

16. Kruss, S.; Hilmer, A.J.; Zhang, J.; Reuel, N.F.; Mu, B.; Strano, M.S. Carbon nanotubes as optical biomedical sensors. *Adv. Drug Deliv. Rev.* **2013**, *65*, 1933–1950.

17. Huang, H.; Zou, M.; Xu, X.; Liu, F.; Li, N.; Wang, X. Near-infrared fluorescence spectroscopy of single-walled carbon nanotubes and its applications. *TrAC Trends Anal. Chem.* **2011**, *30*, 1109–1119.

18. Gohier, A.; Dhar, A.; Gorintin, L.; Bondavalli, P.; Bonnassieux, Y.; Cojocaru, C.S. All-printed infrared sensor based on multiwalled carbon nanotubes. *Appl. Phys. Lett.* **2011**, *98*.

19. Aliev, A.E. Bolometric detector on the basis of single-wall carbon nanotube/polymer composite. *Infrared Phys. Technol.* **2008**, *51*, 541–545.

20. Afrin, R.; Shah, N.A.; Abbas, M.; Amin, M.; Bhatti, A.S. Design and analysis of functional multiwalled carbon nanotubes for infrared sensors. *Sens. Actuators A Phys.* **2013**, *203*, 142–148.

21. Gao, C.; Guo, Z.; Liu, J.-H.; Huang, X.-J. The new age of carbon nanotubes: An updated review of functionalized carbon nanotubes in electrochemical sensors. *Nanoscale* **2012**, *4*, 1948–1963.

22. Vashist, S.K.; Zheng, D.; Al-Rubeaan, K.; Luong, J.H.T.; Sheu, F.-S. Advances in carbon nanotube based electrochemical sensors for bioanalytical applications. *Biotechnol. Adv.* **2011**, *29*, 169–188.

23. Jacobs, C.B.; Peairs, M.J.; Venton, B.J. Review: Carbon nanotube based electrochemical sensors for biomolecules. *Anal. Chim. Acta* **2010**, *662*, 105–127.

24. Ahammad, A.J.S.; Lee, J.J.; Rahman, M.A. Electrochemical Sensors Based on Carbon Nanotubes. *Sensors* **2009**, *9*, 2289–2319.

25. Agüí, L.; Yáñez-Sedeño, P.; Pingarrón, J.M. Role of carbon nanotubes in electroanalytical chemistry: A review. *Anal. Chim. Acta* **2008**, *622*, 11–47.

26. Ye, M.L.; Xu, B.; Zhang, W.D. Voltammetric Behavior of Rutin at a Vertically Aligned Multiwalled Carbon Nanotubes Electrode. *Sens. Lett.* **2013**, *11*, 321–327.

27. Karuwan, C.; Wisitsoraat, A.; Sappat, A.; Jaruwongrungsee, K.; Patthanasettakul, V.; Tuantranont, A. Vertically Aligned Carbon Nanotube Based Electrochemcial Sensor for Salbutamol Detection. *Sens. Lett.* **2010**, *8*, 645–650.

28. Berti, F.; Lozzi, L.; Palchetti, I.; Santucci, S.; Marrazza, G. Aligned carbon nanotube thin films for DNA electrochemical sensing. *Electrochim. Acta* **2009**, *54*, 5035–5041.

29. Wang, J.A.; Zhang, W.D. Sputtering deposition of gold nanoparticles onto vertically aligned carbon nanotubes for electroanalysis of uric acid. *J. Electroanal. Chem.* **2011**, *654*, 79–84.

30. Ye, M.L.; Xu, B.; Zhang, W.D. Sputtering deposition of Pt nanoparticles on vertically aligned multiwalled carbon nanotubes for sensing L-cysteine. *Microchim. Acta* **2011**, *172*, 439–446.

31. Feng, X.; Irle, S.; Witek, H.; Morokuma, K.; Vidic, R.; Borguet, E. Sensitivity of ammonia interaction with single-walled carbon nanotube bundles to the presence of defect sites and functionalities. *J. Am. Chem. Soc.* **2005**, *127*, 10533–10538.

32. Ndiaye, A.; Bonnet, P.; Pauly, A.; Dubois, M.; Brunet, J.; Varenne, C.; Guerin, K.; Lauron, B. Noncovalent Functionalization of Single-Wall Carbon Nanotubes for the Elaboration of Gas Sensor Dedicated to BTX Type Gases: The Case of Toluene. *J. Phys. Chem. C* **2013**, *117*, 20217–20228.

33. Datta, K.; Ghosh, P.; More, M.A.; Shirsat, M.D.; Mulchandani, A. Controlled functionalization of single-walled carbon nanotubes for enhanced ammonia sensing: A comparative study. *J. Phys. D Appl. Phys.* **2012**, *45*.

34. Zhou, Y.; Jiang, Y.D.; Xie, G.Z.; Du, X.S.; Tai, H.L. Gas sensors based on multiple-walled carbon nanotubes-polyethylene oxide films for toluene vapor detection. *Sens. Actuators B: Chem.* **2014**, *191*, 24–30.

35. Cava, C.E.; Salvatierra, R.V.; Alves, D.C.B.; Ferlauto, A.S.; Zarbin, A.J.G.; Roman, L.S. Self-assembled films of multi-wall carbon nanotubes used in gas sensors to increase the sensitivity limit for oxygen detection. *Carbon* **2012**, *50*, 1953–1958.

36. Ahn, K.S.; Kim, J.H.; Lee, K.N.; Kim, C.O.; Hong, J.P. Multi-wall carbon nanotubes as a high-efficiency gas sensor. *J. Korean Phys. Soc.* **2004**, *45*, 158–161.

37. Cui, S.M.; Pu, H.H.; Lu, G.H.; Wen, Z.H.; Mattson, E.C.; Hirschmugl, C.; Gajdardziska-Josifovska, M.; Weinert, M.; Chen, J.H. Fast and Selective Room-Temperature Ammonia Sensors Using Silver Nanocrystal-Functionalized Carbon Nanotubes. *ACS Appl. Mater. Int.* **2012**, *4*, 4898–4904.

38. Tang, Y.; Zhang, Q.H.; Li, Y.G.; Wang, H.Z. Highly selective ammonia sensors based on $Co_{1-x}Ni_xFe_2O_4$/multi-walled carbon nanotubes nanocomposites. *Sens. Actuators B Chem.* **2012**, *169*, 229–234.

39. Varghese, O.K.; Kichambre, P.D.; Gong, D.; Ong, K.G.; Dickey, E.C.; Grimes, C.A. Gas sensing characteristics of multi-wall carbon nanotubes. *Sens. Actuators B Chem.* **2001**, *81*, 32–41.

40. Wilfert, S.; Edelmann, C. Field emitter-based vacuum sensors. *Vacuum* **2012**, *86*, 556–571.

41. Bonard, J.M.; Maier, F.; Stockli, T.; Chatelain, A.; de Heer, W.A.; Salvetat, J.P.; Forro, L. Field emission properties of multiwalled carbon nanotubes. *Ultramicroscopy* **1998**, *73*, 7–15.

42. Guo, P.S.; Chen, T.; Chen, Y.W.; Zhang, Z.J.; Feng, T.; Wang, L.L.; Lin, L.F.; Sun, Z.; Zheng, Z.H. Fabrication of field emission display prototype utilizing printed carbon nanotubes/nanofibers emitters. *Solid State Electron.* **2008**, *52*, 877–881.

43. Nakahara, H.; Kusano, Y.; Kono, T.; Saito, Y. Evaluations of carbon nanotube field emitters for electron microscopy. *Appl. Surf. Sci.* **2009**, *256*, 1214–1217.

44. Zajíčková, L.; Eliáš, M.; Jašek, O.; Kudrle, V.; Frgala, Z.; Matějková, J.; Buršík, J.; Kadlečíková, M. Atmospheric pressure microwave torch

for synthesis of carbon nanotubes. *Plasma Phys. Control. Fusion* **2005**, *47*, B655–B666.

45. Jašek, O.; Eliáš, M.; Zajíčková, L.; Kudrle, V.; Bublan, M.; Matějková, J.; Rek, A.; Buršík, J.; Kadlečíková, M. Carbon nanotubes synthesis in microwave plasma torch at atmospheric pressure. *Mater. Sci. Eng. C* **2006**,*26*, 1189–1193.

46. Jašek, O.; Eliáš, M.; Zajíčková, L.; Kučerová, Z.; Matějková, J.; Rek, A.; Buršík, J. Discussion of important factors in deposition of carbon nanotubes by atmospheric pressure microwave plasma torch. *J. Phys. Chem. Solids* **2007**, *68*, 738–743.

47. Zajíčková, L.; Eliáš, M.; Jašek, O.; Kučerová, Z.; Synek, P.; Matějková, J.; Kadlečíková, M.; Klementová, M.; Buršík, J.; Vojáčková, A. Characterization of Carbon Nanotubes Deposited in Microwave Torch at Atmospheric Pressure. *Plasma Process. Polym.* **2007**, *4*, S245–S249.

48. Zajíčková, L.; Synek, P.; Jašek, O.; Eliáš, M.; David, B.; Buršík, J.; Pizurová, N.; Hanzlíková, R.; Lazar, L. Synthesis of carbon nanotubes and iron oxide nanoparticles in MW plasma torch with Fe(CO)$_5$ in gas feed.*Appl. Surf. Sci.* **2009**, *255*, 5421–5424.

49. Pekárek, J.; Vrba, R.; Prášek, J.; Jašek, O.; Majzlíková, P.; Pekárková, J.; Zajíčková, L. MEMS Carbon Nanotubes Field Emission Pressure Sensor with Simplified Design: Performance and Field Emission Properties Study. *IEEE Sens. J.* **2015**, *15*, 1430–1436.

50. Synek, P.; Jašek, O.; Zajíčková, L.; David, B.; Kudrle, V.; Pizurová, N. Plasmachemical synthesis of maghemite nanoparticles in atmospheric pressure microwave torch. *Mater. Lett.* **2011**, *65*, 982–984.

51. Synek, P.; Jašek, O.; Zajíčková, L. Study of Microwave Torch Plasmachemical Synthesis of Iron Oxide Nanoparticles Focused on the Analysis of Phase Composition. *Plasma Chem. Plasma Process.* **2014**, *34*, 327–341.

52. Bhan, R.K.; Saxena, R.S.; Jalwania, C.R.; Lomash, S.K. Uncooled Infrared Microbolometer Arrays and their Characterisation Techniques. *Def. Sci. J.* **2009**, *59*, 580–589.

53. Majzlíková, P.; Prášek, J.; Eliáš, M.; Jašek, O.; Pekárek, J.; Hubálek, J.; Zajíčková, L. Comparison of different modifications of screen-printed working electrodes of electrochemical sensors using carbon nanotubes and plasma treatment. *Phys. Status Solidi* **2014**, *211*, 2756–2764.

54. Hoa, N.D.; van Quy, N.; Cho, Y.; Kim, D. An ammonia gas sensor based on non-catalytically synthesized carbon nanotubes on an anodic aluminum oxide template. *Sens. Actuators B Chem.* **2007**, *127*, 447–454.

55. Sidek, R.M.; Yusof, F.A.M.; Yasin, F.M.; Wagiran, R.; Ahmadun, F. Electrical response of multi-walled carbon nanotubes to ammonia and carbon dioxide. Proceedings of the 2010 IEEE International Conference on Semiconductor Electronics (ICSE), Melaka, Malaysia, 28–30 June 2010; pp. 263–266.

56. Firouzi, A.; Sobri, S.; Yasin, F.M.; Ahmadun, F. Synthesis of Carbon Nanotubes by Chemical Vapor Deposition and their Application for CO_2 and CH_4 Detection. Proceedings of the 2010 International Conference on Nanotechnology and Biosensors, Hong Kong, China, 28–30 December 2011; Volume 2, pp. 169–172.

57. Han, J.W.; Kim, B.; Li, J.; Meyyappan, M. A carbon nanotube based ammonia sensor on cellulose paper. *RSC Adv.* **2014**, *4*, 549–553.

58. Vikramaditya, T.; Sumithra, K. Effect of Substitutionally Boron-Doped Single-Walled Semiconducting Zigzag Carbon Nanotubes on Ammonia Adsorption. *J. Comput. Chem.* **2014**, *35*, 586–594.

59. Teerapanich, P.; Myint, M.T.Z.; Joseph, C.M.; Hornyak, G.L.; Dutta, J. Development and Improvement of Carbon Nanotube-Based Ammonia Gas Sensors Using Ink-Jet Printed Interdigitated Electrodes. *IEEE Trans. Nanotechnol.* **2013**, *12*, 255–262.

60. Van Hieu, N.; Dung, N.Q.; Tam, P.D.; Trung, T.; Chien, N.D. Thin film polypyrrole/SWCNTs nanocomposites-based NH3 sensor operated at room temperature. *Sens. Actuators B Chem.* **2009**, *140*, 500–507.

61. Cheng, C.Y.; Nakashima, M.; Teii, K. Low threshold field emission from nanocrystalline diamond/carbon nanowall composite films. *Diam. Relat. Mater.* **2012**, *27–28*, 40–44.

62. Obraztsov, A.N.; Zakhidov, A.A.; Volkov, A.P.; Lyashenko, D.A. Non-classical electron field emission from carbon materials. *Diam. Relat. Mater.* **2003**, *12*, 446–449.

Inkjet Printing of Carbon Nanotubes

Ryan P. Tortorich 1 and Jin-Woo Choi [1,2,*]

[1] School of Electrical Engineering and Computer Science, Louisiana State University, Baton Rouge, LA 70803, USA
[2] Center for Advanced Microstructures and Devices, Louisiana State University, Baton Rouge, LA 70803, USA

ABSTRACT

In an attempt to give a brief introduction to carbon nanotube inkjet printing, this review paper discusses the issues that come along with preparing and printing carbon nanotube ink. Carbon nanotube inkjet printing is relatively new, but it has great potential for broad applications in flexible and printable electronics, transparent electrodes, electronic sensors, and so on due to its low cost and the extraordinary properties of carbon nanotubes. In addition to the formulation of carbon nanotube ink and its printing technologies, recent progress and achievements of carbon nanotube inkjet printing are reviewed in detail with brief discussion on the future outlook of the technology.

Keywords: carbon nanotube ink; carbon nanotube patterning; inkjet printing; flexible electronics

INTRODUCTION

Carbon nanotubes (CNTs) have truly become one of the most exciting materials in recent years due to their extraordinary properties. In particular, the electrical properties of carbon nanotubes lend themselves to many applications including use in transistors [1,2], radio-frequency identification (RFID) tags [3], sensors [4,5,6,7,8], photonics [9,10], biological sensing labels [11], and more. One of the most interesting applications of carbon nanotubes is that of transparent electrodes. Considering indium tin oxide (ITO) is the dominant commercial material for transparent electrodes, carbon nanotubes would provide a cheaper alternative. Furthermore, not only can carbon nanotubes assist in reducing the cost of these types of electronic devices, but they can also allow these devices to become flexible.

In order to take advantage of the unique properties of carbon nanotubes, many groups have experimented with various carbon nanotube deposition methods such as dip coating [12], spray coating [2,8,13,14], electrophoretic deposition [15], and others. However, one of the prominent methods of interest today is carbon nanotube printing. There have been demonstrations of screen printing [16], aerosol printing [17,18,19], transfer printing [20], and contact printing [21] to deposit carbon nanotubes on various substrates, but the most favorable form of printing is that of inkjet printing.

Inkjet printing offers unique advantages over other methods of printing. It requires absolutely no prefabrication of templates, allowing for a rapid printing process at low cost. Additionally, due to its precise method of patterning, post-printing steps are not necessary. Furthermore, multiple materials can be deposited simultaneously with the use of multiple ink cartridges, and the amount of deposited material can be controlled with great precision. Finally, due to the nature of inkjet printing technology, multiple layers can be printed on top of one another with great ease. Inkjet printing is currently being used to deposit various types of conductive nanomaterials such as gold [22,23] and silver [24,25]. Although these metals are excellent conductors, carbon nanotubes are cheaper and more versatile in the sense that they can behave as both a semiconductor and a conductor.

Before discussing inkjet printing as it pertains to carbon nanotube printing, it is first necessary to review the various inkjet printing technologies. In general, inkjet printing can be split into two categories, namely continuous and drop-on-demand. As suggested by its name, continuous inkjet printing supplies a continuous stream of ink droplets. These droplets are charged upon leaving the nozzle and are then deflected by voltage plates, where the applied voltage determines whether the droplet will be deposited onto the substrate or recycled through the gutter. Consequently, when the printer is not actually printing anything onto a substrate, a stream of droplets is still being ejected from the nozzle and recycled through the gutter.

While continuous inkjet printers are still used, drop-on-demand inkjet printers are more common. As opposed to a continuous inkjet printer, a drop-on-demand inkjet printer ejects a droplet of ink only when it is told to do so. Therefore, when the printer is not actually printing anything onto a substrate, there are no droplets being ejected from the nozzle. Drop-on-demand inkjet printers can be further split into two categories, namely thermal and piezoelectric. Thermal inkjet printers, sometimes referred to bubble jet printers, contain a thin film resistor in the nozzle. In order to eject a droplet, this thin film resistor is heated by passing current through it. This causes the ink in the nozzle to vaporize, creating a bubble and a large increase in pressure, which forces ink droplets out of the nozzle. Hewlett-Packard, Canon, and Lexmark employ this type of drop-on-demand inkjet printer.

Piezoelectric inkjet printers contain a piezoelectric transducer in the nozzle. When voltage is applied to the piezoelectric transducer, it deforms and causes an increase in pressure, which forces ink droplets out of the nozzle. In terms of consumer printers, Epson employs this type of drop-on-demand inkjet printer. However, many specialized commercial inkjet printers, such as the Fujifilm Dimatix, employ the piezoelectric drop-on-demand technology as well.

Although inkjet printing has its advantages, it also has its obstacles and difficulties. The first step in inkjet printing is formulating ink. There are several issues to consider when mixing ink to be used in an inkjet printer. In general, the ink must maintain a low surface tension as well as a low viscosity. Aside from these properties, incorporating nanomaterials into an ink presents further issues, primarily due to the difficulty of dispersing the nanomaterial within the ink. More specifically, a well-dispersed nanomaterial ink should be free from flocculation of the nanomaterial within the ink. There is a great deal of current research being done on carbon nanotube dispersion, and there have been reports on dispersing carbon nanotubes in water [26,27,28,29,30,31,32,33] as well as organic solvents such as dimethylformamide (DMF) [26,34,35], N-methyl-2-pyrrolidone (NMP) [26,35,36], chloroform [26,37], and others [35].

With the basics of inkjet printing covered, the specifics of carbon nanotube inkjet printing will now be discussed in the following sections. This includes carbon nanotube networks, formulation and preparation of carbon nanotube ink, and key aspects of ongoing research. Along the way, advantages and disadvantages will be discussed for varying methods.

2. CARBON NANOTUBE NETWORK

Before reviewing both the difficulties in formulating carbon nanotube ink and the current research, it is important to first understand how inkjet printing of carbon nanotubes can be used to create conductive traces. When carbon nanotubes, or any one-dimensional nanomaterial, are printed onto a substrate, the solvent evaporates, leaving behind a random network of carbon nanotubes. This network would be analogous to dropping a handful of spaghetti onto a tabletop. Some of the spaghetti might not be in contact with any other spaghetti. In a similar way, some of the carbon nanotubes might be completely isolated without having contact with any other carbon nanotubes. In this case, electrons are confined to a single carbon nanotube. Consequently, isolated carbon nanotubes do not contribute to the conductivity of the printed film. On the other hand, some of the spaghetti may indeed be in contact with other spaghetti, just as some of the carbon nanotubes may be in contact with other carbon nanotubes. This essentially creates an electron pathway. Electrons are capable of traveling from one carbon nanotube to another, ultimately resulting in current, which is the reason for the conductivity of the printed film. Figure 1 demonstrates this concept.

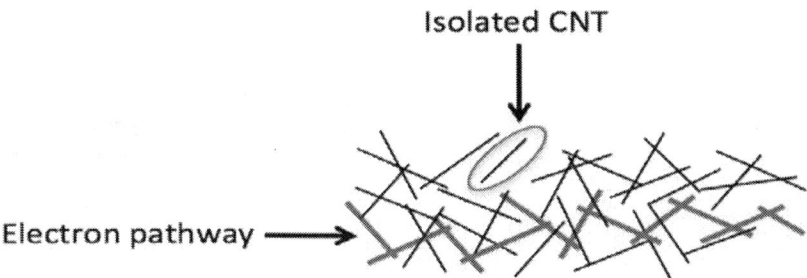

Figure 1. Random carbon nanotube network showing both isolated carbon nanotube and formation of electron pathway via overlapping carbon nanotubes.

As expected, the amount of current is directly related to the number of electron pathways. This suggests that the length of carbon nanotubes plays an important role in the conductivity of a carbon nanotube thin film. Revisiting the aforementioned pasta analogy, if the spaghetti pieces are short, the probability of them touching each other decreases. For a carbon nanotube network, this corresponds to a lower conductivity. On the contrary, if the spaghetti pieces are long, the probability of them touching increases, which corresponds to a higher conductivity in a carbon nanotube network. It will soon be shown that in order to achieve highly conductive films of carbon nanotubes by inkjet printing, it is necessary to print multiple layers of carbon nanotubes. This initially results in a substantial increase in conductivity since each additional layer of carbon nanotubes provides a denser network and produces more electron pathways. However, eventually, the conductivity of the printed film will reach the carbon nanotube bulk conductivity. Figure 2 shows estimated data from two recent reports on carbon nanotube inkjet printing [38,39] as well as our own recent test results. It should be noted that sheet resistance is plotted, which is both more common and useful than conductivity.

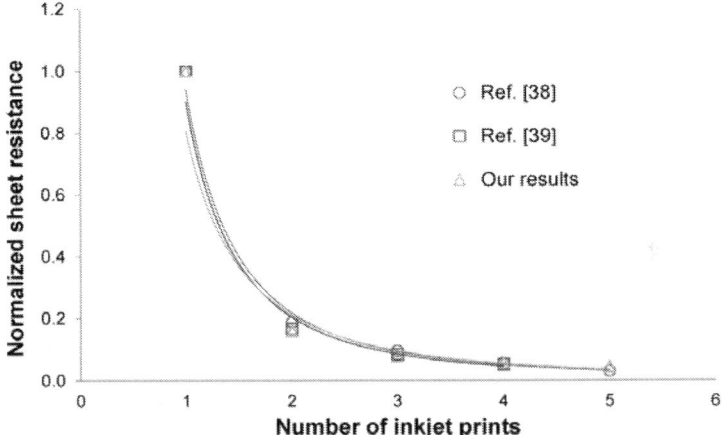

Figure 2. Estimated sheet resistance versus number of prints for two recent reports on carbon nanotube inkjet printing [38,39] and our own recent test results on carbon nanotube inkjet printing. Sheet resistance values are normalized.

The bulk conductivity is determined by many factors. Here again, the length of the carbon nanotubes also plays a major role. As demonstrated by Hecht *et al.*, the conductivity of a carbon nanotube network increases as the length of the carbon nanotubes increases [40]. Additionally, the type of carbon nanotubes used affects the conductivity as well, which includes single-walled or multi-walled, semiconducting or metallic, pristine or functionalized, and other variations. Further, dispersants can even reduce the conductivity of carbon nanotube networks by inhibiting contact between carbon nanotubes [41]. Finally, the drying process can affect the distribution and orientation of carbon nanotubes, which will be discussed in Section 4.4. Nevertheless, this random network of carbon nanotubes is essential in forming a conductive thin film.

3. CARBON NANOTUBE INK

There are many issues that need to be taken into account regarding carbon nanotube ink. First and foremost, carbon nanotube dispersion is a major obstacle. Aside from dispersion, surface tension and viscosity are important characteristics for carbon nanotube ink. Finally, preparation of carbon nanotube ink involves multiple steps to ensure well-dispersed carbon nanotubes and removal of any carbon nanotube bundles.

3.1. Carbon Nanotube Dispersion

Due to the nature of the material, carbon nanotubes are quite difficult to disperse in a liquid. The van der Waals forces between carbon nanotubes can easily cause agglomeration and sedimentation, which is highly undesirable due to the possibility of clogging the inkjet nozzle. As a result, many groups have experimented with various methods of carbon nanotube dispersion through the use of sidewall functionalization, organic solvents, and dispersants in the case of water-based ink.

3.1.1. Functionalized Carbon Nanotube Dispersion

The first method for carbon nanotube dispersion is that of sidewall functionalization. This entails a chemical process whereby molecules are bound to the carbon nanotube sidewalls. One of the most common methods of carbon nanotube functionalization used to enhance dispersion is called carboxylation. In carboxylation, carboxyl groups ($-COOH$) are attached to the sidewalls of carbon nanotubes through a series of chemical steps. Unlike the hydrophobic carbon nanotube sidewalls, these carboxyl groups are hydrophilic, reducing the possibility for carbon nanotube bundling. On the downside, the chemical steps necessary for functionalization tend to introduce defects into the carbon nanotube sidewalls. This effectively decreases the conductivity of the carbon nanotubes. Nevertheless, a few groups have successfully formulated and printed carbon nanotube ink using functionalized carbon nanotubes [41,42,43].

3.1.2. Organic Solvent-Based Carbon Nanotube Dispersion

In order to avoid hindering the conductive nature of carbon nanotubes, other means of dispersion can be used. For example, organic solvents are superb in their ability to disperse carbon nanotubes. There is no need for functionalization of the carbon nanotube sidewalls or addition of other materials to enhance dispersion. Rather, the solvent itself works as a dispersant. The organic solvent molecules adsorb onto the carbon nanotube surface due to a hydrophobic interaction, countering the strong van der Waals forces between the nanotubes [34]. Additionally, due to their inherent low surface tension, there is no need to add a wetting agent to organic solvent-based inks. On the contrary, many organic solvents present some issues for practical use. First, it has been reported that organic solvents have a carbon nanotube concentration limit of approximately 0.1 mg/mL [44]. Organic solvents also tend to be quite volatile, which can cause problems both when the ink is being prepared and when the ink is being used in a cartridge. Unless the cartridge is sealed properly, the solvent will evaporate, leaving behind nothing but the carbon nanotubes and ultimately clogging the nozzle. Another problem encountered when dealing with organic solvents is that of health and environmental effects. If the proper precautions are not taken, there may be some serious consequences. Lastly, organic solvents can be very corrosive to certain polymer materials. As a result, cartridges used for organic solvent-based carbon nanotube ink must be made of materials that resist their corrosive property. This corrosive characteristic also limits the substrate selection for organic solvent-based carbon nanotube inks. Despite the difficulties, many groups have successfully developed and printed carbon nanotube inks using organic solvents such as DMF [3,38,45,46,47,48] and NMP [49].

3.1.3. Water-Based Carbon Nanotube Dispersion
In addition to organic solvent-based carbon nanotube inks, some groups have developed and printed water-based carbon nanotube inks with the use of dispersants rather than functionalization of the carbon nanotubes [39,50,51,52,53,54]. These water-based inks are environmentally friendly, easy to store, and safer to handle. However, water-based inks are much more difficult to develop since carbon nanotubes do not readily disperse in water without the aid of additional dispersants. As the surface of carbon nanotubes is hydrophobic, the nanotubes do not want to be in contact with water. Rather, they bundle together due to the attractive van der Waals forces.

There are a few ways to overcome these strong van der Waals forces. Aside from sidewall functionalization, surfactants and polymers can be used to cover the surface of each carbon nanotube in order to negate the strong van der Waals forces. This is achieved through both physical and chemical means. Surfactants are amphiphilic molecules having a hydrophilic head and a hydrophobic tail. Thus, when a surfactant is introduced into a water-based carbon nanotube ink, the surfactant molecules adsorb onto the surface of each carbon nanotube due to the hydrophobic tail. This essentially forms a barrier around the perimeter of the carbon nanotube, which acts as the physical means to negate the van der Waals forces when carbon nanotubes are in close proximity to each other. Additionally, because the outer layer of the surfactant-covered carbon nanotube consists of the hydrophilic heads, there is a repulsive chemical force between each carbon nanotube. Polymers, on the other hand, are long chains of monomers that wrap around the carbon nanotubes forming a helix. In a similar fashion to surfactants, polymers provide both a physical and a chemical means for overcoming the van der Waals forces. Figure 3 briefly illustrates how surfactants adsorb onto the carbon nanotube surface.

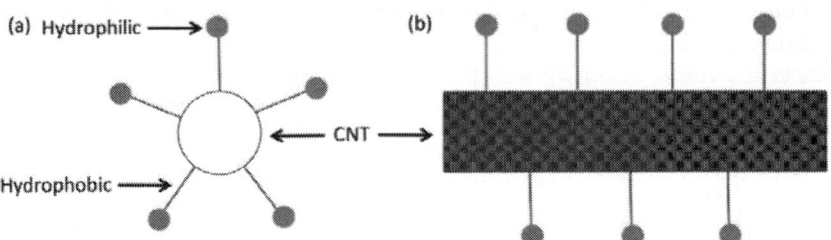

Figure 3. Surfactant-assisted dispersion of carbon nanotubes: (**a**) cross-section of carbon nanotube; and (**b**) side view of carbon nanotube.

3.2. Carbon Nanotube Ink Surface Tension

In order for an ink droplet to be ejected from the nozzle, the ink must maintain a low surface tension. Due to the extremely small volume of ink being ejected from the nozzle (in the pl range), a low surface tension is absolutely necessary. If the surface tension is too high, the ink droplets may remain in the nozzle of the cartridge, which is highly undesirable.

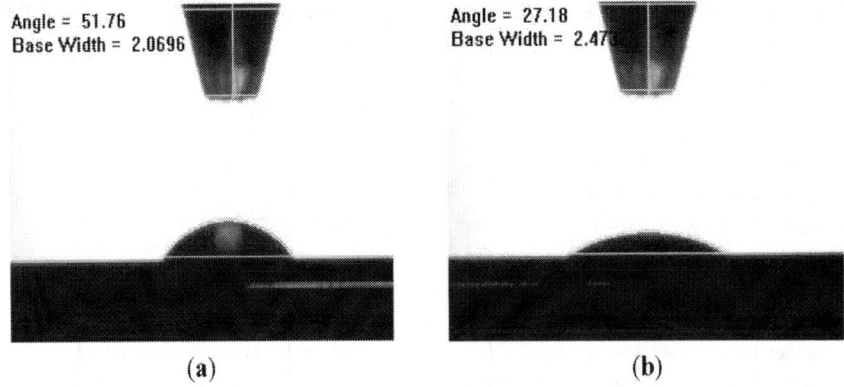

Figure 4. Effect of surfactant on surface tension: (**a**) 3 µL droplet of water without surfactant; and (**b**) 3 µL droplet of water with surfactant. Surfactant clearly decreases the surface tension of the droplet.

As mentioned in Section 3.1.2, organic solvents already have a low surface tension, so they do not require the addition of wetting agents. However, unlike organic solvents, water has a very high surface tension, resulting from the strong cohesive interaction between water molecules. In order to combat this high surface tension, wetting agents are used to lower the surface tension. Typically, surfactants are used as the wetting agent in carbon nanotube inks. In a liquid like water, the surfactant molecules accumulate on the water-air interface due to their amphiphilic structure. This ultimately reduces the cohesive forces between water molecules at the surface, which results in lower surface tension, allowing the water to spread out more on a given surface as shown in Figure 4.

3.3. Carbon Nanotube Ink Preparation

After determining the ingredients and relative concentrations for the carbon nanotube ink, a series of steps are performed in order to obtain useable ink. First, the ink needs to be mixed in order to disperse the carbon nanotubes within the liquid. This can be done in many ways, but the most common approach for dispersing carbon nanotubes is sonication, which uses high frequency vibrations to separate carbon nanotubes within a liquid. Although it works very well, this method also has its drawbacks. In particular, sonication can both shorten carbon nanotubes and cause defects. In the former case, shorter carbon nanotubes reduce the probability of forming an electron pathway in a carbon nanotube network, which can decrease the conductivity of the printed film. In the latter case, defects can negatively affect the inherent conductivity of the carbon nanotubes. Finally, sonication can both physically and chemically affect the solvent and dispersants used in a carbon nanotube ink [44]. Nevertheless, sonication seems to be the primary method of choice for carbon nanotube dispersion.

After dispersing the carbon nanotubes, the ink is centrifuged in order to separate the well-dispersed carbon nanotubes from the bundles or agglomerations, which could clog the printer nozzle. The supernatant solution is then collected and may be centrifuged again. Sometimes, the carbon nanotube ink is also filtered in order to further remove any bundles of carbon nanotubes that could clog the printer nozzle. The filtering step may be performed multiple times to ensure a uniform and well-dispersed carbon nanotube ink. Once the ink is formulated, it is loaded into an inkjet cartridge and ready to be printed.

4. CARBON NANOTUBE INKJET PRINTING

One of the earlier demonstrations of carbon nanotube inkjet printing was reported by Fan et al. in 2005 [39], but one of the more recognized works was reported by Kordás et al. in 2006 [43]. Since then, there have been numerous displays of carbon nanotube inkjet printing, all of which have been successful in producing conductive carbon nanotube films. Rather than providing an exhausting review of each and every demonstration, a few key aspects of current research are discussed in this section.

4.1. Inkjet Printers

As stated previously, there are multiple types of inkjet printers, and all of them have been used for carbon nanotube printing. Consumer inkjet printers are quite cheap and offer familiarity, so there is no need to learn new software or hardware. Nevertheless, these printers are made to print a specific type of ink, so developing useable ink can be a bit more difficult. The new ink must match the original ink in all aspects. Furthermore, in the instance where a new ink clogs the nozzle, some consumer inkjet printers are easier to clean than others. In general, each Hewlett-Packard printer cartridge has its own nozzle, allowing the user to easily remove the cartridge and clean it. On the other hand, the nozzle for Epson printer cartridges is built into the printer itself and cannot be easily removed for cleaning. The most prominent disadvantage for consumer inkjet printers is their overall lack of control. In particular, the drop volume and spacing cannot be adjusted, and the resolution is relatively low. Regardless of these issues, there have been successful demonstrations of printing carbon nanotubes with consumer inkjet printers.

Commercial inkjet printers like the popular Fujifilm Dimatix are specifically made for printing various types of materials. As a result, they have a great deal of control over drop volume and spacing, and they provide better resolution. Although these specialized inkjet printers can be expensive, they seem to be a good choice for carbon nanotube printing due to their superior functionality.

4.2. Sheet Resistance

In order for carbon nanotube films to replace other metallic conductors, they must maintain a comparable sheet resistance. Some groups were able to achieve a sheet resistance below 1 kΩ/$\square$ using multiple layers of carbon nanotubes, the lowest being 78 Ω/$\square$ with a total of 200 prints demonstrated by Chen *et al.* [53]. Although this is a very low sheet resistance, performing 200 prints is not ideal. Taking the print number into account, the lowest recorded sheet resistance is 760 Ω/$\square$ with a total of 12 prints [50].

One key factor that can play a major role in sheet resistance is that of dispersants. As mentioned in Section 2, dispersants can reduce the conductivity of carbon nanotube thin films. When dispersants are used in carbon nanotube ink, they form a physical barrier around the carbon nanotubes. In a carbon nanotube network, these dispersants can inhibit the contact between carbon nanotubes, possibly resulting in a very large decrease in conductivity. In order to diminish this effect, the dispersant concentration can be decreased. Consequently, there may be a reduction in the amount of dispersant that covers each nanotube, allowing for better contact between carbon nanotubes. On the contrary, decreasing the dispersant concentration can also result in a lower concentration of carbon nanotubes, which subsequently decreases the conductivity. Another possible way to prevent dispersants from reducing the conductivity is by simply removing them. Many dispersants are soluble in water and other liquids. By placing the substrate into one of these solvents, the dispersants may detach from the carbon nanotubes and dissolve into the liquid. It should be noted that during this process, some carbon nanotubes might detach from the substrate as well and disperse in the solvent, which can significantly reduce the conductivity.

4.3. Transparency

Aside from sheet resistance, in order for carbon nanotube films to replace transparent electrodes like indium tin oxide, they must maintain a comparable transparency. This involves a delicate balance because increasing the conductivity through multiple prints directly affects the transparency. As more and more carbon nanotubes are deposited onto a given substrate, the film becomes less and less transparent. Although Chen et al. were able to achieve a very low sheet resistance, the transmittance was only 10% [53]. To the authors' knowledge, there has not been a report of carbon nanotube inkjet printing that demonstrates both good sheet resistance and good transparency. However, Mustonen et al. did accomplish this task using a composite ink made of carboxyl functionalized single-walled carbon nanotubes (SWCNT-COOHs) and poly(3,4-ethylenedioxythiophene)-poly(styrenesulfonate) (PEDOT-PSS) [55]. With 25 prints, the conductive film reached a sheet resistance of 1 kΩ/$\square$ and a transmittance of 70%.

4.4. The Coffee Stain Effect

The well-known phenomenon denoted as the coffee stain effect occurs as a droplet of ink dries on the substrate. During drying, carbon nanotubes are pushed to the perimeter of the droplet due to an internal flux [56]. Denneulin et al. even demonstrated that carbon nanotubes orient themselves in specific directions at the perimeter of a drying ink droplet [57]. In order to overcome this, Denneulin et al. used a SWCNT-

COOH/PEDOT-PSS composite ink. Other methods for limiting the coffee stain effect include heating the substrate and treating the substrate surface, which can both accelerate the drying process [38]. This helps to prevent flocculation of carbon nanotubes, allowing for a more uniform distribution and ultimately a more conductive film.

Seeing that carbon nanotube inkjet printing is quite involved, Table 1 on the following page provides a side-by-side comparison of recent reports on carbon nanotube inkjet printing.

Table 1. Comprehensive comparison of recently reported carbon nanotube printing.

Reference Number	Cited Papers	Printer	Solvent	Dispersant and Concentration	CNT Type and Concentration	Preparation	Best Sheet Resistance	Notable Feature
[43]	[39]	Canon	Water	Functionalized	MWCNT-COOH 0.26 mg/mL	Sonication Stirring Centrifuge	40 kΩ/□ 90 prints	One of the first reported
[42]	[43,55]	Dimatix	Water	Functionalized	SWCNT-COOH 0.1 mg/mL	Sonication Centrifuge	Not reported	FET-like behavior
[41]	[42,43,49]	Dimatix	Water	Functionalized	SWCNT-COOH (carboxylic acid) SWCNT-CONH$_2$ (amide) SWCNT-PEG (polyethylene glycol) SWCNT-PABS (polyaminobenzene sulfonic acid) 0.13 mg/mL	Sonication Centrifuge	Estimated 2 kΩ/□ (for COOH and PABS) 14 prints (Assumed)	Fully inkjet printed FET
[46]	-	MicroJet	DMF	n/a	SWCNT 20 µg/mL	Centrifuge	Not reported	Gas sensing
[47]	-	MicroJet	DMF	n/a	SWCNT 0.01 mg/mL	Sonication	Not reported	Field emission display
[3]	[38,46]	Dimatix	DMF	n/a	SWCNT 0.4 mg/mL	Sonication	Estimated 150 Ω/□ 25 prints	RFID and gas detection
[38]	[43]	MicroJet	DMF	n/a	SWCNT 0.02 mg/mL	Centrifuge	Estimated 333 Ω/□ 8 prints	Uniform CNT network
[45]	[48,58]	MicroJet	DMF	n/a	SWCNT 0.001 µg/mL or 0.04 mg/mL (Assumed)	Sonication Filtering	Not reported	Doping of CNT Films
[48]	[42,43,49]	MicroJet	DMF	n/a	SWCNT 0.001 µg/mL or 0.04 µg/mL	Sonication Centrifuge Filtering	Not reported	Fully inkjet printed FET
[49]	[43]	Microdrop Autodrop	NMP	n/a	SWCNT 0.003 mg/mL	Sonication Centrifuge Filtering	Not reported	Use of CNT as active layer in TFT
[39]	-	Not reported	Water	Special dispersant S27000	MWCNT 3 mg/mL	Centrifuge Sonication	11.6 k Ω/□ 4 prints	One of the first reported
[52]	-	HP	Water	Gellan gum or xanthan gum <1 mg/mL	SWCNT or MWCNT Concentration not reported	Sonication	Not reported	Water vapor detection
[53]	[38,43,59]	Epson	Water	SDS 10 mg/mL	SWCNT 0.2 mg/mL	Sonication Centrifuge	78 Ω/□ 200 prints	Supercapacitors
[50]	[39,41,43,51,57,60]	Epson	Water	Combination of 3 different dispersants 150 mg/mL	MWCNT 0.15 mg/mL	Mixing Ball-milling Centrifuge	760 Ω/□ 12 prints	Low sheet resistance
[51]	[38,39,43,47,49,52,59]	Microfab	Water	Solsperse® 46000 5 mg/mL Byk 348 1 mg/mL	MWCNT 10 mg/mL	Sonication	Not reported	Electroluminescent device

Table 1 includes information such as the printer used, the ink ingredients and preparation, and the sheet resistance if reported. For consistency, concentrations that were reported as a weight percent were converted to a mass per volume value (*i.e.*, µg/mL or mg/mL). Also, the table has been organized into three sections based on how the carbon nanotubes were dispersed, namely functionalization, organic solvent, or water with a dispersant.

5. CONCLUSIONS AND FUTURE OUTLOOK

Although carbon nanotube inkjet printing is relatively new, it seems to be a very promising method for deposition. Of course, there are a few obstacles to overcome before inkjet printing will become a commercial method for depositing carbon nanotubes, but it will not take long. With ongoing research in the area of carbon nanotube dispersion, stable carbon nanotube inks will soon be available. Furthermore, commercial inkjet printers like the Fujifilm Dimatix offer better control and resolution than general office inkjet printers. Takagi *et al.* have even demonstrated a method for further enhancing inkjet printing resolution by substrate surface modification [61]. In terms of applications, carbon nanotube inkjet printing can be used to fabricate transistors [41,48,49], sensors [3,46], electroluminescent devices [51], and more. Also, given the current progress, carbon nanotubes seem to be a potential candidate for next generation printable, flexible, and transparent electrodes.

REFERENCES

1. Zhou, Y.; Gaur, A.; Hur, S.-H.; Kocabas, C.; Meitl, M.A.; Shim, M.; Rogers, J.A. P-channel, n-channel thin film transistors and p–n diodes based on single wall carbon nanotube networks. *Nano Lett.* **2004**, *4*, 2031–2035.

2. Artukovic, E.; Kaempgen, M.; Hecht, D.S.; Roth, S.; Grüner, G. Transparent and flexible carbon nanotube transistors. *Nano Lett.* **2005**, *5*, 757–760.

3. Yang, L.; Zhang, R.; Staiculescu, D.; Wong, C.P.; Tentzeris, M.M. A novel conformal RFID-enabled module utilizing inkjet-printed antennas and carbon nanotubes for gas-detection applications. *IEEE Antennas Wirel. Propag. Lett.* **2009**, *8*, 653–656.

4. Li, J.; Lu, Y.; Ye, Q.; Cinke, M.; Han, J.; Meyyappan, M. Carbon nanotube sensors for gas and organic vapor detection. *Nano Lett.* **2003**, *3*, 929–933.

5.   Chopra, S.; McGuire, K.; Gothard, N.; Rao, A.M.; Pham, A. Selective gas detection using a carbon nanotube sensor. *Appl. Phys. Lett.* **2003**, *83*, 2280–2282.

6.   Zhang, T.; Mubeen, S.; Myung, N.V.; Deshusses, M.A. Recent progress in carbon nanotube-based gas sensors. *Nanotechnology* **2008**, *19*, 332001.

7.   Fu, D.; Okimoto, H.; Lee, C.W.; Takenobu, T.; Iwasa, Y.; Kataura, H.; Li, L.-J. Ultrasensitive detection of DNA molecules with high on/off single-walled carbon nanotube network. *Adv. Mater.* **2010**, *22*, 4867–4871.

8.   Scardaci, V.; Coull, R.; Coleman, J.N.; Byrne, L.; Scott, G. Carbon Nanotube Network Based Sensors. In Proceedings of 2012 12th IEEE Conference on Nanotechnology (IEEE-NANO), Birmingham, UK, 20–23 August 2012; pp. 1–3.

9.   Avouris, P.; Freitag, M.; Perebeinos, V. Carbon-nanotube photonics and optoelectronics. *Nat. Photonics* **2008**, *2*, 341–350.

10.   Choi, W.B.; Chung, D.S.; Kang, J.H.; Kim, H.Y.; Jin, Y.W.; Han, I.T.; Lee, Y.H.; Jung, J.E.; Lee, N.S.; Park, G.S.; *et al.* Fully sealed, high-brightness carbon-nanotube field-emission display. *Appl. Phys. Lett.* **1999**, *75*, 3129–3131.

11.   Abera, A.; Choi, J.-W. Quantitative lateral flow immunosensor using carbon nanotubes as label. *Anal. Methods* **2010**, *2*, 1819–1822.

12.   Ng, M.H.A.; Hartadi, L.T.; Tan, H.; Poa, C.H.P. Efficient coating of transparent and conductive carbon nanotube thin films on plastic substrates. *Nanotechnology* **2008**, *19*, 205703.

13.   Geng, H.-Z.; Kim, K.K.; So, K.P.; Lee, Y.S.; Chang, Y.; Lee, Y.H. Effect of acid treatment on carbon nanotube-based flexible transparent conducting films. *J. Am. Chem. Soc.* **2007**, *129*, 7758–7759.

14.   Schrage, C.; Kaskel, S. Flexible and transparent SWCNT electrodes for alternating current electroluminescence devices. *ACS Appl. Mater. Interfaces* **2009**, *1*, 1640–1644.

15.   Sarkar, A.; Daniels-Race, T. Electrophoretic deposition of carbon nanotubes on 3-amino-popyl-triethoxysilane (APTES) surface functionalized silicon substrates. *Nanomaterials* **2013**, *3*, 272–288.

16.   Li, J.; Lei, W.; Zhang, X.; Zhou, X.; Wang, Q.; Zhang, Y.; Wang, B. Field emission characteristic of screen-printed carbon nanotube cathode. *Appl. Surf. Sci.* **2003**, *220*, 96–104.

17. Jones, C.S.; Lu, X.; Renn, M.; Stroder, M.; Shih, W.-S. Aerosol-jet-printed, high-speed, flexible thin-film transistor made using single-walled carbon nanotube solution. *Microelectron. Eng.* **2010**, *87*, 434–437.

18. Vaillancourt, J.; Zhang, H.; Vasinajindakaw, P.; Xia, H.; Lu, X.; Han, X.; Janzen, D.C.; Shih, W.-S.; Jones, C.S.; Stroder, M.; *et al*. All ink-jet-printed carbon nanotube thin-film transistor on a polyimide substrate with an ultrahigh operating frequency of over 5 GHz. *Appl. Phys. Lett.* **2008**, *93*, 243301.

19. Ha, M.; Xia, Y.; Green, A.A.; Zhang, W.; Renn, M.J.; Kim, C.H.; Hersam, M.C.; Frisbie, C.D. Printed, sub-3V digital circuits on plastic from aqueous carbon nanotube inks. *ACS Nano* **2010**, *4*, 4388–4395.

20. Zhou, Y.; Hu, L.; Grüner, G. A method of printing carbon nanotube thin films. *Appl. Phys. Lett.* **2006**, *88*, 123109.

21. Liu, C.-X.; Choi, J.-W. Patterning conductive PDMS nanocomposite in an elastomer using microcontact printing. *J. Micromech. Microeng.* **2009**, *19*, 085019.

22. Chow, E.; Herrmann, J.; Barton, C.S.; Raguse, B.; Wieczorek, L. Inkjet-printed gold nanoparticle chemiresistors: Influence of film morphology and ionic strength on the detection of organics dissolved in aqueous solution. *Anal. Chim. Acta* **2009**, *632*, 135–142.

23. Zhao, N.; Chiesa, M.; Sirringhaus, H.; Li, Y.; Wu, Y.; Ong, B. Self-aligned inkjet printing of highly conducting gold electrodes with submicron resolution. *J. Appl. Phys.* **2007**, *101*, 064513.

24. Kim, D.; Moon, J. Highly conductive ink jet printed films of nanosilver particles for printable electronics.*Electrochem. Solid State Lett.* **2005**, *8*, J30–J33.

25. Lee, S.-H.; Shin, K.-Y.; Hwang, J.Y.; Kang, K.T.; Kang, H.S. Silver inkjet printing with control of surface energy and substrate temperature. *J. Micromech. Microeng.* **2008**, *18*, 075014.

26. Ham, H.T.; Choi, Y.S.; Chung, I.J. An explanation of dispersion states of single-walled carbon nanotubes in solvents and aqueous surfactant solutions using solubility parameters. *J. Colloid Interface Sci.* **2005**, *286*, 216–223.

27.   O'Connell, M.J.; Bachilo, S.M.; Huffman, C.B.; Moore, V.C.; Strano, M.S.; Haroz, E.H.; Rialon, K.L.; Boul, P.J.; Noon, W.H.; Kittrell, C.; *et al.* Band gap fluorescence from individual single-walled carbon nanotubes.*Science* **2002**, *297*, 593–596.

28.   Liu, J.; Rinzler, A.G.; Dai, H.; Hafner, J.H.; Bradley, R.K.; Boul, P.J.; Lu, A.; Iverson, T.; Shelimov, K.; Huffman, C.B.; *et al.* Fullerene pipes. *Science* **1998**, *280*, 1253–1256.

29.   Islam, M.F.; Rojas, E.; Bergey, D.M.; Johnson, A.T.; Yodh, A.G. High weight fraction surfactant solubilization of single-wall carbon nanotubes in water. *Nano Lett.* **2003**, *3*, 269–273.

30.   Moore, V.C.; Strano, M.S.; Haroz, E.H.; Hauge, R.H.; Smalley, R.E.; Schmidt, J.; Talmon, Y. Individually suspended single-walled carbon nanotubes in various surfactants. *Nano Lett.* **2003**, *3*, 1379–1382. Zhang, X.; Liu, T.; Sreekumar, T.V.; Kumar, S.; Moore, V.C.; Hauge, R.H.; Smalley, R.E. Poly(vinyl alcohol)/SWNT composite film. *Nano Lett.* **2003**, *3*, 1285–1288.

31.   O'Connell, M.J.; Boul, P.; Ericson, L.M.; Huffman, C.; Wang, Y.; Haroz, E.; Kuper, C.; Tour, J.; Ausman, K.D.; Smalley, R.E. Reversible water-solubilization of single-walled carbon nanotubes by polymer wrapping. *Chem. Phys. Lett.* **2001**, *342*, 265–271.

32.   Zheng, M.; Jagota, A.; Semke, E.D.; Diner, B.A.; Mclean, R.S.; Lustig, S.R.; Richardson, R.E.; Tassi, N.G. DNA-assisted dispersion and separation of carbon nanotubes. *Nat. Mater.* **2003**, *2*, 338–342.

33.   Nguyen, T.T.; Nguyen, S.U.; Phuong, D.T.; Nguyen, D.C.; Mai, A.T. Dispersion of denatured carbon nanotubes by using a dimethylformamide solution. *Adv. Nat. Sci. Nanosci. Nanotechnol.* **2011**, *2*, 035015.

34.   Ausman, K.D.; Piner, R.; Lourie, O.; Ruoff, R.S.; Korobov, M. Organic solvent dispersions of single-walled carbon nanotubes: Toward solutions of pristine nanotubes. *J. Phys. Chem. B* **2000**, *104*, 8911–8915.

35.   Hasan, T.; Scardaci, V.; Tan, P.; Rozhin, A.G.; Milne, W.I.; Ferrari, A.C. Stabilization and "debundling" of single-wall carbon nanotube dispersions in n-methyl-2-pyrrolidone (NMP) by polyvinylpyrrolidone (PVP).*J. Phys. Chem. C* **2007**, *111*, 12594–12602.

36. Liu, C.-X.; Choi, J.-W. Improved dispersion of carbon nanotubes in polymers at high concentrations.*Nanomaterials* **2012**, *2*, 329–347.

37. Song, J.-W.; Kim, J.; Yoon, Y.-H.; Choi, B.-S.; Kim, J.-H.; Han, C.-S. Inkjet printing of single-walled carbon nanotubes and electrical characterization of the line pattern. *Nanotechnology* **2008**, *19*, 095702.

38. Fan, Z.; Wei, T.; Luo, G.; Wei, F. Fabrication and characterization of multi-walled carbon nanotubes-based ink. *J. Mater. Sci.* **2005**, *40*, 5075–5077.

39. Hecht, D.; Hu, L.; Grüner, G. Conductivity scaling with bundle length and diameter in single walled carbon nanotube networks. *Appl. Phys. Lett.* **2006**, *89*, 133112.

40. Gracia-Espino, E.; Sala, G.; Pino, F.; Halonen, N.; Luomahaara, J.; Mäklin, J.; Tóth, G.; Kordás, K.; Jantunen, H.; Terrones, M.; *et al.* Electrical transport and field-effect transistors using inkjet-printed SWCNT films having different functional side groups. *ACS Nano* **2010**, *4*, 3318–3324.

41. Mustonen, T.; Mäklin, J.; Kordás, K.; Halonen, N.; Tóth, G.; Saukko, S.; Vähäkangas, J.; Jantunen, H.; Kar, S.; Ajayan, P.M.; *et al.* Controlled ohmic and nonlinear electrical transport in inkjet-printed single-wall carbon nanotube films. *Phys. Rev. B* **2008**, *77*, 125430.

42. Kordás, K.; Mustonen, T.; Tóth, G.; Jantunen, H.; Lajunen, M.; Soldano, C.; Talapatra, S.; Kar, S.; Vajtai, R.; Ajayan, P.M. Inkjet printing of electrically conductive patterns of carbon nanotubes. *Small* **2006**, *2*, 1021–1025.

43. Hecht, D.S.; Hu, L.; Irvin, G. Emerging transparent electrodes based on thin films of carbon nanotubes, graphene, and metallic nanostructures. *Adv. Mater.* **2011**, *23*, 1482–1513.

44. Matsuzaki, S.; Nobusa, Y.; Shimizu, R.; Yanagi, K.; Kataura, H.; Takenobu, T. Continuous electron doping of single-walled carbon nanotube films using inkjet technique. *Jpn. J. Appl. Phys.* **2012**, *51*, 06FD18:1–06FD18:3.

45. Yun, J.-H.; Chang-Soo, H.; Kim, J.; Song, J.-W.; Shin, D.-H.; Park, Y.-G. Fabrication of Carbon Nanotube Sensor Device by Inkjet Printing. In Proceedings of 3rd IEEE International Conference on Nano/Micro Engineered and Molecular Systems, Sanya, China, 6–9 January 2008; 2008; pp. 506–509.

46. Song, J.-W.; Kim, Y.-S.; Yoon, Y.-H.; Lee, E.-S.; Han, C.-S.; Cho, Y.; Kim, D.; Kim, J.; Lee, N.; Ko, Y.-G.; *et al.* The production of transparent carbon nanotube field emitters using inkjet printing. *Physica E* **2009**, *41*, 1513–1516.

47. Okimoto, H.; Takenobu, T.; Yanagi, K.; Miyata, Y.; Shimotani, H.; Kataura, H.; Iwasa, Y. Tunable carbon nanotube thin-film transistors produced exclusively via inkjet printing. *Adv. Mater.* **2010**, *22*, 3981–3986.

48. Beecher, P.; Servati, P.; Rozhin, A.; Colli, A.; Scardaci, V.; Pisana, S.; Hasan, T.; Flewitt, A.J.; Robertson, J.; Hsieh, G.W.; *et al.* Ink-jet printing of carbon nanotube thin film transistors. *J. Appl. Phys.* **2007**, *102*, 043710.

49. Kwon, O.-S.; Kim, H.; Ko, H.; Lee, J.; Lee, B.; Jung, C.-H.; Choi, J.-H.; Shin, K. Fabrication and characterization of inkjet-printed carbon nanotube electrode patterns on paper. *Carbon* **2013**, *58*, 116–127.

50. Azoubel, S.; Shemesh, S.; Magdassi, S. Flexible electroluminescent device with inkjet-printed carbon nanotube electrodes. *Nanotechnology* **2012**, *23*, 344003.

51. In het Panhuis, M.; Heurtematte, A.; Small, W.R.; Paunov, V.N. Inkjet printed water sensitive transparent films from natural gum–carbon nanotube composites. *Soft Matter* **2007**, *3*, 840–843.

52. Chen, P.; Chen, H.; Qiu, J.; Zhou, C. Inkjet printing of single-walled carbon nanotube/RuO2 nanowire supercapacitors on cloth fabrics and flexible substrates. *Nano Res.* **2010**, *3*, 594–603.

53. Noh, J.; Jung, M.; Jung, K.; Lee, G.; Lim, S.; Kim, D.; Kim, S.; Tour, J.M.; Cho, G. Integrable single walled carbon nanotube (SWNT) network based thin film transistors using roll-to-roll gravure and inkjet. *Org. Electron.* **2011**, *12*, 2185–2191.

54. Mustonen, T.; Kordás, K.; Saukko, S.; Tóth, G.; Penttilä, J.S.; Helistö, P.; Seppä, H.; Jantunen, H. Inkjet printing of transparent and conductive patterns of single-walled carbon nanotubes and PEDOT-PSS composites. *Phys. Status Solidi B* **2007**, *244*, 4336–4340.

55. Deegan, R.D.; Bakajin, O.; Dupont, T.F.; Huber, G.; Nagel, S.R.; Witten, T.A. Capillary flow as the cause of ring stains from dried liquid drops. *Nature* **1997**, *389*, 827–829.

56. Denneulin, A.; Bras, J.; Carcone, F.; Neuman, C.; Blayo, A. Impact of ink formulation on carbon nanotube network organization within inkjet printed conductive films. *Carbon* **2011**, *49*, 2603–2614.

57. Nobusa, Y.; Yomogida, Y.; Matsuzaki, S.; Yanagi, K.; Kataura, H.; Takenobu, T. Inkjet printing of single-walled carbon nanotube thin-film transistors patterned by surface modification. *Appl. Phys. Lett.* **2011**, *99*, 183106.

58. Small, W.R.; in het Panhuis, M. Inkjet printing of transparent, electrically conducting single-walled carbon-nanotube composites. *Small* **2007**, *3*, 1500–1503.

59. Denneulin, A.; Bras, J.; Blayo, A.; Khelifi, B.; Roussel-Dherbey, F.; Neuman, C. The influence of carbon nanotubes in inkjet printing of conductive polymer suspensions. *Nanotechnology* **2009**, *20*, 385701.

60. Takagi, Y.; Nobusa, Y.; Gocho, S.; Kudou, H.; Yanagi, K.; Kataura, H.; Takenobu, T. Inkjet printing of aligned single-walled carbon-nanotube thin films. *Appl. Phys. Lett.* **2013**, *102*, 143107.

Magnetic Carbon Nanotubes: Synthesis, Characterization and Anisotropic Electrical Properties

Il Tae Kim[1] and Rina Tannenbaum[a]

[1] Georgia Institute of Technology, United States

1. INTRODUCTION

Carbon nanotubes (CNTs) have been the focus of extensive research in recent years due to their exceptional mechanical, thermal, and electrical properties (Treacy et al., 1996; Lourie et al., 1998; Yu et al., 2000; Lukic et al., 2005). As a result of their nanoscale dimensions and high surface area, CNTs could also be considered as efficient templates for the assembly and tethering of nanoparticles on their surface (Grzelczak et al., 2006). The decoration of CNTs with various compounds and various structures could increase their surface functionality and the tunability of their properties, such as their electrical and magnetic characteristics (Korneva et al., 2005; Kuang et al., 2006). Recent reports described the attachment of various inorganic nanoparticles to either the external surface of the CNTs, or to the internal surface of the CNT cavity, through several experimental methods (Han et al., 2004; Qu et al., 2006). In this context, it is important to note that the control of the size of these tethered nanoparticles is of primary importance for the purpose of tailoring the physical and chemical properties of these hierarchical materials.

Iron oxide nanoparticles, such as magnetite and maghemite, have been of technological and scientific interest due to their unique electrical and magnetic properties. These nanoparticles can be used in such diverse fields as high-density information storage and electronic devices (Sun et al., 2000; Pu et al., 2005; Yi et al., 2006; Jia et al., 2007; Wan et al., 2007). Maghemite, γ-Fe_2O_3, is the allotropic form of magnetite, Fe_3O_4 (Rockenberger et al., 1999; Pileni et al., 2003; Sun et al., 2004). These two iron oxides are crystallographically isomorphous. The main difference is the presence of ferric ions only in γ-Fe_2O_3, and both ferrous and ferric ions in Fe_3O_4. As a result, while the magnetic properties of Fe_3O_4 are superior, γ-Fe_2O_3 is more stable, since the iron cannot be further oxidized under ambient conditions. This renders γ-Fe_2O_3 nanoparticles easier to work with, especially in the presence of organic solvents and organic ligands, and consequently, they have been widely used for magnetic storage in a variety of fields such as floppy disks and cassette tapes. However, maghemite-CNT nanohybrid materials have not been studied as extensively as magnetite-CNT nanohybrid materials, with the exception of several few examples (Sun et al., 2005 ; Youn et al., 2009).

The alignment of CNTs in a variety of matrices can be used to reinforce, intensify, and enhance some of the properties of the resulting systems, as well as introduce various degrees of anisotropy into the properties of the desired nanomaterials (Kimura et al., 2002; Garmestani et al., 2003). The alignment of CNTs in a suspension under a magnetic field requires that the energy produced by the torque acting on a magnetically-anisotropic segment exceeds the thermal energy of that particular segment, such that: $\delta U \sim B^2 n \delta \chi > kT$, where B is the field strength, n is the number of carbon atoms in the segment, and $\delta \chi$ is the magnetic anisotropy (Fisher et al., 2003). However, due to the low magnetic susceptibility of CNTs, their alignment by the application of an external magnetic field requires a relatively high magnetic field (Camponeschi et al., 2007). This drawback could be eliminated by enhancing the magnetic susceptibility of carbon nanotubes via the tethering of magnetic nanoparticles onto their surface. In zero field, the magnetic moments of the maghemite nanoparticles randomly point in different directions, resulting in a vanishing net magnetization. However, if a sufficient homogeneous magnetic field is applied, the magnetic moments of the nanoparticles align in parallel, and the resulting dipolar interactions are sufficiently large to overcome thermal motion and to reorient the magnetic CNTs.

In this chapter, we describe and report a convenient approach for the decoration of CNTs with near-monodisperse maghemite nanoparticles by employing a novel and simple modified sol-gel process (in-situ process) with an iron salt as precursor, followed by calcination. The resulting hybrid nanomaterials are superparamagnetic at room temperature and are conducive to facile alignment under relatively low magnetic fields. Subsquently, the nanohybrid materials, i.e. the magnetized carbon nanotubes, were incorporated into a polymer matrix and aligned by the application of a magnetic field, forming polymer composites with an aligned filler phase. It is therefore expected that the composites formed in this manner would exhibit anisotropic mechanical and electrical properties that would depend on and correlate with the parallel and perpendicular direction to the magnetic field that has been applied and under which the alignment has taken place.

2. EXPERIMENTAL DETAILS

2.1. Synthesis of maghemite-mwcnt nanohybrid materials

Pure-MWCNTs were first dispersed in a solution mixture of concentrated H_2SO_4 and HNO_3 with the volume ratio of 3:1. The suspension was ultra-sonicated for 3 hrs at room temperature. After that, the concentration of the suspension was diluted up to 50% and filtered with a PTFE membrane (0.45 µm pore size) with the aid of a vacuum pump. Carboxylated MWCNT (MWCNT-COOH) was washed with de-ionized water several times to reach neutral pH and dried under vacuum at 50º C overnight. The synthesis of maghemite-MWCNT was performed by first adding 0.65 g $Fe(NO_3)_3$ $9H_2O$ to 20 ml of absolute ethanol (100% purity) and stirring until the $Fe(NO_3)_3$ $9H_2O$ was dissolved completely. Subsequently, this iron salt solution was added to a suspension of oxidized MWCNTs with a mass ratio of 4:1 ($Fe(NO_3)_3$ $9H_2O$: MWCNTs mass ratio of 4:1), stirred, and sonicated for 3 hrs. Twenty ml of 1.2 mM of NaDDBS were added to the solution and stirred for 30 min. Then, 1.2 ml of propylene oxide was added as a gelation agent and stirred for 30 min. The mixture was then placed in a Fisher Scientific iso-temperature oven for drying for 3 days at 100º C. The resulting powder products were washed with ethanol several times and dried at 50º C. The calcination of these powders was performed in a furnace under argon atmosphere at both 500º C and 600º C for 2 hrs. The overall strategy for the preparation of MWCNT/γ-Fe_2O_3 is shown in Figure 1 (Kim et al., 2010).

2.2. Fabrication of polymer nanocomposites with aligned feature

Various weight percents of magnetic multi-walled carbon nanotubes (m-MWCNTs) were dispersed in a small amount of ethanol with sonication for 1 hr. Epoxy resin (PR2032) was added to the suspension and mixed with a mechanical stirrer for 30 min in order to obtain optimal dispersion. After that, the nanocomposite solution was sonicated to evaporate entire solvent at 50º C. The curing agent (PH3660) was added into the solution, mixed, and degassed under vacuum. The solution was immediately poured into a mold, and a 0.3 T magnetic field was applied for 1 hr at room temperature, for 1 hr at 60º C, and for another 1 hr at 60º C without a magnetic field. The nanocomposite was post-cured at 60º C for 6 hrs in the iso-temperature oven (Kim et al., 2011).

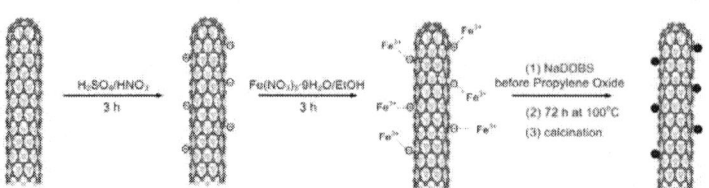

Figure 1. Schematic representation for the preparation of nanohybrid materials, MWCNT/γ-Fe₂O₃ via a modified sol-gel technique (Reprinted with permission from Kim et al., *J. Phys. Chem. C* 2010, 114, 15, 6944-6951, Copyright 2010 ACS).

2.3. Characterization

The dried samples were ground into a fine powder using a ceramic mortar and pestle. Tiny amounts of samples were rarified with KBr powder, ground, and pressed in a KBr pellet with a punch and die. A Nicolet Nexus 870 spectrometer scanned the range from 4000 to 400 cm^{-1} with a resolution of 2 cm^{-1} and data spacing of 0.964 cm^{-1}. XRD measurements were performed using an X'pert Pro Alpha-1 (wavelength of 1.54 Å). XRD peaks were collected from 2θ = 0º to 90º with a step size of 0.02º. XPS scans of powder samples were taken using a Surface Science Laboratories SSX-100 ESCA spectrometer using monochromatic Al Kα radiation (1486.6 eV). Raman spectra were recorded in the range of 200-2000 cm^{-1} at ambient temperature using a WITEC Spectra Pro 2300I spectrometer equipped with an Ar-ion laser, which provided a laser beam of 514 nm wavelength. The magnetic properties of MWCNTs were measured using a 5.5 T Quantum Design Superconducting Quantum Interface Device (SQUID) magnetometer. The alignment of the sample was conducted by a magnet (GMW-5403) at 0.3 T. The morphology and aligned feature of as-prepared samples were also characterized using SEM (LEO 1530). TEM samples were prepared by placing a droplet of solution onto a TEM grid, and for the observation of aligned features, samples were micro-tomed into 100 nm thick slices using a diamond knife and placed on a TEM grid. These samples were analyzed using a Hitachi HF2000, 200 kV transmission electron microscopy. The electrical conductivity data of as-prepared composites were collected using impedance analyzer (Solartron Instruments SI 1260 with dielectric interface 1296) for the frequency range 0.1 Hz ~ 1 MHz. All the data were collected under an AC voltage of 0.1 V. Contact was achieved by silver painting the two ends of the samples, and then using coaxial probers on a probe station attached to the impedance analyzer (Peng et al., 2008).

3. DECORATION OF CARBON NANOTUBES WITH MAGNETIC NANOPARTICLES AND THE CHARACTERISTICS OF THE RESULTING HYBRID NANOSTRUCTURES

A variety of methods to form nanohybrid materials on the surface of CNTs have been reported. Correa-Duarte group (Correa-Duarte et al., 2005) coated CNTs with iron oxide nanoparticles (magnetite/maghemite) via a layer by layer (LBL) assembly technique and aligned CNT chains in relatively small external magnetic fields. Subsequently, the resulting magnetic CNT structures could be used as building blocks for the fabrication of nanocomposite materials. Cai group (Wan et al., 2007) decorated CNTs with magnetite nanoparticles in liquid polyols. As a result, these nanoparticles could have significant potential for application in the fields of sensors. In addition, Gao group (Jia et al., 2007) initiated the self-assembly of magnetite particles along MWCNTs via a hydrothermal process. The resulting materials feature nanoparticle beads along the CNT surface, rendering this as an appropriate material to be used as a functional device.

The maghemite-CNT nanocomposite systems also have been reported even though research has not been studied as extensively as magnetite-CNT system. Liu group (Sun et al., 2005) decorated MWCNTs with maghemite via the pyrolysis of ferrocene at different temperatures. This product is expected to provide an efficient way for the large-scale fabrication of magnetic CNT composites. Jung group (Youn et al., 2009) decorated single-wall CNTs (SWCNTs) with iron oxide nanoparticles along the nanotube via a magneto-evaporation method. The nanotubes were aligned vertically on ITO surfaces, suggesting the possibility of rendering this process adequate and cost-effective for mass production. The method described in this work consisted of the use of an iron-oleate complex, oleic acid, and truncated SWCNTs to create iron oxide nanoparticles. The research also demonstrated the anisotropic properties of vertically aligned SWCNTs in a nanocmposite by comparing current densities of the aligned and non-aligned CNTs.

Keeping pace with these researches' streaming, we have developed the MWCNT/γ-Fe$_2$O$_3$ nanohybrid materials. As a first step, the MWCNTs were carboxylated in order to introduce negative charges on their surface, which in turn will interact with Fe (III) ions present in a strong acid solution. This process was also coupled with sonication to ensure dispersion of the MWCNTs in the suspension. The x-ray photoelectron spectroscopy (XPS) wide-survey (Fig. 2a)and high resolution spectra (Fig. 2b)reveal not only the presence of carbon-carbon bonding of MWCNTs at 285 eV binding energy but also the formation of a carbonyl moiety consistent with carboxylated groups at 288 eV binding energy. Nucleation sites for the iron oxide were generated at the CNT surface due to the electrostatic interaction between Fe (III) ions and the carboxylate surface groups of acid-treated CNTs. In this system, the occurrence of gelation was inhibited by the addition of a surface active molecule, sodium dodecylbenzenesulfonate (NaDDBS), before the addition of propylene oxide, which is a gel promoter. The surfactant interfered in the growth stage of the iron oxide nanoparticles (gel phase) and prevented the formation of a gel. This occurred because the NaDDBS molecules had already coordinated to the iron (III) centers due to the attraction between the negatively-charged hydrophilic head of the surfactant and the positively-charged iron (Matarredona et al., 2003; Camponeschi et al., 2008). Therefore, due to the presence of the NaDDBS molecules, no aggregates of γ-Fe_2O_3 were formed but rather the nanoparticles remained individually isolated and dispersed along the length of the CNTs.

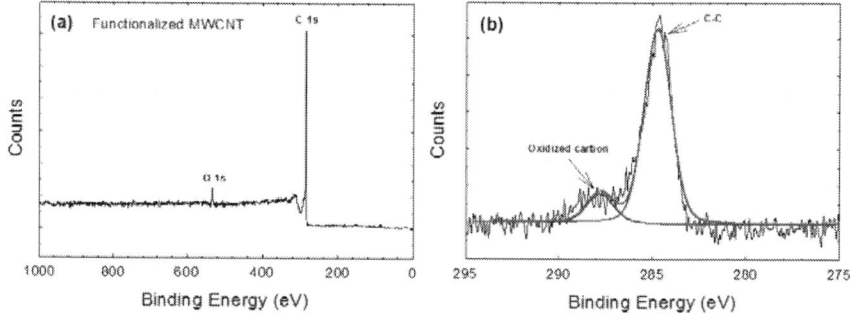

Figure 2. a) The XPS survey spectrum of functionalized MWCNTs. (b) The high-resolution XPS spectrum of C1s. (Adapted with permission from Kim et al., *J. Phys. Chem. C* 2010, 114, 15, 6944-6951, Copyright 2010 ACS).

X-ray diffraction patterns of MWCNT containing iron oxide nanoparticles calcinated at different temperatures with the initial $Fe(NO_3)_3$ $9H_2O$: MWCNTs mass ratio of 4:1 and 2:1 demonstrate the high crystalline nature of the nanoparticles as shown in Figure 3. The diffraction peak at $2\theta = 26^\circ$ can be confidently indexed as the (002) reflection of the MWCNTs, similar to that of pure MWCNTs. The other peaks in the range of $20^\circ < 2\theta < 80^\circ$ correspond to the (220), (311), (400), (422), (511), (440), and (533) reflections of maghemite (γ-Fe_2O_3) and/or magnetite (Fe_3O_4). When the mass ratio of $Fe(NO_3)_3$ $9H_2O$ and MWCNTs increases from 2:1 to 4:1, the intensity of the carbon (002) reflection decreases. Also, when calcination temperature increases from 500° C to 600° C, the crystal structure of the product becomes better-defined. Because XRD patterns of maghemite and magnetite are practically identical (Sun et al., 2005), x-ray diffraction alone cannot be used to distinguish between the two phases. Therefore, we employed additional experimental techniques to discern between these two phases.

The FTIR spectrum of the product of this modified sol-gel process shows the presence of well-crystallized iron oxide nanoparticles after calcination at 600º C as shown in Figure 4. Maghemite (γ-Fe_2O_3) has an inverse spinel structure and therefore, it can be seen as an iron-deficient form of magnetite. If the powder is not heat-treated, a weak peak from 800 to 400 cm^{-1} is shown. This is evidence of an amorphous iron oxide phase with minimal long-range order typical of maghemite or magnetite. However, after calcination, IR bands show strong peaks at 576 and 460 cm^{-1}, which correspond to a partial vacancy ordering in the octahedral positions in the maghemite crystal structure (White et al., 1967; de Faria et al., 1997; Millan et al., 2007).

X-ray photoelectron spectroscopy (XPS) as well as Raman spectroscopy confirmed that the iron oxide nanoparticles formed were indeed maghemite and not magnetite. After the formation of oxidized MWCNTs decorated with iron oxide nanoparticles followed by calcination at 600º C, Figure 5 shows XPS characteristic iron peaks in addition to carbon and oxygen. The position of the Fe (2p3/2) and Fe (2p1/2) peaks were marked at 711.3 and 724.4 eV, respectively, which are in good agreement with the values reported for γ-Fe_2O_3 in the literature (Hyeon et al., 2001; Sun et al., 2005). Therefore, this suggests the formation of γ-Fe_2O_3 in our samples. Raman spectroscopy can also effectively distinguish between maghemite and magnetite nanoparticles. The strong peak at ~1350 cm^{-1} can be assigned to the D band of MWCNTs, while another dominant peak at ~1576 cm^{-1} can be ascribed the G band of MWCNTs as shown in Figure 6 (Jorio et al., 2003). In contrast to magnetite, the maghemite bands are not well-defined, but rather consist of several broad peaks around 350, 500, and 700 cm^{-1}, which are unique to these species and are absent in other types of iron oxide nanoparticles (de Faria et al., 1997). This supports the conclusion that the nanoparticles bound at the walls of the MWCNTs are maghemite and not magnetite.

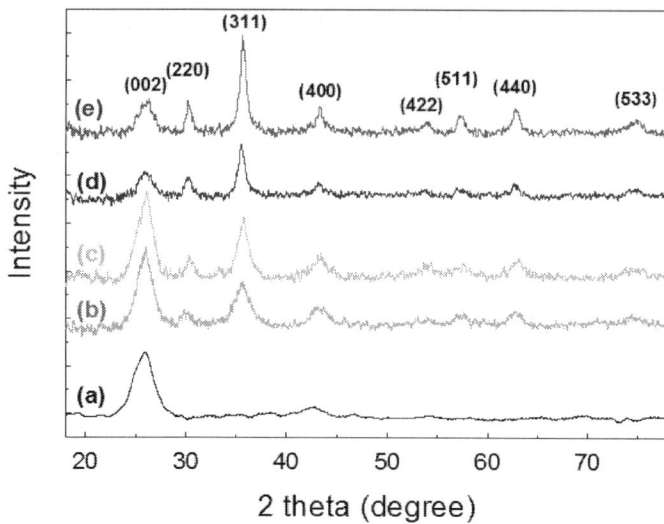

Figure 3. XRD patterns of MWCNT/γ-Fe₂O₃ nanostructures fabricated with two different mass ratios of Fe(NO₃)₃ 9H₂O and MWCNTs: (a) MWCNT; (b) 2:1 at 500º C; (c) 2:1 at 600º C; (d) 4:1 at 500º C; (e) 4:1 at 600º C (Reprinted with permission from Kim et al., *J. Phys. Chem. C* 2010, 114, 15, 6944-6951, Copyright 2010 ACS).

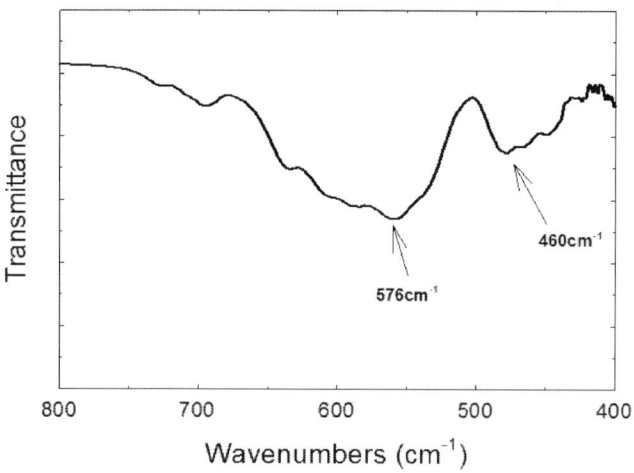

Figure 4. FTIR spectrum of MWCNT/γ-Fe₂O₃ after calcination at 600º C (Reprinted with permission fromKim et al., *J. Phys. Chem. C* 2010, 114, 15, 6944-6951, Copyright 2010 ACS).

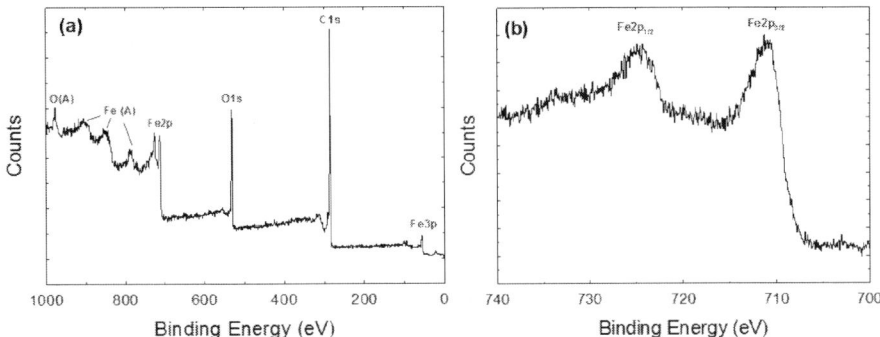

Figure 5. a) The XPS survey spectrum of MWCNT/γ-Fe₂O₃. (b) The high-resolution XPS spectrum of Fe 2p bands (Adapted with permission from Kim et al., *J. Phys. Chem. C* 2010, 114, 15, 6944-6951, Copyright 2010 ACS).

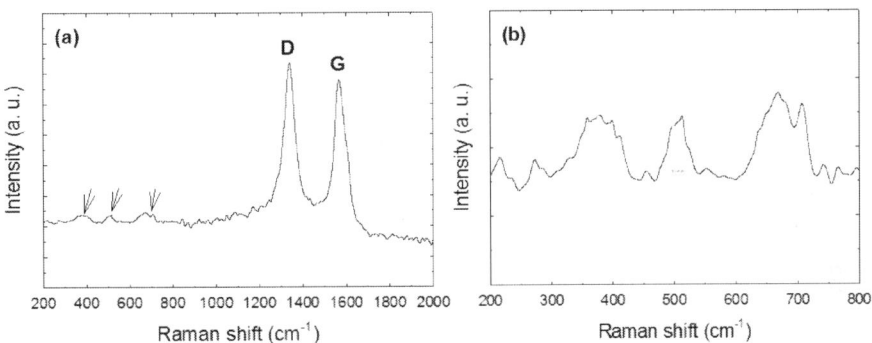

Figure 6. a) The Raman spectrum of MWCNT/γ-Fe₂O₃ nanostructure prepared at 600º C with the mass ratio of 4:1. (b) The detailed Raman spectrum of the same sample in the 200-800 cm⁻¹ spectral range (Reprinted with permission from Kim et al., *J. Phys. Chem. C* 2010, 114, 15, 6944-6951, Copyright 2010 ACS).

Scanning electron microscopy (SEM) and transmission electron microscopy (TEM) images of MWCNTs/γ-Fe₂O₃ confirmed that γ-Fe₂O₃ was attached to the walls of the MWCNTs as shown inFigure 7. The high-resolution transmission electron microscopy (HRTEM) image of a nanoparticle (Figure 7(b)) illustrates the maghemite interlayer spacing of the (311) lattice plane of approximately 0.25 nm (Hyeon et al., 2001). Furthermore, the inset image of Figure 7(b) shows the electron diffraction patterns of maghemite, indicating the high crystallinity of the maghemite nanoparticles. At a mass ratio of 4:1 between the $Fe(NO_3)_3$ $9H_2O$ precursor and the MWCNTs, the particle size increased with increasing temperature from 500º C to 600º C, and the average sizes were 10.1 nm and 10.8 nm, respectively as shown in Figure 7(c) and (d). Similarly, when the mass ratio of $Fe(NO_3)_39H_2O$ precursor and MWCNT was 2:1, the average particle sizes as a result of the increased temperature were 7.9 nm and 8.4 nm, respectively (Figure 7(e) and 7(f)), which also slightly increased with increasing temperature. This result indicated that both a higher mass ratio between the $Fe(NO_3)_39H_2O$

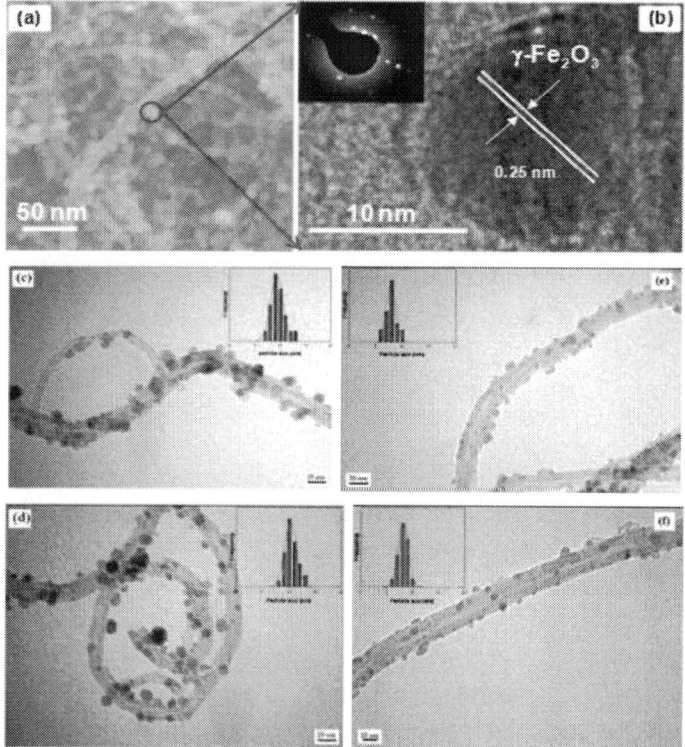

Figure 7. a) SEM image of MWCNT/γ-Fe_2O_3 hybrid structures prepared with 4:1 mass ratio of iron salt and MWCNT; (b) High resolution TEM image of maghemite. Inset shows diffractions of a single maghemite nanoparticle. TEM images of MWCNT/γ-Fe_2O_3 prepared with 4:1 mass ratio of iron salt and MWCNT: (c) High-resolution image prepared at 500º C; (d) High magnification image prepared at 600º C. TEM images of MWCNT/γ-Fe_2O_3 prepared with 2:1 mass ratio; (e) High magnification image prepared at 500º C; (f) High magnification image prepared at 600º C (Adapted with permission from Kim et al., *J. Phys. Chem. C* 2010, 114, 15, 6944-6951, Copyright 2010 ACS, andKim et al., *Carbon* 2011, 49, 1, 54-61, Copyright 2011 Elsevier).

precursor and the MWCNT and increasing temperature led to larger nanoparticles, and therefore, we can conclude that particle size could be controlled by the precursor to MWCNT mass ratio and temperature.

Chemical analysis using EDS during the TEM analysis showed the presence of Fe, O, and C in the maghemite-MWCNT system as shown in Figure 8, and the calculated atomic ratio of Fe and O was close to 2:3, which suggested the formation of γ-Fe_2O_3.

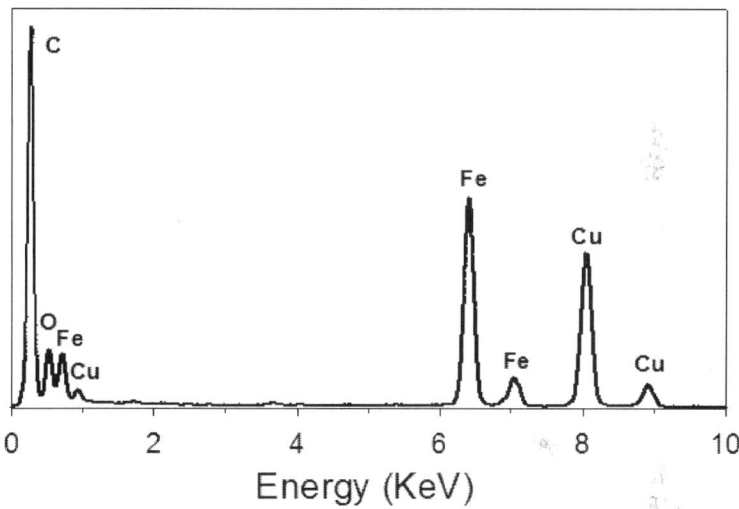

Figure 8. Energy dispersion spectrum (EDS) of the MWCNT/γ-Fe_2O_3 hybrid material (Adapted with permission from Kim et al., *J. Phys. Chem. C* 2010, 114, 15, 6944-6951, Copyright 2010 ACS).

The magnetic properties of the as-prepared MWCNTs/γ-Fe_2O_3 nanocomposites were measured using Superconducting Quantum Interference Device (SQUID) magnetometer. The magnetization hysteresis loops were measured in fields between ±50 kOe at room temperature as shown in Figure 9(a). The saturation magnetization (M_s) of the samples obtained is below 2 emu/g, which is considerably smaller than that of bulk iron (M_s = 222 emu/g) as shown in Table 1. Coercivity is below 10 Oe, which is larger than that of bulk iron (H_c = 1 Oe). The conclusion drawn from the measurement of magnetic properties is that both samples, having different ratios between Fe(NO_3)_3 9H_2O precursor and MWCNT, exhibit superparamagnetic behavior at room temperature. This should be mainly attributed to the small size of γ-Fe_2O_3 nanoparticles that were formed in the presence of MWCNTs (Pascal et al., 1999). This result is in good accordance with the TEM observation of the small sizes of the maghemite nanoparticles mentioned above.

The magnetic attraction of our sample was also tested by placing a magnet near a vial containing the maghemite-MWCNT nanostructures as shown in Figure 9(c) and 9(d). Our samples can be easily dispersed in solution and form a stable suspension. When a magnet approaches the vial, magnetic carbon nanotubes are attracted toward the magnet. This phenomenon illustrates that the maghemite nanoparticles that are anchored on the surface of the MWCNTs impart to the composite material a magnetic response similar to that observed with magnetite.

This novel method for the magnetization of carbon nanotubes through the tethering of magnetic iron oxide nanoparticles with controlled size and site distribution would open up a slew of new opportunities for applications in which the alignment of CNTs is not only desired, but is actually required. While many groups have studied strategies to align MWCNT/Fe_3O_4 nanostructures under external magnetic fields due to their strong magnetic properties, very little attention has been devoted to MWCNT/γ-Fe_2O_3 conjugate nanomaterials. Therefore, we would like to show that this latter system also exhibits similar interesting properties and can constitute a facile gateway to MWCNT alignment processes under tight morphological control and relatively low magnetic fields, resulting in enhanced anisotropic electrical conductivity behavior, in the following sections.

Table 1. Magnetic properties as a function of both different mass ratio of $Fe(NO_3)_3$ $9H_2O$ and MWCNT and different calcination temperatures.

Magnetic properties	Calcination temperature (ºC)	2:1	4:1
M_s (emu/g)	500	0.3	2.0
	600	0.2	1.4
H_c (Oe)	500	4.8	2.8
	600	6.3	9.6

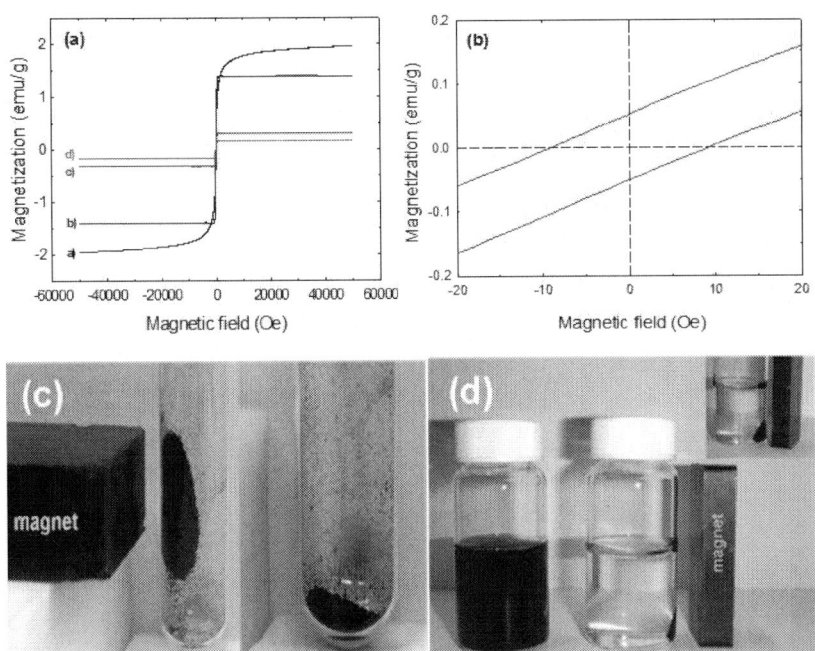

Figure 9. a) Magnetization vs. applied magnetic field for the magnetic carbon nanotubes prepared at different mass ratios and temperatures: 4:1 mass ratio of $Fe(NO_3)_3$ $9H_2O$ and MWCNT at a) 500º C, b) 600º C, and 2:1 mass ratio of $Fe(NO_3)_3$ $9H_2O$ and MWCNT at c) 500º C, d) 600º C. (b) The enlarged hysteresis loop of the MWCNT/γ-Fe_2O_3 structures formed from a 4:1 mass ratio of $Fe(NO_3)_3$ $9H_2O$ and MWCNT calcinated at 600º C. The photographs of magnetic carbon nanotubes (c) in the presence (left image) and in the absence (right image) of a magnet and (d) suspended in ethanol in the absence (left image) and in the presence (right image) of an externally-placed magnet (Adapted with permission from Kim et al., *J. Phys. Chem. C* 2010, 114, 15, 6944-6951, Copyright 2010 ACS, and Kim et al., *Carbon* 2011, 49, 1, 54-61, Copyright 2011 Elsevier).

4. ALIGNMENT STRATEGIES OF CARBON NANOTUBES IN POLYMER MATRICES

Alignments of CNTs by electric, shear induced field, and magnetic field were reported previously by several groups (Chen et al., 2001; Nagahara et al., 2002). Bauhofer group (Martin et al., 2005) successfully demonstrated the application of AC electric fields allowing both the alignment of carbon nanofibers in epoxy resin and their connection into a network. Zhu group (Zhu et al., 2009) studied electric field aligned MWCNT/epoxy nanocomposites with a sample size of up to several centimetres using fast UV polymerization, showing significant anisotropic properties for storage modulus and electrical conductivity.

For the characterization of aligned composite systems using shear induced field, we probed the effects of shear flow on the alignment of dispersed SWCNTs in polymer solutions as a previous study (Camponeschi et al., 2006). The sample solutions were placed in the 8.5 mm gap between the outer cylinder and the spindle, as shown Figure 10. In turn, the spindle was allowed to rotate for one week at several different angular velocities ranging from 12 to 100 rpm. TEM samples were taken in situ from the solutions flowing in circular motion in the gap between the outer cylinder and inner cylinder as shown in Figure 10(b).

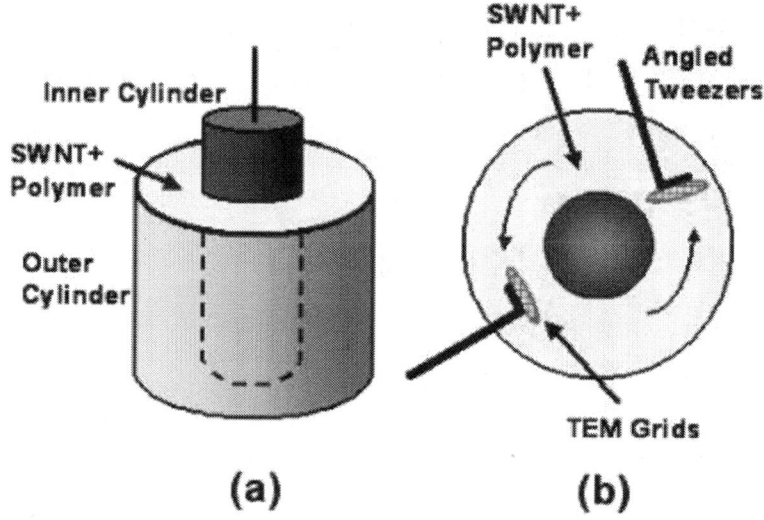

(a)   **(b)**

Figure 10. a) Concentric cylinder arrangement in the Brookfield viscometer. (b) TEM sample retrieval and preparation (Reprinted with permission from Camponeschi et al., *Langmuir* 2006, 22, 4, 1858-1862. Copyright 2006 ACS).

In this experimental set up, for systems in which effective dispersion of the carbon nanotubes was achieved by the combined action of both NaDDBS and Carboxymethylcellulose (CMC). The only system in which tube alignment was observed was for the NaDDBS/CMC/SWCNT solution that was subjected to shear stresses at the highest angular velocity used in the experiments as shown in Figure 11.

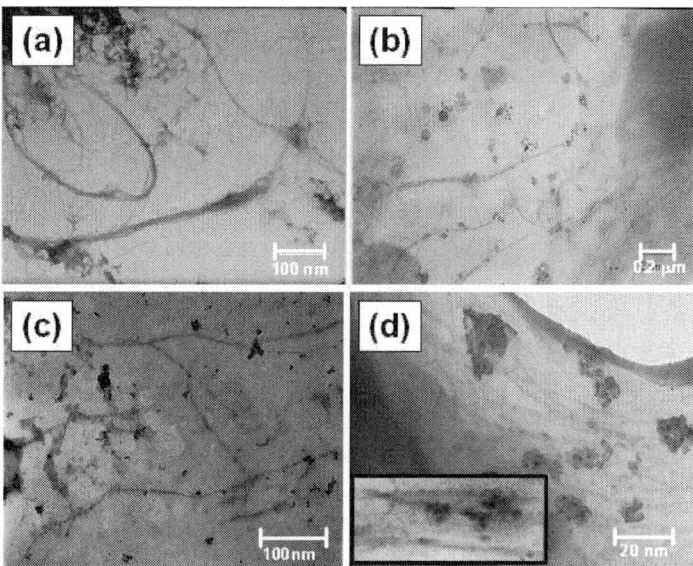

Figure 11. Oriented carbon nanotubes dispersed with NaDDBS and CMC and subjected to shear flow at 100 rpm. The inset image is a 4-fold magnification of the larger image showing the local orientation of the surface modified SWCNT (Reprinted with permission from Camponeschi et al., *Langmuir* 2006, 22, 4, 1858-1862. Copyright 2006 ACS).

A high magnetic field is an efficient and direct ways to align carbon nanotubes. Tanimoto group have found that a high magnetic field of 7 T aligns arc-grown MWCNTs (Fujiwara et al., 2001). They dried a MWCNT dispersion in methanol under a constant magnetic field and observed the MWCNTs alignment parallel to the field. This result was explained by the difference between the diamagnetic susceptibilities parallel $(\chi_{//})$ and perpendicular $(\chi_{\perp})$ to the tube axis; if $|\chi_{\perp}|$ is larger than $|\chi_{//}|$, a MWCNT tends to align parallel to the magnetic field by overcoming thermal energy (Ajiki et al., 1993; Fujiwara et al., 2001). More recently, Steinert and Dean (Steinert & Dean, 2009) obtained solution cast PET-carbon nanotube composite films by applying a magnetic field, resulting in increased conductivity with the increase of the applied magnetic field. Furthermore, in our previous study (Camponeschi et al., 2007), we prepared magnetically aligned carbon nanotube composite systems; thus, carbon nanotubes were aligned parallel to the direction of magnetic field, resulting in enhanced mechanical properties. However, due to the low magnetic susceptibility of carbon nanotubes, their alignment by the application of an external magnetic field requires a relatively high magnetic field. This draw-back could be solved by enhancing the magnetic susceptibility of carbon nanotubes by tethering magnetic nanoparticles on their surface, as we developed MWCNT/γ-Fe$_2$O$_3$ hybrid materials.

The samples for SEM were prepared by dispersing as-prepared nanostructures in water solution with surfactant, sonicating for 30 min, and then depositing the samples onto silicon wafer under an external field. Figure 12 shows the SEM images of magnetic carbon nanotubes. When a droplet of dispersed hybrid materials in a water solution was dried under the magnetic field, the surface-modified MWCNT were aligned easily as shown in Figure 12(a). However, when the nanocomposite solution was dried without applying magnetic field, the surface-modified MWCNT did not exhibit alignment features.

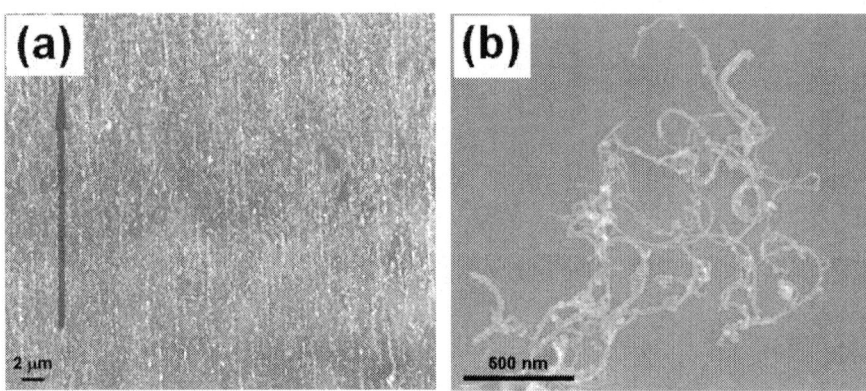

Figure 12. a) SEM image of aligned magnetic carbon nanotube hybrid materials parallel to the direction of magnetic field. (b) SEM image of magnetic carbon nanotube hybrid materials that were not subjected to a magnetic field (Reprinted with permission from Kim et al., *Carbon* 2011, 49, 1, 54-61.2011. Copyright 2011 Elsevier).

The TEM images of composites in which surface-modified MWCNT (m-MWCNT) and unmodified MWCNT were embedded in epoxy matrices are shown in Figure 13(a) through 13(d). We first compared the alignment features of the MWCNT/epoxy nanocomposite and the m-MWCNT/epoxy nanocomposite systems, under the same experimental conditions, i.e. the same strength of the externally-applied magnetic field (0.3 T). Figure 13(a)and 13(b), representing MWCNT/epoxy composites with 0.5 wt% MWCNT and 1.0 wt% MWCNT, respectively, did not reveal any alignment features of filler phase in the polymer matrix under the externally-applied magnetic field. However, in the case of the m-MWCNT/epoxy nanocomposite systems also having 0.5 wt% m-MWCNT and 1.0 wt% m-MWCNT and shown in Figure 13(c)and 13(d), respectively, it is obvious that the m-MWCNTs embedded in the epoxy matrix have indeed aligned parallel to the direction of magnetic field (0.3 T). Comparing the alignment features of aligned m-MWCNT hybrid materials and aligned m-MWCNT/epoxy composites (Figure 12(a), 13(c), and 13(d)), it becomes evident that the m-MWCNT hybrid materials in the absence of a polymer matrix show better alignment, fact which could be attributed to the viscosity of the polymer matrix during processing. Therefore, we can conclude that the m-MWCNT hybrids can be aligned under a relatively weak magnetic field even when embedded in a polymer matrix. This alignment is expected to directly affect the anisotropic conductivity of the resulting epoxy composites, as will be shown in the subsequent section (Figure 14 and 15). The bundling of the m-MWCNTs in the polymer matrix, as observed in the inset in Figure 13(c), may be attributed to the anisotropic nature of the dipolar interactions of the iron oxide nanoparticles near the ends of the carbon nanotubes, i.e. the near-linear stacking of the north and south poles of the m-MWCNT in the polymer matrix, resulting in their observed end-to-top connectivity (Butter et al., 2003; Correa-Duarte et al., 2005).

5. ANISOTROPIC ELECTRICAL CONDUCTIVITY OF COMPOSITE SYSTEM

The electric conductivities of the m-MWCNT/epoxy composites were measured at a series of different frequencies, from 0.1 Hz to 1 MHz. The real and imaginary parts of the impedance (Z' and Z'') were collected, and the magnitude of the AC conductivity (σ) was calculated using equations:

$$Z = Z' + iZ''$$

$$\sigma = \frac{1}{\sqrt{Z'^2 + Z''^2}} \frac{L}{A}$$

(1)

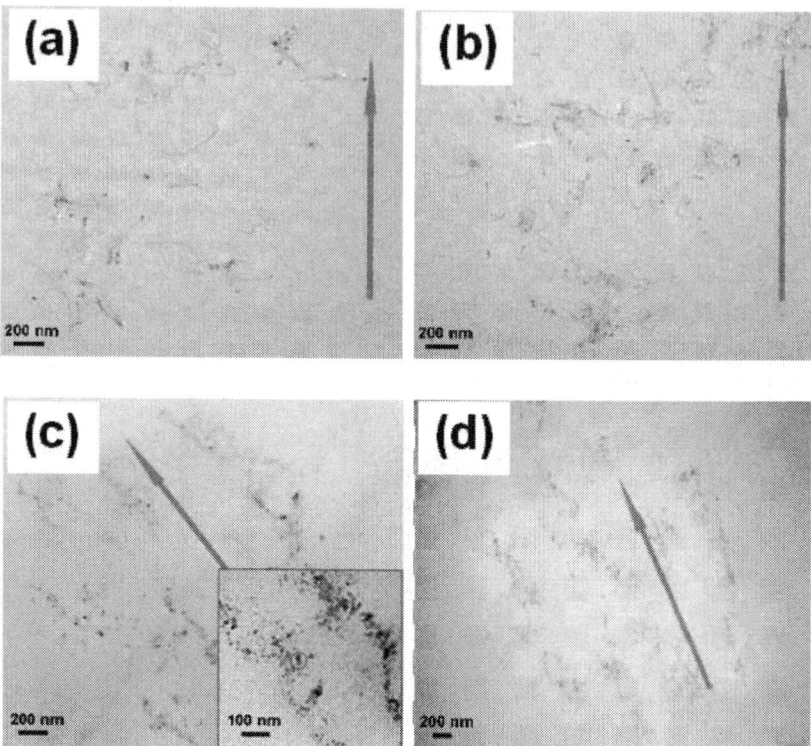

Figure 13. a) TEM image of MWCNT/epoxy composites with 0.5 wt% filler loading. (b) TEM image of MWCNT/epoxy composites with 1.0 wt% filler loading. (c) TEM images of m-MWCNT/epoxy composites with 0.5 wt% filler loading. Inset shows the end-to-top connectivity between two m-MWCNTs under an external magnetic field. (d) TEM image of m-MWCNT/epoxy composites with 1.0 wt% filler loading (Adapted with permission from Kim et al., *Carbon* 2011, 49, 1, 54-61. Copyright 2011 Elsevier).

where, i is the imaginary unit, L is the path length along the measurement direction, and A is the electrode cross-sectional area. Figure 14 shows various conductivities of a series of m-MWCNT/epoxy nanocomposite samples containing various degrees of content of m-MWCNT in the polymer matrix. At the same magnetic field (0.3 T), the conductivity increased with increasing m-MWCNT content in composites. In the case of 0.1 wt% filler content, the nanocomposite exhibited dielectric behavior because the low mass-fraction of m-MWCNT made it difficult to form interconnected m-MWCNT networks that would have facilitated electron flow. However, for m-MWCNT contents of 0.5

wt% and higher, the samples exhibited increased conductivity as a function of increased mass-fraction of the m-MWCNT.

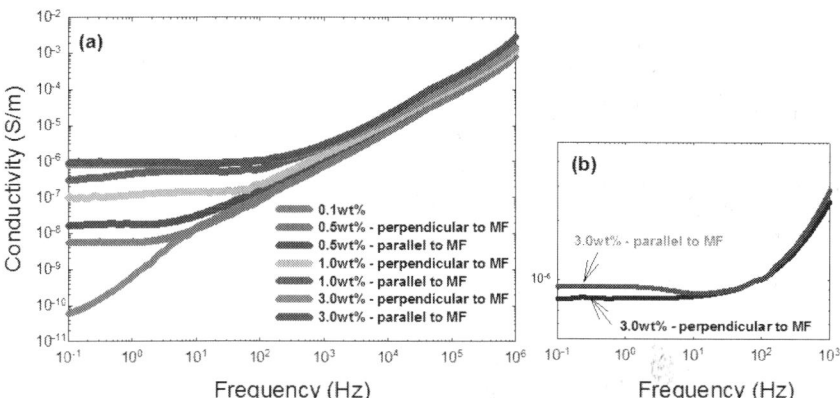

Figure 14. a) The conductivity of m-MWCNT/epoxy composites as a function of frequency for different mass loading of m-MWCNT as measured in the direction parallel to the magnetic field and perpendicular to the magnetic field. (b) The magnified region of a nanocomposite with a 3.0 wt% filler loading (Adapted with permission from Kim et al., *Carbon* 2011, 49, 1, 54-61. Copyright 2011 Elsevier).

Percolation theory predicts a critical concentration or percolation threshold where the material converts from a capacitor to a conductor (Weber et al., 1997; Ounaies et al., 2003). In order to determine the percolation threshold of the aligned system, the volume conductivity data could be fitted to a power law in terms of volume fraction of m-MWCNT.

$$\sigma_c \propto \left(v - v_c\right)^t$$

where σ_c is the composite conductivity, v is the m-MWCNT volume fraction in the composite, v_c is the critical volume fraction, and t is the critical exponent. We assumed that the density of m-MWCNT is the same as that of unmodified-MWCNT (2.1 g/cm³), since both mass and volume of m-MWCNT increase similarly. The inset in Figure 15 shows the plot of σ_c as a function of $v-v_c$ for the parallel measurements. The linear fit to the data generated a straight line with v_c= 0.2 vol% (corresponding to 0.4 wt%), which gives a good fit.

When we compared the results of samples in which conductivity was measured in the direction of the m-MWCNT alignment (parallel to the magnetic field) and perpendicular to the m-MWCNT alignment (perpendicular direction to the magnetic field) for the same mass fraction of m-MWCNT, we observed that the conductivity measured parallel to the magnetic field was higher than that measured perpendicular to the magnetic field, indicating a cooperative effect due to the alignment of the m-MWCNTs in the polymer matrix, as was previously shown in Figure 13(c)and 13(d). Figure 15 shows the variation of the conductivities extracted from the plateau region at low frequency as a function of m-MWCNT mass fractions in the epoxy nanocomposite for both the parallel and perpendicular directions with respect to the magnetic field. The measured conductivities are summarized in Table 2.

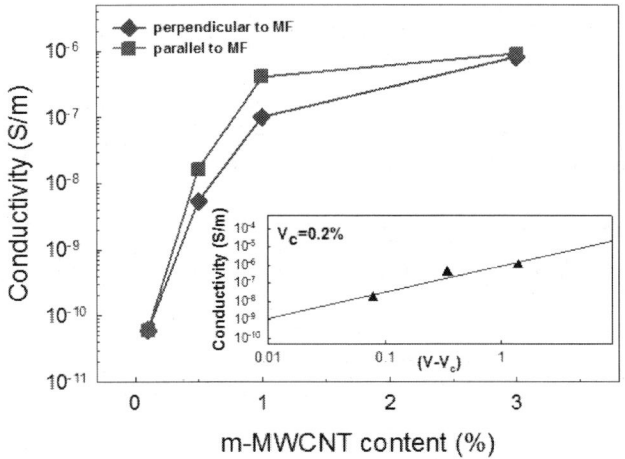

Figure 15. The conductivity of m-MWCNT/epoxy composites as a function of different mass loading of m-MWCNT measured in the direction parallel to the magnetic field and perpendicular to it. Inset shows percolation equation fit to the experimental conductivity data obtained parallel to the direction of the magnetic field (Adapted with permission from Kim et al., *Carbon* 2011, 49, 1, 54-61. Copyright 2011 Elsevier).

We would like to note that for 3.0 wt% m-MWCNT sample, even though the conductivity in the parallel direction was somewhat larger than that in the perpendicular direction (see Figure 14(b)), the values obtained were, nevertheless, quite similar. This is most likely due to the following factors: (a) We assumes that the viscosity of the composite solution containing 3.0 wt% m-MWCNT is higher than for other compositions as evidenced by the superior alignment of m-MWCNT without polymer matrix to that of m-MWCNT/epoxy composites, as discussed in a previous section. By introducing higher mass fractions of the carbon nanotubes into the polymer solution, the viscosity of the system could be further increased, fact which could then handicap with the alignment process. Therefore, we can conclude that when the magnetic field was applied to the 3.0 wt% m-MWCNT sample, the alignment of the decorated carbon nanotubes was not as effective as in the less concentrated samples, and hence, the differences between the conductivities in the parallel and the perpendicular directions were not as pronounced, mainly due to the higher viscosity of the solution. (b) In addition, the conductivity of 3.0 wt% m-MWCNT sample (measured in either direction) was not much higher than the conductivity of the 1.0 wt% m-MWCNT sample (seeFigure 15). Tethered iron oxide (maghemite) nanoparticle has high resistivity (Mei et al., 1987). Hence, the higher viscosity of the 3.0 wt% m-MWCNT sample may lead to the formation of iron oxide rich regions, resulting in a decrease of the conductivity.

Table 2. The conductivity of m-MWCNT/epoxy composites in the directions that were parallel and perpendicular to the externally-applied magnetic field as a function of m-MWCNT content.

m-MWCNT content (wt%)	Conductivity (S/m)		Conductivity ratio of parallel and perpendicular
	Parallel to MF	Perpendicular to MF	
3.0	1.0×10^{-6}	8.5×10^{-7}	1.2
1.0	4.1×10^{-7}	1.0×10^{-7}	4.1
0.5	1.6×10^{-8}	5.3×10^{-9}	3.0
0.1	6.0×10^{-11}		1.0

6. ANISOTROPIC RESPONSE M-CNT-EPOXY COMPOSITE SYSTEM TO COMPRESSION

The initial goal of using a magnetic field on carbon nanotube composites was to promote the alignment of the carbon nanotubes, which would improve the mechanical properties of the composite, particularly in the direction of the alignment. The effects on the glass transition temperature should be relatively simple to predict since they are well documented (Akima et al., 2006; Bliznyuk et al., 2006;Dou et al., 2006; Lanticse et al., 2006; Park et al., 2006). It is expected that increasing the extent of alignment and orientation of the carbon nanotubes and epoxy matrix chains (Al-Haik et al., 2004) will result in an increase in the glass transition temperature of the composite (Ajayan et al., 1994; Akima et al., 2006; Bliznyuk et al., 2006; Dou et al., 2006; Lanticse et al.,

2006; Park et al., 2006). Table 3summarizes the T_g values for the various samples tested.

Table 3. The glass transition temperature of the m-CNT/epoxy nanocomposites measured in samples subjected to various external magnetic fields.

Fe:CNT	Magnetic field (Tesla)	$T_g \pm 5.2$ (°C)
0:0	0	54.9
0:0	0.4	63.8
0:0	0.8	68.7
2:1	0	41.6
2:1	0.4	65.6
2:1	0.8	75.1
4:1	0	45.5
4:1	0.4	68.9
4:1	0.8	86.0

The glass transition temperature of the epoxy matrix increased with increasing magnetic field and implies that there is some molecular orientation/alignment occurring in the epoxy (Garmestani et al., 2003). When the magnetic CNTs are introduced into the matrix, the glass transition temperature of the nanocomposite decreased, probably due to the plasticizing effect of the NaDDBS molecules associated with the Fe_2O_3 nanoparticles tethered to the surface of the CNTs. However, in the samples in which the magnetic CNTs were aligned when the nanocomposite was subjected to an external magnetic field, the general trend showed that an increase in the extent of alignment caused an increase in T_g, as expected.

Preliminary compression data that illustrate the effect of a magnetic field on the modulus of the epoxy matrix naocomposites are shown in Figure 16.

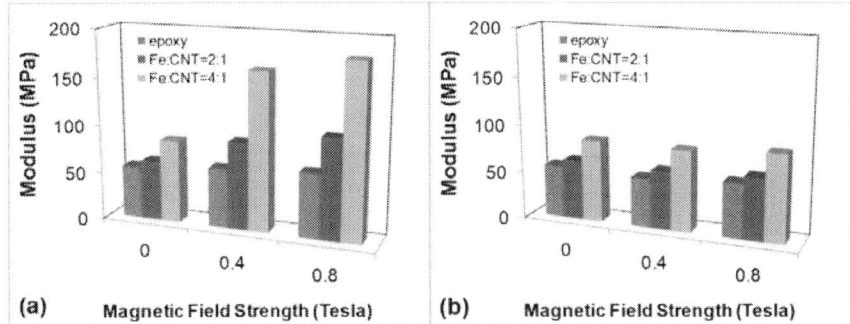

Figure 16. The moduli of the m-CNT/epoxy nanocomposites that were subjected to an external magnetic field measured at room temperature.(a) Measured in the direction that is parallel to the applied magnetic field;(b) Measured in the direction that is perpendicular to the applied magnetic field.

The modulus was measured at room temperature both in the direction that is parallel to the applied magnetic field (Figure 16a)and in the direction that is perpendicular to the applied magnetic field (Figures 16b). The presence of the externally-applied magnetic field had little effect on the pure epoxy. It has been shown in previous work that the orientation and possible alignment of epoxy chains is indeed possible, but only at very high magnetid fields (Garmestani et al., 2003). Hence, the moduli of pure epoxy are constant, irrespective of the magnitude of the magnetic field applied on the samples and of the direction of the measurement. Conversely, the moduli observed for the epoxy filled with the m-CNTs indeed increase with the increase in the magnetic field, mainly for measurements conducted parallel to the direction of the magnetic field. Moreover, higher concentrations of the iron oxide nanoparticles tethered to the surface of the CNTs result in considerably higher moduli, particularliy in the direction parallel to the magnetic field, probably due to an increase in the susceptibility of the m-CNTs to the applied magnetic field.

7. SUMMARY AND OUTLOOK

In this chapter, we have demonstrated on CNT-inorganic hybrid system, especially, CNT/γ-Fe$_2$O$_3$hybrid materials. We developed the synthesis method of MWCNT/γ-Fe$_2$O$_3$ nanostructures via an easy and novel modified sol-gel process. Our study shows that NaDDBS molecules are intimately involved in inhibiting the formation of an iron oxide gel. As a result, well-defined and well-dispersed maghemite nanoparticles can be obtained. In addition, the particle size of these nanoparticles could be precisely modulated by changing the temperature and the mass ratio of the $Fe(NO_3)_3$ $9H_2O$ precursor and MWCNTs. Finally, tethered γ-Fe_2O_3 magnetic nanoparticles on the surface of MWCNTs imparted superparamagnetic properties to the composite material.

Due to the acquired magnetic property of the m-MWCNTs, they could be aligned either alone or embedded in a polymer matrix by the application of only a relatively weak magnetic field. Conductivity measurements performed on m-MWCNT/epoxy composites showed that the conductivity of the m-MWCNT/epoxy composites increased with increasing m-MWCNT contents with low percolation threshold (~0.4–0.5 wt% m-MWCNT loading). Moreover, the conductivity measured in the direction parallel to the magnetic field was higher than that measured in the direction perpendicular to it. However, the alignment of a nanocomposite sample having a loading of 3.0 wt% m-MWCNT was not as effective as samples with lower nanofiller content because of the higher solution viscosity in the more concentrated samples. This hurdle could, in principle, be overcome by either applying a stronger magnetic field or selecting other polymer matrices with low solution viscosity.

In summary, our facile magnetic functionalization method could be effectively applied for the development of conductive films, composites with conductive polymers, and bio-based composites with aligned features. Furthermore, we suggest that this maghemite-CNT hybrid material may be used for biomedical applications such as drug delivery or special medical applications such as cancer diagnosis in the not-so-distant future (Sincai et al., 2001; Sousa et al., 2001).

8. ACKNOWLEDGEMENTS

This work was supported in part by grants from NSF, Division of Engineering, award No. ECCS-0535382 and by the Air Force/Bolling AFB/DC MURI, award No. F49620-02-1-0382. Il Tae Kim was supported by a Paper Science and Engineering (PSE) Graduate Fellowships from the Institute of Paper Science and Technology (IPST) at the Georgia Institute of Technology. The authors are indebted to Drs. Erin Camponeschi, Hamid Garmestani, Karl Jacob and Allen Tannenbaum for their invaluable contributions and stimulating input.

REFERENCES

1. P. M. Ajayan, O. Stephan, C. Colliex, D. Trauth, 1994 Aligned carbon nanotube by cutting a polymer resin-nanotube composite. Science, 265, 1212

2. N. Akima, Y. Iwasa, S. Brown, A. M. Barbour, J. Cao, J. L. Musfeldt, H. Matsui, N. Toyota, M. Shiraishi, H. Shimoda, O. Zhou, 2006 Strong

anisotropy in the far-infrared absorption spectra of stretch-aligned single-walled carbon nanotubes. Adv. Mater., 18, 9, 1166 1169.

3. H. Ajiki, T. Ando, 1993 Magenetic-properties of carbon nanotubes, J. Phys. Soc. Jpn., 62, 7, 2470 2480 .

4. M. S. Al-Haik, H. Garmestani, D. S. Li, M. Y. Hussaini, S. S. Sablin, R. Tannenbaum, K. Dahmen, 2004 Mechanical properties of magnetically oriented epoxy. J. Polym. Sci. Polym. Phys., 42, 1586 1600.

5. V. N. Bliznyuk, S. Singamaneni, R. L. Sanford, D. Chiappetta, B. Crooker, P. V. Shibaev, 2006 Matrix mediated alignment of single wall carbon nanotubes in polymer composite films. Polymer, 47, 11, 3915 3921.

6. K. Butter, P. H. Bomans, P. M. Frederik, G. J. Vroege, A. P. Philipse, 2003 Direct observatioin of dipolar chains in ferrofluids in zero field using cryogenic electron microscopy, J. Phys.: Conds. Matter., 15, 15, S1451 S1470 .

7. E. Camponeschi, B. Florkowski, R. Vance, G. Garrett, H. Garmestani, R. Tannenbaum, 2006 Uniform directional alignment of single-walled carbon nanotubes in viscous polymer flow, Langmuir, 22, 4, 1858 1862.

8. E. Camponeschi, R. Vance, M. Al-Haik, H. Garmestani, R. Tannenbaum, 2007 Properties of carbon nanotube-polymer composites aligned in a magnetic field, Carbon, 45, 10, 2037 2046 .

9. E. Camponeschi, J. Walker, H. Garmestani, R. Tannenbaum, 2008 Surfactant effects on the particle size of iron (III) oxides formed by sol-gel synthesis, J. Non-Cryst. Solids, 351, 34, 4063 4069.

10. X. Q. Chen, T. Saito, H. Yamada, K. Matsushige, 2001 Aligning single-wall carbon nanotubes with an alternating-current electric field, Appl. Phys. Lett., 78, 23, 3714 3716 .

11. M. A. Correa-Duarte, M. Grzelczak, V. Salgueirino-Maceira, M. . Giersig, L. M. Liz-Marzan, M. Farle, K. Sierazdki, R. Diaz, 2005 Alignment of carbon nanotubes under low magnetic fields through attachment of magnetic nanoparticles, J. Phys. Chem. B, 109, 41, 19060 19063 .

12. D. L. A. de Faria, S. V. Silva, M. T. de Oliveira, 1997 Raman microspectroscopy of some iron oxides and oxyhydroxides, J. Raman Spectrosc., 28, 11, 873 878 .

13. S. X. Dou, W. K. Yeoh, O. Shcherbakova, D. Wexler, Y. Li, S. M. Ren, P. Munroe, S. K. Chen, S. K. Tan, B. A. Glowacki, Manus. Mac-Driscoll, J.

P., 2006 Alignment of carbon nanotube additives for improved performance of magnesium diboride superconductors. Adv. Mater., 18, 6, 785 788.

14. J. E. Fischer, W. Zhou, J. Vavro, M. C. Llaguno, C. Guthy, R. Haggenmueller, 2003 Magnetically aligned single wall carbon nanotube films: Preferred orientation and anisotropic transport properties, J. Appl. Phys., 93, 4, 2157 2163.

15. M. Fujiwara, E. Oki, M. Hamada, Y. Tanimoto, I. Mukouda, Y. Shimomura, 2001 Magnetic orientation and magnetic properties of a single carbon nanotube, J. Phys. Chem. A, 105, 18, 4383 4386.

16. H. Garmestani, M. S. Al-Haik, K. Dahmen, R. Tannenbaum, D. Li, S. S. Sablin, M. Y. Hussaini, 2003 Polymer-mediated alignment of carbon nanotubes under high magnetic fields, Adv. Mater., 15, 22, 1918 1921 .

17. M. Grzelczak, M. A. Correa-Duarte, L. M. Liz-Marzan, 2006 Carbon nanotubes encapsulated in wormlike hollow silica shells, Small, 2, 10, 1174 1177 .

18. L. Han, W. Wu, F. L. Kirk, J. Luo, M. M. Maye, N. N. Kariuki, Y. H. Li, C. Wang, C. J. Zhong, 2004 A direct route toward assembly of nanoparticle-carbon nanotube composite materials, Langmuir, 20, 14, 6019 6025 .

19. T. Hyeon, S. S. Lee, J. Park, Y. Chung, N. H. Bin, 2001 Synthesis of highly crystalline and monodisperse maghemite nanocrystallites without a size-selection process, J. Am. Chem. Soc., 123, 51, 12798 12801 .

20. B. Jia, L. Gao, J. Sun, 2007 Self-assembly of magnetite beads along multiwalled carbon nanotubes via a simple hydrothermal process, Carbon, 45, 7, 1476 1481.

21. A. Jorio, M. A. Pimenta, A. G. Souza, R. Saito, G. Dresselhaus, M. S. Dresselhaus, J. New, Phys, 5139 139.

22. Kim, I. T. , G. Nunnery, K. Jacob, J. Schwartz, X. Liu, R. Tannenbaum, 2010 Synthesis, characterization, and alignment of magnetic carbon nanotubes tethered with maghemite nanoparticles, J. Phys. Chem. C, 114, 15, 6944 6951 .

23. T. Kim, A. Tannenbaum, R. Tannenbaum, 2011 Anisotropic conductivity of magnetic carbon nanotubes embedded in epoxy matrices, Carbon, 49, 1, 54 61 .

24. T. Kimura, H. Ago, M. Tobita, S. Ohshima, M. Kyotani, M. Yumura, 2002 Polymer composites of carbon nanotubes aligned by a magnetic field, Adv. Mater., 14, 19, 1380 1383 .

25. G. Korneva, H. Ye, Y. Gogotsi, D. Halverson, G. Friedman, J. C. Bradley, K. Kornev, 2005 Carbon nanotubes loaded with magnetic particles, Nano Lett., 5, 5, 879 884 .

26. Q. Kuang, S. F. Li, Z. X. Xie, S. C. Lin, X. H. Zhang, S. Y. Xie, R. B. Huang, L. S. Zheng, 2006 Controllable fabrication of SnO2-coated nanotubes by chemical vapor multiwalled carbon deposition, Carbon, 44, 7, 1166 1172 .

27. L. J. Lanticse, Y. Tanabe, K. Matsui, Y. Kaburagi, K. Suda, M. Hoteida, M. Endo, E. Yasuda, 2006 Shear-induced preferential alignment of carbon nanotubes resulted in anisotropic electrical conductivity of polymer composites. Carbon, 44, 14, 3078 3086.

28. O. Lourie, D. M. Cox, H. D. Wagner, 1998 Buckling and collapse of embedded carbon nanotubes, Phys. Rev. Lett., 81, 8, 1638 1641.

29. B. Lukic, J. W. Seo, R. R. Bacsa, S. Delpeux, F. Beguin, G. Bister, A. Fonseca, J. B. Nagy, A. Kis, S. Jeney, A. J. Kulik, L. Forro, 2005 Catalytically grown carbon nanotubes of small diameter have a high Young's modulus, Nano Lett., 5, 10, 2074 2077 .

30. C. A. Martin, J. K. W. Sandler, A. H. Windle, M. K. Schwarz, W. Bauhofer, K. Schulte, M. S. P. Shaffer, 2005 Electric field-induced aligned multi-wall carbon nanotube networks in epoxy composites, Polymer, 46, 3, 877 886 .

31. O. Matarredona, H. Rhoads, Z. R. Li, J. H. Harwell, L. Balzano, D. E. Resasco, 2003 Dispersion of single-walled carbon nanotubes in aqueous solutions of the anionic surfactant NaDDBS, J. Phys. Chem. B, 107, 48, 13357 13367.

32. Y. Mei, Z. J. Zhou, H. L. Luo, 1987 Electrical-resistivity of rf-sputtered iron-oxide thin-films, J. Appl. Phys., 61, 8, 4388 4389 .

33. A. Millan, F. Palacio, A. Falqui, E. Snoeck, V. Serin, A. Bhattacharjee, V. Ksenofontov, P. Gutlich, I. Gilbert, 2007 Maghemite polymer nanocomposites with modulated magnetic properties, Acta Mater. 55, 6, 2201 2209.

34. L. A. Nagahara, I. Amlani, J. Lewenstein, R. K. Tsui, 2002 Directed placement of suspended carbon nanotubes for nanometer-scale assembly, Appl. Phys. Lett., 80, 20, 3826 3828 .

35. Z. Ounaies, C. Park, K. E. Wise, E. J. Siochi, J. S. Harrison, 2003 Electrical properties of single wall carbon nanotube reinforced polyimide composites, Compos. Sci. Technol., 63, 11, 1637 1646.

36. C. Park, J. Wilkinson, S. Banda, Z. Ounaies, K. E. Wise, G. Sauti, P. T. Lillehei, J. S. Harrison, 2006 Aligned single-wall carbon nanotube polymer composites using an electric field. J. Poly. Sci. B: Poly. Phys., 44, 12, 1751 1762.

37. C. Pascal, J. L. Pascal, F. Favier, M. L. E. Moubtassim, C. Payen, 1999 Electrochemical synthesis for the control of gamma-Fe2O3 nanoparticle size. Morphology, microstructure, and magnetic behavior, Chem. Mater., 11, 1, 141 147.

38. C. Q. Peng, Y. S. Thio, R. A. Gerhardt, 2008 Conductive paper fabricated by layer-by-layer assembly of polyelectrolytes and ITO nanoparticles, Nanotechnology, 19, 50, 505603.

39. M. P. Pileni, 2003 The role of soft colloidal templates in controlling the size and shape of inorganic nanocrystals, Nature Mater., 2, 3, 145 150.

40. H. T. Pu, F. J. Jiang, 2005 Towards high sedimentation stability: magnetorheological fluids based on CNT/Fe3O4 nanocomposites, Nanotechnology, 16, 9, 1486 1489.

41. L. T. Qu, L. Dai, E. Osawa, 2006 Shape/size-controlled syntheses of metal nanoparticles for site-selective modification of carbon nanotubes, J. Am. Chem. Soc., 128, 16, 5523 5532 .

42. Rockenberger, E. C. Scher, A. P. Alivisatos, 1999 A new nonhydrolytic single-precursor approach to surfactant-capped nanocrystals of transition metal oxides, J. Am. Chem. Soc., 121, 49, 11595 11596 .

43. M. Sincai, D. Ganga, D. Bica, L. Vekas, 2001 The antitumor effect of locoregional magnetic cobalt ferrite in dog mammary adenocarcinoma, J. Magn. Magn. Mater., 225, 1-2, 235 240 .

44. M. H. Sousa, J. C. Rubim, P. G. Sobrinho, F. A. Tourinho, 2001 Biocompatible magnetic fluid precursors based on aspartic and glutamic acid modified maghemite nanostructures, J. Magn. Magn. Mater., 225, 1-2, 67 72 .

45. B. W. Steinert, D. R. Dean, 2009 Magenetic field alignment and electrical properties of solution cast PET-carbon nanotube composite films, Polymer, 50, 3, 898 904 .

46. S. H. Sun, C. B. Murray, D. Weller, L. Folks, A. Moser, 2000 Monodisperse FePt nanoparticles and ferromagnetic FePt nanocrystal superlattices, Science, 287, 5460, 1989 1992 .

47. S. H. Sun, H. Zeng, D. B. Robinson, S. Raoux, P. M. Rice, S. X. Wang, G. X. Li, 2004 Monodisperse MFeO4 (M = Fe, Co, Mn) nanoparticles, J. Am. Chem. Soc., 126, 1, 273 279 .

48. Z. Sun, Z. Liu, Y. Wang, B. Han, J. Du, J. Zhang, 2005 Fabrication and characterization of magnetic carbon nanotube composites, J. Mater. Chem., 15, 42, 4497 4501.

49. Z. Y. Sun, H. Q. Yuan, Z. M. Liu, B. X. Han, X. R. Zhang, 2005 A highly efficient chemical sensor material for H2S: alpha-Fe2O3 nanotubes fabricated using carbon nanotube templates, Adv. Mater., 17, 24, 2993 2997.

50. M. M. J. Treacy, T. W. Ebbesen, J. M. Gibsoj, 1996 Exceptionally high Young's modulus observed for individual carbon nanotubes, Nature, 381, 6584, 678 680 .

51. D. K. Yi, S. S. Lee, J. Y. Ying, 2006 Synthesis and application of magnetic nanocomposite catalysts, Chem. Mater., 18, 10, 2459 2461 .

52. S. C. Youn, D. Jung, Y. K. Ko, Y. W. Jin, J. M. Kim, H. Jung, 2009 Vertical alignment of carbon nanotubes using the magneto-evaporation method, J. Am. Chem. Soc., 131, 2, 742 748 .

53. M. F. Yu, B. S. Files, S. Arepalli, R. Ruoff, 2000 Tensile loading of ropes of single wall carbon nanotubes and their mechanical properties, Phys. Rev. Lett., 84, 24, 5552 5555 .

54. Wan, W. Cai, J. Feng, X. Meng, E. Liu, 2007 In situ decoration of carbon nanotubes with nearly monodisperse magnetite nanoparticles in liquid polyols, J. Mater. Chem., 17, 12, 1188 1192.

55. Weber, M. R. Kamal, 1997 Estimation of the volume resistivity of electrically conductive composites, Polym. Compos., 18, 6, 711 725 .

56. W. B. White, B. A. Deangeli, 1967 Interpretation of vibrational spectra of spinels, Spectrochim. Acta, Part A, A 23, 4, 985 995.

57. Y. F. Zhu, C. , W. Zhang, R. P. Zhang, N. Koratkar, J. Liang, 2009 Alignment of multiwalled carbon nanotubes in bulk epoxy composites via electric field, J. Appl. Phys., 105, 5, 054319.

Low-Energy Irradiation Damage in Single-Wall Carbon Nanotubes

Satoru Suzuki[1]

[1] NTT Basic Research Laboratories, NTT Corporation, 3-1, Morinosato Wakamiya, Atsugi, Kanagawa, Japan

1. INTRODUCTION

Single-wall carbon nanotubes (SWCNTs) are one of the most promising materials for future nano-electronics, because of their unique quasi-one-dimensional structures and excellent electric and mechanical properties. They also have very high chemical stability, owing to their robust sp^2-bonding carbon network (graphene) with no dangling bonds. Because of the structural robustness, low-energy (typically 10 eV-20 keV) electron and photon irradiation in a vacuum had been generally assumed not to cause damage to SWCNTs when the energy is smaller than the knock-on threshold. In fact, analytical tools that use low-energy electrons or photons, such as scanning electron microscopy (SEM), had been commonly used for characterization of SWCNTs without serious concerns.

In 2004, however, we reported that electron irradiation in a SEM caused severe damage (low-energy irradiation damage) in SWCNTs produced by both thermal chemical vapor deposition and laser ablation methods (Suzuki et al., 2004b). Other techniques using low-energy electrons and vacuum-ultraviolet (VUV) light or soft x-rays (especially high-brilliance synchrotron radiation light), such as low-energy electron microscopy (LEEM) and photoemission spectroscopy, also inevitably damage SWCNTs. Therefore, paying attention to the low-energy irradiation damage is practically important for those who study SWCNTs. For example, when we measure the Raman and photoluminescence (PL) spectra and electric properties and take SEM images of the same SWCNTs, the SEM observations should be done last. Doing the high-resolution SEM observation first would inevitably cause severe damage and tremendously affects the following measurements.

The low-energy irradiation damage and its defect characteristics are also physically interesting. In this chapter, we will review the physical and chemical property changes induced by the damage, and the defect properties, which are significantly different from those of other types of damage. We will examine the defect-induced metal-semiconductor transition of the room-temperature electric properties and discuss its mechanism. We will also summarize other types of damage, which are often confused with the low-energy irradiation damage, focusing on the differences between them.

Before continuing to the main text, I must briefly explain how I compare spectra obtained form the same SWCNT sample. In many studies of the physical or chemical treatment of SWCNTs and graphene, spectra are often normalized to the maximum peak height. In contrast, when I show irradiation- and annealing-induced changes of Raman and PL spectra, the spectra are never normalized. That is, I obtain the spectra under the same condition to the best of my ability and directly compare the raw spectra. This methodology has been applied in all of our related reports, unless otherwise mentioned. With arbitrary spectral normalization, we would no longer be able to discuss the reversibility of the damage and recovery, which is a very important characteristic of low-energy irradiation damage.

2. WHAT IS LOW-ENERGY IRRADIATION DAMAGE?

We define low-energy irradiation damage as damage solely caused by irradiation of low-energy particles, where low-energy means that the energy is much smaller than the threshold energy of knock-on damage. Thus, the mechanism of the damage is completely different from knock-on damage. Moreover, we discriminate low-energy irradiation damage and secondary damage caused by the irradiation, such as damage by radicals. Irradiation by both electrons and photons irradiation was found to damage SWCNTs. However, other particles, such as atoms and ions, or quasi-particles such as plasmons, may also cause the damage.

3. PROPERTY CHANGES CAUSED BY LOW-ENERGY IRRADIATION DAMAGE

3.1. Raman and pl spectra

In a Raman spectrum, a SWCNT shows the so-called G band (tangential mode) and disorder-induced D band, which are characteristic of a graphene sheet. The D band is ideally inactive and its appearance is evidence of symmetry breaking. The intensity ratio of the G and D bands is often utilized as an indicator of the degree of crystallinity. Another very important mode of SWCNTs is the radial breathing mode (RBM), which is often used for diameter evaluation. For a general review of Raman spectroscopy of CNTs, see (Dresselhaus et al., 2005), for example. Like other types of damage, low-energy irradiation damage generally decreases the G band and RBM intensities (There are some exceptions at the edges of the resonance window, as discussed below) and the G/D intensity ratio and increases the D band intensity, as shown in Figs. 1(a) and (b). Generally, the decrease of intensity is more prominent for the RBM than for the G band. Considering that the detectable Raman intensity from individual SWCNTs is owing to the resonance enhancement effect, the disappearance of Raman spectra is probably due to a reduction of the resonance enhancement. The initially divergent joint density of states, which is a characteristic of one-dimensional systems, would be considerably broadened by the formation of defects. Low-energy irradiation damage causes almost no broadening or almost no shift of the Raman peaks including the D band (Suzuki et al., 2010), although significant D band broadening due to gas-phase reaction has been observed (Yang et al., 2006. Zhang et al., 2006).

When SWCNTs are moderately damaged (or considerably recover from severe damage), originally hidden non-resonant RBM peaks sometimes appear. At the excitation wavelength of 785 nm, metallic and semiconducting SWCNTs are usually observed at about 150-160 and 200-240 cm^{-1}, respectively. In Fig. 2(a), however, the moderately damaged (partially recovered) SWCNTs show a sharp peak at 182 cm^{-1} in the off-resonance region. The metallic SWCNTs at 156 cm^{-1}, which were initially not strongly excited in this sample, also became more prominent in the moderately damaged sample. Further damage extinguishes these peaks again, as also shown in the figure. Similar off-resonant RBM peaks are also often observed in doped SWCNTs grown from boron- and nitrogen-containing feedstocks, as shown in Fig. 2(b) (Suzuki & Hibino, 2011). I think that the defects slightly shift the absorption energy or broaden the absorption edge and this makes the originally off-resonant peak resonant. Similarly, complicated behavior of the RBM intensity with increasing damage is observed at the edge of the resonance window (Suzuki & Kobayashi, 2007a).

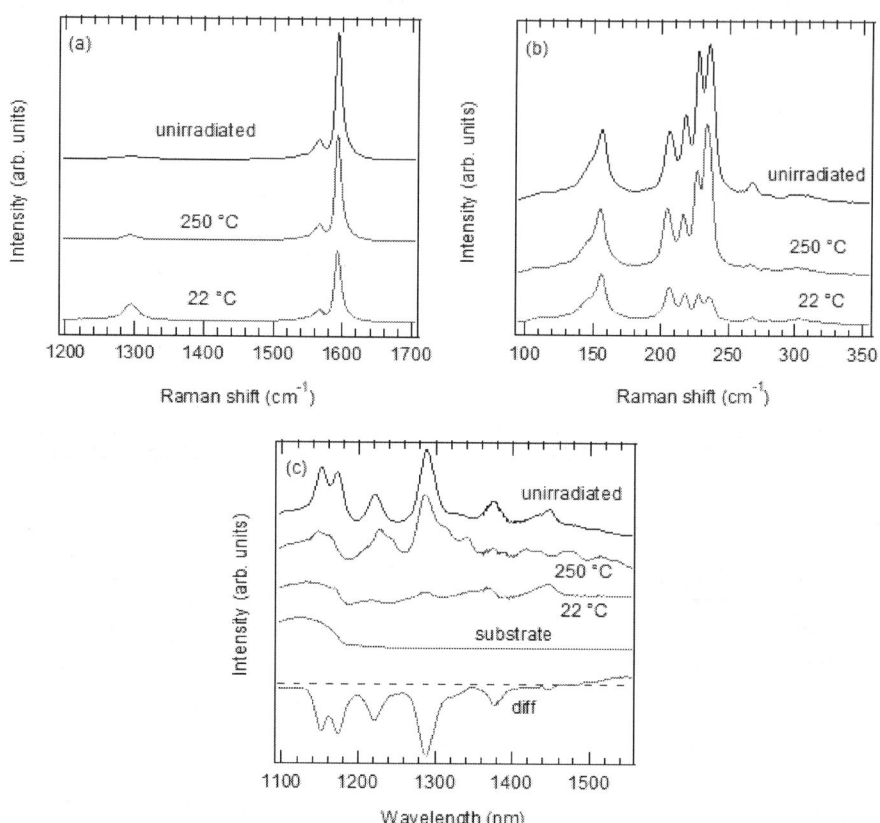

Figure 1. a) G and D band, and (b) RBM regions of Raman spectra, and (c) PL spectra of unirradiated SWCNTs and of SWCNTs irradiated at 250 and 22º C. The irradiated electron energy and dose were 20 keV and 5.7×10^{16} cm^{-2}. The excitation wavelength was 785 nm.

These results mean that the Kataura plot is modified by the defects.

The PL peak intensity of suspended semiconducting SWCNTs is more sensitively decreased than the Raman peak intensity, as shown in Fig. 1(c). In addition, broad spectral intensity newly appears at the longer wavelength side when the extent of the damage is moderate.

Severe damage finally extinguishes all spectral intensities in Raman (including the D band (Suzuki et al., 2005a)) and PL spectra (Suzuki & Kobayashi, 2007b). However, note that, in marked contrast to the damage caused by knock-on collisions and by radicals, the low-energy irradiation damage itself never eliminates a SWCNT.

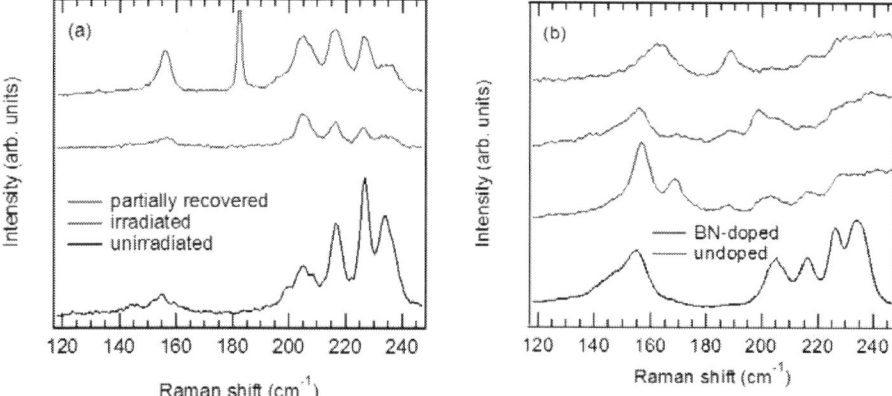

Figure 2. A) RBM spectra of unirradiated and electron-irradiated SWCNTs and partially recovered SWCNTs. The electron energy and irradiation dose were 20 keV and 6.3×10^{16} cm^{-2}, respectively. The irradiated and considerably damaged SWCNTs were partially recovered by annealing in Ar atmosphere at 350º C. The wavenuber range of 160-200 cm^{-1} is the off-resonance region and a peak is rarely observed there, initially. (b) RBM spectra of undoped and BN-doped SWCNTs. The doped SWCNTs also often exhibit peaks in the off-resonance region. The excitation wavelength was 785 nm.

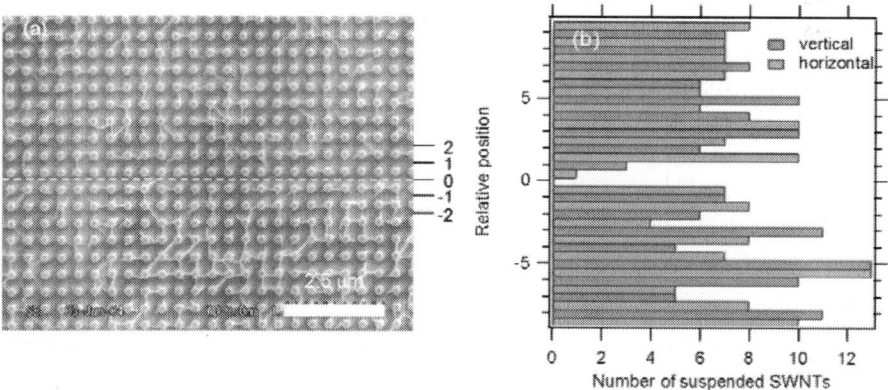

Figure 3. a) SEM image of the SWNT sample after eliminating the electron-irradiated SWCNTs. The irradiation was done along the dashed line. The electron energy and the local dose were 1 keV, and 1.5×10^{19} cm^{-2}, respectively. Note that the elimination was done by selective combustion in air, not by the irradiation itself. (b) Position dependence of the number of SWCNTs suspended between neighboring pillars in (a). Here, diagonally suspended SWCNTs were neglected.

3.2. Chemical Stability

The low-energy irradiation itself does not cut a SWCNT. However, the damage significantly decreases the chemical tolerance of SWCNTs, because the irradiation-induced defects make the sidewall chemically active. Therefore, we can selectively eliminate the irradiated SWCNTs by heating in air, as shown in Fig. 3 (Suzuki et al., 2005a). A part of the SWCNT sample was intensively irradiated in a SEM using the line scan mode along the dashed line. Then, the sample was heated in air at 420º C for 30 m. The irradiated SWCNTs were selectively eliminated by combustion due to the reduced chemical tolerance. Note that a irradiation dose that is too high often has an entirely opposite effect, because the irradiation-induced contaminants on the SWCNT surfaces protect the SWCNTs from oxygen. As shown in (b), the irradiation effects almost completely disappear about 600 nm from the irradiation line. Such a high spatial resolution can be easily obtained using a convergent electron beam.

There have been many attempts to functionalize SWCNTs by using other molecules or metal particles. In many cases, defects are intentionally created to functionalize the sidewall, which is originally inert (Yan et al., 2005). Low-energy irradiation damage could also be applied for spatially selective functionalization with electron beam lithography.

3.3. Electric Properties

The electric properties are much more sensitively changed by low-energy irradiation damage than Raman and PL spectra. Moderate irradiation can convert a metallic field effect transistor (FET) into semiconducting. I will discuss this remarkable phenomenon in sec. 5. Here, I focus on the intensive irradiation effects I have studied by in-situ electric measurements during electron irradiation in a SEM equipped with piezo-actuated micro-probes for electric measurements (Suzuki, 2011). The device used here consists of two SWCNTs (A branch is seen between the electrodes) suspended between the drain and source electrodes (height: 300 nm), as shown in Fig. 4(a). The high-magnification SEM image was taken after all experiments had been completed. Otherwise, the conductivity of the SWCNTs would almost vanish. Fig. 4(b) shows the results of in-situ electric measurements during irradiation. The whole SWCNTs were first irradiated by an electron beam using the normal SEM observation mode. The SEM observation gradually decreased the conductivity.

Then, at ~42.76 s, they were intensively irradiated by using the line scan mode. This irradiation decreased the conductivity by two orders of magnitude in only a few seconds. As shown in Fig. 4(c), avery abrupt current decrease occurred at least within the initial 44 ms, which is the time resolution of the measurements. The gate voltage characteristics of the device before and after the irradiation are shown in Fig. 4(d). The irradiation decreased the two-probe conductivity by four to five orders of magnitude in the whole gate voltage range. Considering that the initial resistivity of the device would be dominated by the contact resistance between the SWNTs and electrodes, the intrinsic conductivity decrease would be much larger. Thus, intensive irradiation finally makes a SWCNT almost insulating. Similar results had been observed in previous works by another group (Marquardt et al., 2008. andVijayaraghavan et al., 2010) and in our early work (Suzuki and Kobayashi., 2005), in which conventional on-substrate SWCNT devices were used. Here, I would like to get remind the readers again that even intensive irradiation does not cut a SWCNT. In fact, a SWCNT can be completely recovered by annealing, as shown later in sec. 4.2.

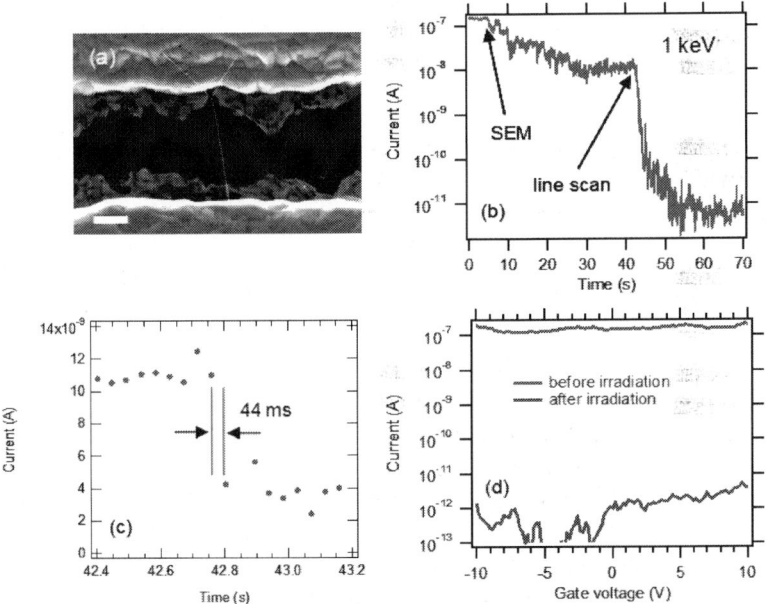

Figure 4. a) SEM image of a suspended SWNT device obtained after experiments. The substrate acts as a back-gate electrode. Scale bar: 200 nm. (b) Drain current during SEM observation and line scans. The drain voltage was set to 0.1 V, and the substrate (back-gate) was grounded. The electron energy was 1 keV. During the SEM observation, the irradiation dose rate was 1.2×10^{13} cm^{-2}s^{-1} on average in the observation area. The dose rate during the lines scan was ~6×10^{16} cm^{-2}s^{-1}. (c) Time evolution of the drain current just before and after the line scan. The data collection was done using the sampling mode of a semiconductor parameter analyzer (Agilent 4156º C) and the time interval is not always the same. (d) Gate characteristics of the device before and after the electron irradiation. The drain voltage was 0.1 eV.

4. CHARACTERISTICS OF LOW-ENERGY IRRADIATION DAMAGE

4.1. Energy dependence

The low-energy irradiation damage has been observed in an energy range of several electron-volts to 25 keV. One important characteristic of the damage is that, in general, a lower energy is more destructive than a higher energy (Suzuki et al., 2004b), as shown in Fig. 5(a) (except for the energies below ~20 eV, where the optical absorption and electron energy loss spectra strongly reflect the specific electronic density of states of graphene). This can be well understood by the fact that the interaction (the cross section of electronic excitation) between a SWCNT and an incident electron (photon) is generally larger at a smaller energy. A high-energy electron or photon easily penetrates a SWCNT without any electric excitation. An exception is when the photon energy is tuned to near the C 1s absorption edge, for which severer damage was observed than at the energy below the absorption threshold. However, the resonance effect does not seem to be very prominent, as will be shown in Fig. 11(b).

The existence of the threshold energy is also expected for the low-energy irradiation damage. As seen in the RBM spectra shown in Fig. 5(b), distinct spectral intensity decreases are observed at ~6 eV or larger for semiconducting SWCNTs with diameters of about 1.2-1.0 nm at 200-240 cm^{-1} (Suzuki & Kobayashi, 2008). Decreases in the G/D ratio are also clearly observed at 6 eV or higher, as shown inFig. 5(c). Notably, the damage in metallic SWCNTs with diameters of ~1.6 nm at 140-160 cm^{-1} is not clear up to 8 eV. This is because of the diameter dependence of low-energy irradiation damage, as discussed in sec. 4.3. In electron (hole) tunneling injection studies using a scanning tunneling microscope (STM), slightly lower threshold energy of ~ 4 eV has been observed (Yamada et a., 2009). The reason for the discrepancy may be that the threshold energy depends on the diameter, chirality, or detailed defect structures. Anyway the threshold energy of the low-energy irradiation damage seems to be several electron volts.

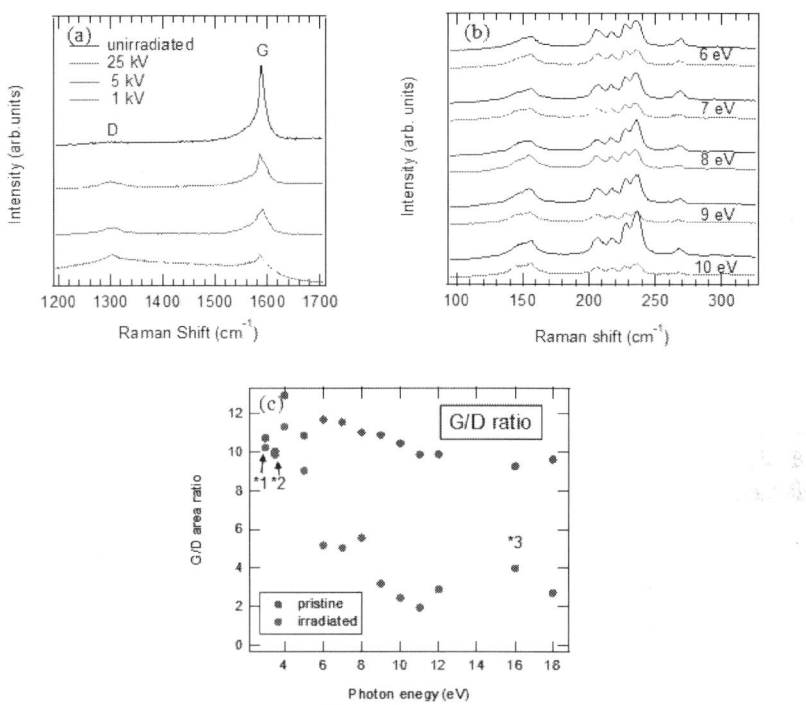

Figure 5. a) G and D band spectra of pristine and electron-irradiated SWCNTs, showing the acceleration voltage dependence of electron irradiation damage. The irradiation dose was 2.7×10^{18} cm^{-2}. (b) RBM spectra of pristine SWCNTs (black lines) and SWCNTs illuminated by 6-10 eV photons (red). Because the spectra were originally slightly sample-dependent, spectra before and after the irradiation are shown for each photon energy. (c) G/D area ratio of pristine SWCNTs and SWCNTs irradiated at various photon energies. In (b) and (c), the photon dose was 5×10^{17} cm^{-2} except for ^{*2} and ^{*3} in (c). The excitation wavelength was 785 nm. ^{*1}) higher order light cut by a pyrex window. ^{*2}) very intense unmonochromatized light through a pyrex window (hv≤3.5 eV). ^{*3}) The photon dose was 2.5×10^{17} cm^{-2}.

4.2. Reversible damage and recovery

Probably the most important characteristic of the low-energy irradiation damage is the reversibility of the damage and recovery. Figure 6 shows Raman spectra of SWCNTs before and after VUV light irradiation and of SWCNTs annealed at 300º, 600º, 800º, 900º C. The irradiation caused severe damage and drastically decreased the G/D ratio and the RBM intensities. All the RBM peaks above 200 cm^{-1}corresponding to a diameter less than ~1.2 nm almost completely disappeared. However, the annealing at 300º, 600º, and 800º C gradually recovered the spectra, and at 900º C, all the peaks including the once disappeared peaks are almost fully recovered. I would like to point out once again that the spectra were not normalized at all. Thus, the results reveal that not only the spectral shape but also the spectral intensity itself is almost fully recovers by annealing.

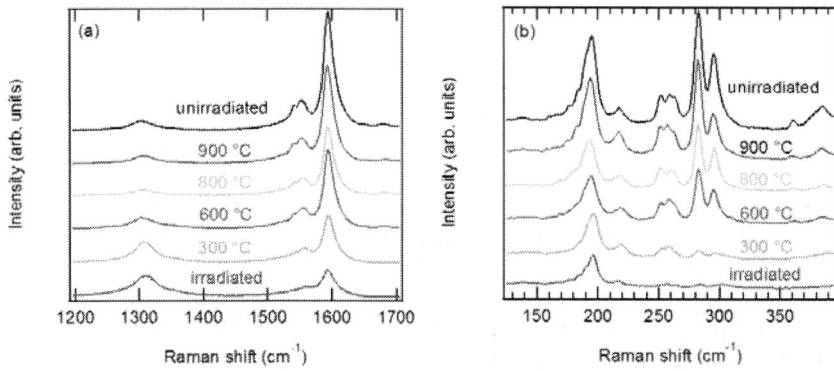

Figure 6. A) G and D band, and (b) RBM spectra of unirradiated, photon-irradiated, and annealed SWCNTs. The SWCNTs were irradiated by unmonochromatized synchrotron radiation light (hv≤1 keV) up to a dose of 8×10^{20} cm^{-2}. The excitation wavelength for the Raman measurements was 633 nm.

The reversible damage and recovery is also observed in the electric properties. As mentioned in sec. 3.3, the low-energy irradiation damage almost extinguishes the electric conductivity. However, the extinguished conductivity is also fully recoverd by annealing, as shown in Fig. 7(a). An originally metallic SWCNT device was intensively irradiated in a SEM using the line scan mode. The irradiation extinguished the conductivity and made the SWCNT almost insulating. However, the conductivity was fully recovered by annealing in a vacuum at 300º C. Moreover, the reversibility can be observed repeatedly, as shown in Fig. 7(b). The complete reversibility of the electric properties has also been observed by another group, although they attribute the conductivity decrease and recovery to substrate charging and its release (Marquardt et al., 2008. and Vijayaraghavan et al., 2010). I will discuss this issue later in sec. 7.4.

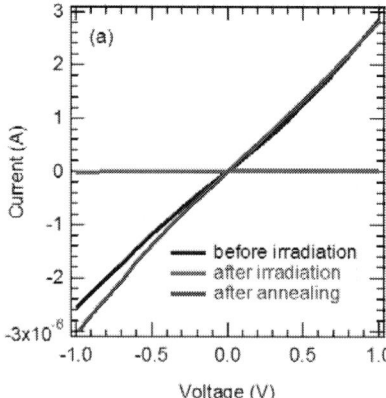 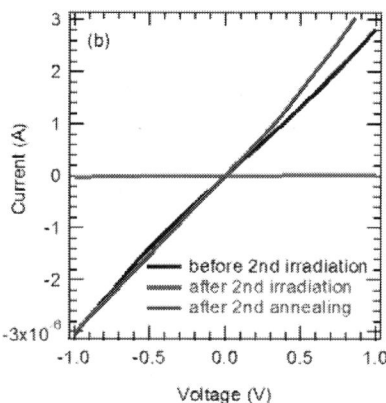

Figure 7. a) Electric properties of a SWCNT device before and after the first electron irradiation and after annealing. The SWCNT was intensively irradiated by electrons of 20 keV up to a dose of $1\times10^{20}\mathrm{cm}^{-2}$. Then, the irradiated device was annealed at 300º C in Ar atmosphere for 30 min. (b) The electric properties of the same SWCNT device before and after the second irradiation and after the second annealing. The SWCNT was intensively irradiated again by 20-keV electrons up to a dose of 1.7×10^{20} cm^{-2}. After the irradiation, the SWCNT was annealed at 300º C for 30 min and fully recovered again.

The reversibility of the damage and recovery indicates that the damage is not accompanied by a reduction of carbon atoms and that the number of carbon atoms is preserved. Recently, Mera et al. directly measured ion desorption from SWCNTs under soft X-ray illumination (Mera et al, 2010). They also excluded emission of carbon atoms from the SWCNTs.

4.3. Diameter dependence

Another important characteristic of the low-energy irradiation damage is that strong diameter dependence is observed when the irradiation is done at room temperature or above (Suzuki and Kobayashi, 2006). For example, in the RBM and PL spectra shown in Figs. 1, 2, 5, and 6, we can clearly see the diameter dependence of the damage; that is, thinner SWCNTs are more severely damaged. Especially, in Fig. 1(c), we can see a large difference in the extent of the damage due to very small diameter difference. The SWNTs observed at about 1151 and 1172 nm can be assigned to (12,1) and (11,3) tubes having diameters of 0.995 and 1.014 nm, respectively. These two peaks were considerably weakened by the photon irradiation at 250º C. On the other hand, the occurrence of the damage was not obvious for thicker SWNTs after the

same irradiation dose at 250º C. The peaks at 1224 nm is assigned to (10,5)
tubes having a diameter of 1.050 nm. The diameter difference between the
considerably damaged (11,3) and hardly damaged (10,5) tubes is only 0.037
nm. It is very interesting that such a small diameter difference results in the
distinctly different sensitivity to the irradiation.

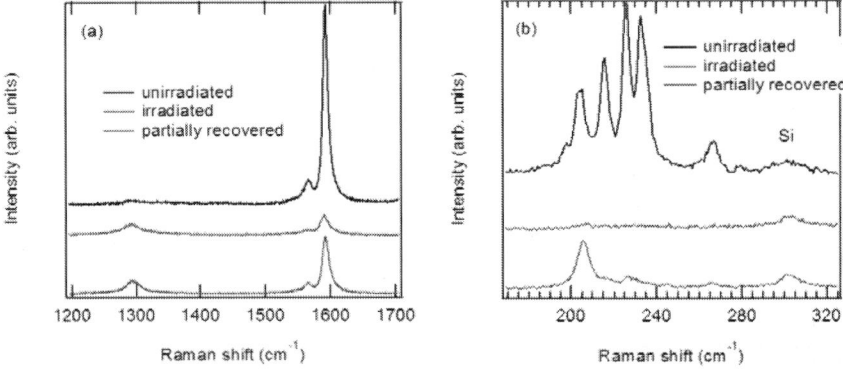

Figure 8. a) G and D band and (b) RBM spectra of pristine, irradiated, and
partially recovered SWCNTs. The SWCNTs were irradiated by 1-keV electrons
up to a dose of 8×10^{17} cm^{-2}. Then, the damaged SWCNTs were partially
recovered by annealing at 400º C in Ar atmosphere for 30 min. The excitation
wavelength was 785 nm. In (b), the hump at ~300 cm^{-1} is from the Si substrate.

The diameter dependence is observed in the recovery process, as shown in Fig.
8. The electron irradiation largely decreased the G/D ratio (a) and once almost
completely extinguished all of the RBM peaks (b). The sample was partially
recovered by annealing at 400º C. Then, only the peak at about 205 cm^{-1},
corresponding to the thickest SWNTs among the initially observed ones,
significantly recovered.

The diameter dependence of damage is more or less also observed in knock-on
damage (Krasheninnikov & Nordlund, 2010) and damage by radicals (Yang et
al., 2006, Zhang et al., 2006b). However, the diameter dependence of low-
energy irradiation damage is more prominent, as mentioned above. The
damage caused by knock-on collision and radicals also occurs in thick MWCNTs
and graphite, but the low-energy irradiation damage has not. Also noteworthy
is that the diameter dependence is not prominent when SWCNTs are irradiated
at low temperature, as shown in the next section.

4.4. Temperature dependence of the damage

Severer damage is observed at lower temperatures (Suzuki & Kobayashi, 2007a). As shown in Fig. 1, the irradiation at 250º C results in much less damage than at 22º C. The temperature dependence is seen at lower temperatures, and less damage is observed at -27º C than at -267º C (6 K), as shown in Fig. 9. These results suggest that low-energy irradiation-induced defects can be healed even at -27º C. In fact, the electric conductivity of irradiated SWCNTs gradually recovers at room temperature, as shown in Fig. 10 (Suzuki & Kobayashi, 2007). The temperature dependence of the damage is completely opposite to that observed in gas phase reactions (An et al., 2002, Zhang et al., 2006a).

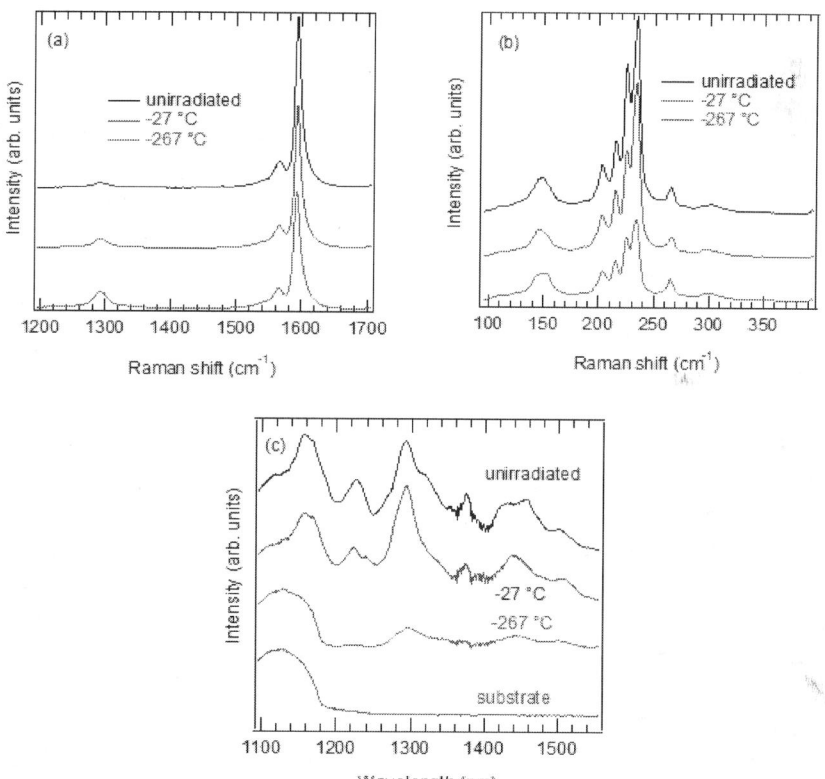

Figure 9. a) G and D band, (b) RBM, and (c) PL spectra of pristine SWCNTs and of SWCNTs irradiated at -27º, and -267º C. The SWCNTs were irradiated by 40-eV photons up to a dose of 7.2×10^{17} cm^{-2}. The excitation wavelength was 785 nm. In (b), the hump at ~300 cm^{-1} is from the Si substrate.

Fig. 9(b) also shows that the diameter dependence of the damage is less prominent at -27º C, meaning that the recovery of all the observed SWCNTs is almost completely forbidden at this temperature, regardless of the diameter. In other words, the less damage in a thicker SWCNT observed at room temperature and above is a consequence of the fact that a defect created in the thicker SWCNT can be more quickly healed by the thermal energy at the irradiation temperature. The diameter dependence can mainly be ascribed to the diameter dependence of the defect healing rather than to that of defect creation.

4.5. Activation energy of defect healing

The recovery of the damage at room temperature or below suggests that the activation energy of the defect healing is quite small. We have proposed a simple method for determining the activation energy of defect healing in SWCNTs (Suzuki et al., 2010). For example, recovery curves of the G/D ratio can be used for the analysis. In Fig. 11(a), we show Raman spectra of SWCNTs before and after irradiation and after 2, 14, and 64 min annealing at 240º C. The annealing gradually recovered the irradiated SWCNTs. From these measurements, we obtained recovery curves of the G/D ratio at several temperatures, as shown in Fig. 11(b). The reason for the relatively small values of the G/D ratio is that we adopted the area ratio rather than the peak height ratio, in order to decrease static errors.

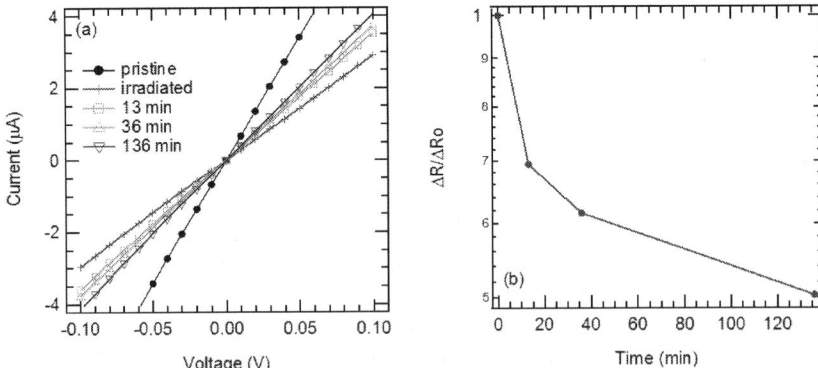

Figure 10. a) Current-voltage characteristics of a SWCNT device before and just after electron irradiation and 13, 36, 136 min after the irradiation. The electron energy and irradiation dose were 20 keV and 1.8×10^{15} cm^{-2}, respectively. **(b)** Time evolution of the irradiation-induced resistance (ΔR) at 22º C. The initial value of the irradiation-induced resistance (ΔRo) was 19.8 kΩ.

The activation energy E_a of defect healing is given by

$$E_a = \frac{k}{\left(T_1^{-1} - T_2^{-1}\right)} \ln \frac{t_1}{t_2}$$

(1)

where t_1 and t_2 are annealing times during which the G/D ratio R increases from R_1 to R_2. Here, it is sufficient that the G/D ratio is just a monotonic function of the defect density (We do not have to assume a specific relation between the G/D ratio and defect density, such as that the G/D ratio is inversely proportional to the defect density). The activation energy seems to depend on the extent of the damage. In the region of $2.8 \leq R \leq 3.0$ (heavily damaged), we obtained an activation energy value of 1.4 ±0.2 eV, as shown in Fig. 11(c), from the recovery curves at 140º and 120º C and eq. (1), whereas at $6 \leq R \leq 6.5$ (lightly damaged), a smaller value of 0.7 ±0.2 eV was obtained. I would like to mention that these values may be affected by gas absorption at defect sites, because the SWCNTs were once exposed to air after the irradiation. Anyway, the values are small enough for the defects to be healed at moderate temperatures.

Interestingly, although a partial recovery of low-energy irradiation damage at room temperature is easily observed in the electric properties (Fig. 10), it has not been observed in Raman spectra. Once we obtain the activation energy value, we can estimate the recovery curve at a given temperature T_3. Using the recovery curve R_1 (R_2) at T_1 (T_2), the recovery curve at T_3 is given by

$$R_3\left(t_3\right) = R_{1,2}\left(t_3 \exp\left[\frac{E_a}{k}\left(T_{1,2}^{-1} - T_3^{-1}\right)\right]\right).$$

(2)

Fig. 11(d) shows the recovery curves at 20º C simulated from the experimental recovery curves at 140º and 120º C and eq. (2). The two independently obtained curves are almost consistent. Note that the unit of the horizontal axes is "year". Recovery of the G/D ratio from 2.8 to 3.0 at 20º C would take about 230 years. Similarly, the recovery from 6.0 to 6.5 at 20º C was estimated to take about 7 years (Suzuki et al., 2010). Thus, the recovery would be much too slow to observe at room temperature in usual experiments. The very long recovery time at room temperature is a consequence of the relatively slow recovery at elevated temperatures in Raman spectra. On the other hand, the recovery of the electric properties is much more rapid. Annealing at 300º C for 30 min often results in recovery of conductivity of several orders of magnitude, as already shown in Fig. 7.

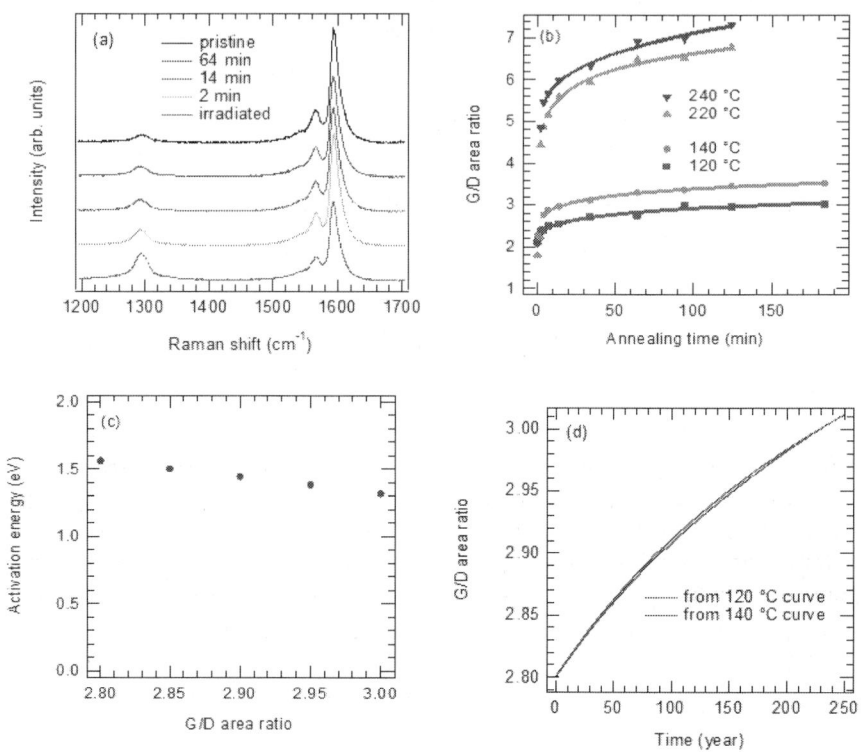

Figure 11. a) Raman spectra of SWCNTs before and after electron irradiation and after annealing at 240º C. The electron energy and dose were 20 keV, and 8×10¹⁶ cm⁻², respectively. The excitation wavelength was 785 nm. (b) Recovery curves of the G/D area ratio otbained at several annealing temperatures. (c) Activation energy of the defect healing obtained from the recovery curves at 140º and 120º C. (d) Simulated recovery curves of the G/D ratio at room temperature (20º C) obtained from the recovery curves at 120º and 140º C in (b), respectively.

Though we evaluate the activation energy from the G/D ratio here, any other quantity that is a monotonic function of the defect density can basically be used for the analyses. This method can also be used to analyze other kinds of defects and desorption barriers of chemisorbed atoms or molecules on SWCNTs.

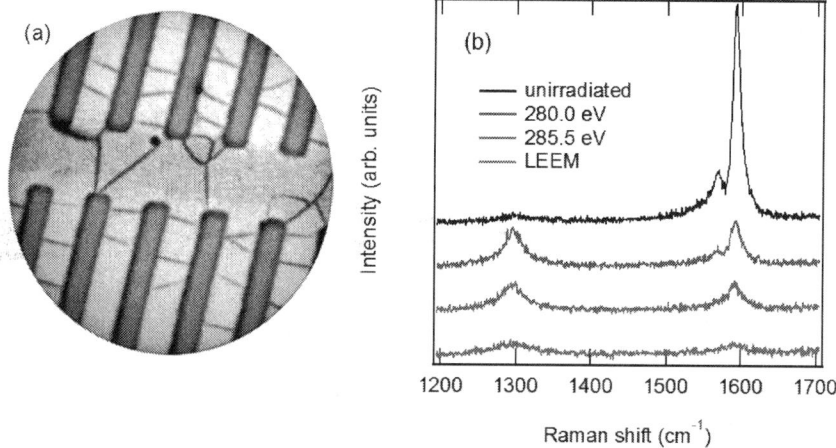

Figure 12. a) LEEM image of suspended SWNTs on a patterned Si substrate. (b) G and D band Raman spectra of unirradiated and soft X-ray-illuminated (280. 0 and 285.5 eV) SWCNTs and electron-irradiated (~20 eV) SWCNTs. The excitation wavelength was 785 nm. The soft X-ray illumination was done at BL-27SU at SPring-8, Hyogo, Japan. The electron irradiation was a consequence of LEEM observation. The soft X-ray and electron irradiations were done after thorough degassing.

4.6. Occurrence of the damage in an ultra-high vacuum

As I mentioned in sec. 2, the low-energy irradiation damage is caused by the irradiation itself. We found that the damage does not depend on the remnant gas pressure at $\sim 10^{-4}$ Pa or below (Suzuki et al., 2008). The low-energy irradiation damage has been observed in a surface-science-grade ultra-high vacuum (UHV) of $\sim 1\times 10^{-8}$ Pa by VUV light illumination (Suzuki & Kobayashi, 2006a. Suzuki & Kobayashi, 2007a, Mera et al., 2009. Mera et al., 2010.), electron beam irradiation (Arima et al., 2009.), and electron (hole) injection from a STM tip (Berthe et al., 2007. Yamada et al., 2009). The damage occurs as ever when SWCNTs are thoroughly degassed in a UHV before irradiation. Other examples of occurrence of the damages in UHV surface analysis systems are shown in Fig. 12(b). The SWCNTs were irradiated by electrons during LEEM observation [Fig. 12(a)] or by soft X-rays at a photoemission spectroscopy beamline attached to a synchrotron radiation ring. The damage to SWCNTs is especially severe in LEEM observation using very low-energy electrons of several tens electron volts, due to the energy dependence of the damage (sec. 4.1). In an UHV, no irradiation-induced change is observed even in high-energy-resolution photoemission spectroscopy (Suzuki et al., 2004a), indicating that chemical reactions with gas molecules are negligible. Nevertheless, very severe damage is observed in Raman spectra.

4.7. Structure dependence

A low-energy electron and photon can easily dissociate a small molecule (for example, photodissociation). On the other hand, such low-energy irradiation damage (or structural change) is not commonly observed inside the bulk of a metal or semiconductor. Actually, it has not been reported for graphite. Very interestingly, even among CNTs, the damage has been reported for SWCNTs but not for MWCNTs. An electron irradiation experiment in an SEM has shown that the irradiation causes no reduction of the conductivity of MWCNTs with a diameter of ~10 nm (Bachtold et al., 1998. Hobara et al., 2004). The irradiation conditions used in those studies (4 C cm^{-2} of 20-keV and 20 C cm^{-2} of 10-keV electrons) roughly correspond to 10 to 1000 fold of a value that can cause a SWCNT conductivity decrease of a few orders of magnitude (Suzuki, 2011). Thus, the damage seems to be specific to SWCNTs or thin CNTs with a diameter of ~1 nm. Even among SWCNTs, the extent of the damage strongly depends on the diameter: Thinner SWCNTs are more severely damaged, as discussed in section 4.3.

The diameter dependence of the damage may explain the occurrence of the damage in SWCNTs and its absence in MWCNTs and graphite. Considering that the damage strongly depends on the diameter among SWCNTs, it would be possible that a MWCNT of 10-nm diameter is no longer damaged by low-energy irradiation at room temperature. If the occurrence and the absence of the damage originate in the diameter difference, we can expect that strain in the sidewall plays an essential role in the defect formation or its stabilization. Alternatively, it is interesting to view the occurrence and absence of the damage in terms of dimensionality. Graphite, in which the damage does not occur, is a three-dimensional material, and a SWCNT, in which the damage occurs, is a one-dimensional material. Notably, it has been well established that structural changes occur in zero-dimensional fullerenes by photon and electron irradiation (Zhao et al., 1994. Onoe et al., 2003). This is generally described as "polymerization" instead of damage, because the irradiation causes chemical bonds to form between neighboring fullerenes. The structural change can be reversibly restored by annealing, exactly like the low-energy irradiation damage of a SWCNT. The electronic states, which spread in the whole crystal in a bulk material, should be localized in low-dimensional materials or nanomaterials, and the degrees of freedom of atomic movement should become larger. Thus, in low-dimensional materials or nanomaterials, local structural change would easily occur with low-energy irradiation and the defect structure would be stabilized (See also sec. 6).

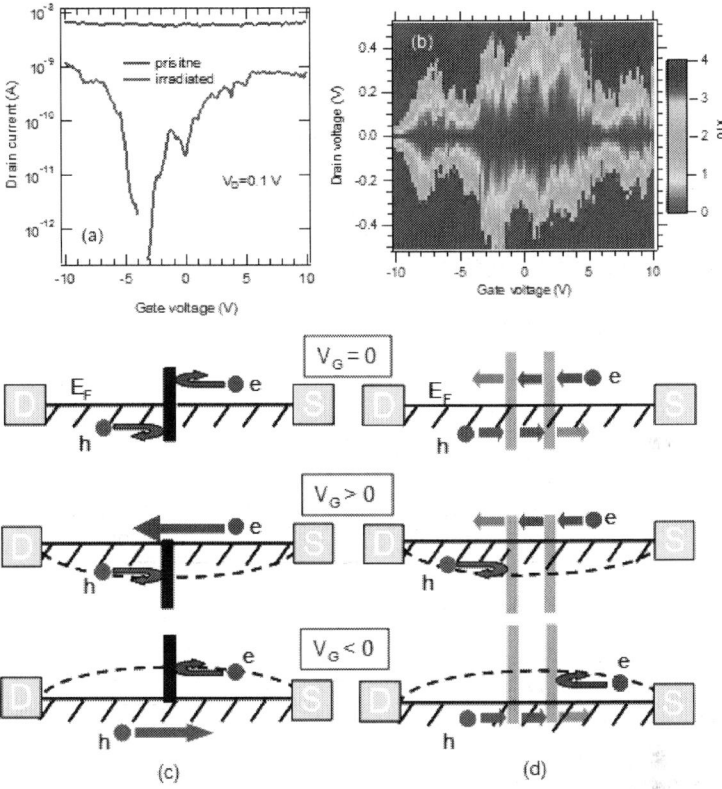

Figure 13. a) Room-temperature gate voltage characteristics of a SWCNT device before and after electron irardiation. The SWCNT was once scanned by an electron beam of 100 pA. The electron energy and scan speed were 20 keV and 400 nm/s, respectively. (b) Room-temperature Coulomb diamond characteristics of a SWCNT device before and after electron irradiation. The SWCNT was once scanned by an electron beam of 500 pA. The electron energy and scan speed were 20 keV and 400 nm/s, respectively. (c) Schematic explanations of the defect-induced semiconductng properites. (d) Schematic explanations of the defect-induced Coulomb oscillation properites.

In terms of the relation between the damage and structure, it is very interesting to explore whether the damage occurs in graphene, which is a two-dimensional material and can be considered to be a SWCNT of infinite diameter. Zhou et al. reported that soft x-ray illumination damages graphene, on the basis of their C 1s x-ray absorption and Raman spectroscopy results (Zhou et al., 2009). Very interestingly, the illumination effects increased with a decreasing number of layers of exfoliated graphene and were negligible even for monolayer epitaxial graphene on SiC, which has a relatively strong interaction with the substrate. These results suggest that low dimensionality is strongly related to the low-energy irradiation damage.

5. METAL-SEMICONDUCTOR TRANSITION OF A SWCNT-FET INDUCED BY DEFECTS

As mentioned in sec. 3.3, intensive irradiation finally makes a SWCNT almost insulating. However, when the damage is moderate, a metal-semiconductor transition of the electric properties is often observed. In our early study, we irradiated the whole device in a SEM and observed the conversion of the electric properties at 28 K (Vijayaraghavan et al., 2005). Further irradiation caused an increase of the nominal band gap observed in the low-temperature electric properties. More recently, we succeeded in converting the room-temperature device characteristics from metallic to semiconducting by local irradiation using an electron beam lithography system (Suzuki et al., 2008). Before the irradiation, the device characteristics were almost gate-independent, which is a common feature of a metallic SWCNT. A part of a metallic SWCNT was once scanned by an electron beam. Then, the room-temperature gate characteristics of the device were converted to semiconducting, as shown in Fig. 12(a). After the irradiation, ambipolar semiconducting gate characteristics were clearly observed.

Room-temperature Coulomb oscillations have been observed when defects form a small dot in a SWCNT channel (Matsumoto et al., 2003). The low-energy irradiation damage can also be used to fabricate such small dots intentionally. As shown in Fig. 12(b), after irradiation, multi-dot Coulomb oscillation properties are sometimes observed at room temperature.

A schematic model of a possible mechanism for the irradiation-induced semiconducting properties is shown in Fig. 12(c) (Kanzaki et al., 2007. Suzuki et al., 2008). The temperature dependence of device characteristics after irradiation shows that an energy barrier for carriers is formed in the SWCNT channel. The barrier height observed in the electric properties reaches about 0.6 eV, when irradiation-induced semiconducting properties are observed at room temperature. Recently, a STM study more directly showed that a local band gap is actually formed in a metallic SWCNT by a carrier injection-induced defect (Yamada et al., 2009). This suggests that the defect-induced local band gap opening is the origin of the barrier. The carrier transport is inhibited by the barrier at the gate voltage of around 0 V. The device still turns on at large gate voltage. This can be reasonably explained in terms of gate-induced band bending in a metallic SWCNT. The density of states near the Fermi level of a metallic SWCNT is very small. Owing to the small density of states near the Fermi level, we can bend the band by applying gate voltage and reduce the effective barrier height for an electron. At sufficiently large gate voltage, the device will turn on. Thus, the metal-semiconductor transition is explained by the defect-induced barrier formation and gate-induced band bending. When Coulomb oscillation is observed, the defects seem to act as tunneling barriers, as schematically shown in Fig. 12(d). In this case, tiny multi-dots divided by the defects seem to be formed in the vicinity of the irradiated part.

The defect-induced conversion of the electric properties seems to be caused by defects formed by other methods. In fact, conversion of the electric properties from metallic to semiconducting also occurs when defects are induced by plasma treatment of metallic SWCNT-FETs. More interestingly, the defect-induced semiconducting electric properties well explain the fact that the ratio of "semiconducting" SWCNTs that act as FETs has been reported to strongly depend on the growth method (Suzuki et al., 2008, Mizutani et al., 2009). The plasma-enhanced CVD method has been reported to produce preferentially semiconducting SWCNTs, and the ratio of semiconducting SWCNTs has been reported to reach about 90 % (Li et al., 2004. Ohnaka et al., 2006) or even 97 % (Mizutani et al., 2009). On the other hand, for the laser ablation method, which generally produces high-quality SWCNTs, the semiconducting SWCNT ratio was evaluated to be quite small, about 30 % (Li et al., 2004). I think that the growth method dependence of the "semiconducting" SWCNT ratio is mainly due to the growth method dependence of defect density. Distinguishing whether the electronic structure is metallic or semiconducting by electric measurements may be inconclusive, especially when the SWCNT shows "semiconducting" properties.

6. MECHANISM OF THE LOW-ENERGY IRRADIATION DAMAGE

The low-energy irradiation damage can be caused by 10-eV photons, which have very small momentum. This indicates that the momentum of an incident particle would have no essential role in the damage, which is in remarkable contrast to knock-on damage. Thus, the defect formation would be due to bond breaking, which follows an electronic excitation by the energy of the incident particle. An energy of ~10 eV is still high enough to cut C-C bonds.

Thus, it is reasonable that low-energy irradiation creates a defect with finite probability if the defect structure is stable and the lifetime is long enough. A simple example of this kind of structural change is photodissociation of a molecule. A bond breaking following electronic excitation can easily dissociate a small molecule In a bulk crystal, on the other hand, even the breaking of several bonds would result in immediate re-bonding without any structural change because an atom has very little freedom of displacement due to the existence of surrounding atoms. The situation in a SWCNT is one between a molecule and bulk. More than one bond breaking would be necessary to stabilize the defect. Among related carbon materials, low-energy electron and photon irradiation-induced structural change (polymerization) is known to occur for fullerenes. On the other hand, the damage has not been reported for graphite or MWCNTs, as discussed in sec. 4.7.

Interestingly, Yamada et al. (Yamada et al., 2009) have proposed, on the basis of thier STM results, that carrier injection first creates primary defects whose lifetime is very short ($<$50 ms). Most of them are quickly annihilated and the structure is restored. However, in rare cases, a primary defect fails to recover and a stable defect is created. The quantum efficiency of the primary defect formation was evaluated to be 2×10^{-10} at a bias voltage of 3.5 V near the defect creation threshold.

The detailed atomic structure of a low-energy irradiation-induced defect is not clear at present. A detectable change has not been observed even with microscopy techniques, such as SEM. This is one of the main reasons that the low-energy irradiation had not been recognized for such a long time, although SEM had been commonly used for characterizing SWCNTs since their discovery. Our previous TEM observation showed that the tube wall is not clearly destroyed regardless of severe damage (Suzuki et al., 2005b). The Stone-Wales defect, which is formed by a C-C bond rotation, is consistent with the conservation of the number of carbon atoms. However, the stable structure seems to contradict the relatively small activation energy and healing at a moderate temperature or even at room temperature or below. Another possible defect is a vacancy in the tube wall with a migratory C adatom on the surface. The observed activation energies (0.7-1.4 eV, sec. 4.5) are very close to the C adatom migration energies, which are theoretically predicted to depend on the SWCNT diameter and to be 0.6 to 1.3 eV (Krasheninnikov et al., 2004). However, a high-resolution TEM observation has shown that annihilation of the vacancy and migratory adatom is governed by the recombination barrier rather than by the adatom migration barrier itself (Hashimoto et al., 2004). The existence of such a vacancy-adatom defect (with the adatom bounded in the vicinity of the vacancy) was also strongly suggested by a scanning tunneling microscopy study (Lee et al., 2005). The vacancy-adatom defect is simply formed by breaking two bonds of a C atom. The recombination barrier of the vacancy-adatom defect in SWCNTs has been calculated to be ~1-2 eV (Okada, 2007), which is rather close to the observed activation energies. Determining the presice defect structure is a future issue.

7. OTHER TYPES OF DAMAGE AND IRRADIATION-INDUCED PHENOMENA

The low-energy irradiation damage is often confused with other types of damage and irradiation-induced phenomena. Here, I would like to summarize differences in the defect characteristics of low-energy irradiation damage and other damages.

7.1. Knock-on damage

Knock-on damage is caused by ballistic ejection of an atom from a solid by an incident particle. Thus, the damage is accompanied by a loss of SWCNT mass. The displacement energy (kinetic energy at which an atom can escape from the solid.) in a SWCNT is considered to depend on the diameter and to be 15-20 eV. However, when the incident particle is an electron whose mass is much smaller than a carbon atom, the threshold energy becomes ~80 keV from the energy and momentum conservation laws. The threshold energy for a photon will be much larger. Thus, the energy at which knock-on damage is observed is much larger than that at which the low-energy irradiation damage is normally observed. Knock-on damage occurs in MWCNTs and graphite. For a recent review focusing on carbon and other nanomaterials, see (Krasheninnikov, 2010).

7.2. Contaminant effects

Low-energy irradiation often causes hydrocarbon contaminants to adhere to the sample surface. The contaminant adhesion is prominent in a conventional SEM, in which a UHV is usually unavailable. At the electron energy where severe damage is observed (1 keV or smaller), severe contaminant adhesion is also observed. This is because the interaction both between electrons and a SWCNT and between electron and hydrocarbon gases are strong at such a low energy. However, effects caused by the damage and by the contaminant adhesion can be easily distinguished by annealing. The contaminants do not sublimate even at ~900º C, whereas the low-energy irradiation damage can be recovered by annealing at moderate temperatures, as discussed in sec. 4. In our experiments, it is rather difficult to detect contaminant effects because the spectra or electric properties of pristine SWCNTs and SWCNTs recovered from the damage are almost identical (see Figs. 6 and 7), although I do not deny that the contaminants cause some so-called environmental effects (Ohno, 2010).

7.3. Damage by radicals

Irradiation in remnant gases can also cause radical-induced etching, which is the opposite of the contaminant adhesion. For example, the cutting of MWCNTs has been clearly demonstrated under gas atmosphere formed by intentional gas bleeding (Yuzvinsky et al., 2005). In this way, the damage by radicals is generally accompanied by etching, which eventually cuts and eliminates CNTs. Thus, this damage can not be fully recovered. In some literatures, reversible chemisorption and desorption on SWCNT or graphene has been suggested. However, spectra before chemisorption and after desorption by annealing are often compared after arbitrary normalization. This damage also occurs in thick MWCNTs and graphite, although some diameter dependence of the damage is observed (Yang et al., 2006, Zhang et al., 2006b). Metallic SWCNTs are preferentially damaged by radicals (Yang et al., 2006. Zhang et al., 2006b), although such preference has not been observed for low-energy irradiation damage. Moreover, severer damage is observed at higher temperatures due to more activated chemical reactions (An et al., 2002. Zhang, 2006a). This is entirely opposite to the low-energy irradiation damage (sec. 4.4). Of course, the radical effect becomes severer at higher pressures and negligible in a UHV. Reader should recall that low-energy irradiation damage occurs in a UHV after thorough degassing, although no indication of chemical reaction is observed (sec. 4.6).

In a standard SEM in which the adhesion of contaminants occurs, the radical effects seem to be less important. Otherwise, the contaminants would not adhere due to etching. According to my experience, intensive irradiation in a conventional SEM can not cut or eliminate even a SWCNT (Fig. 7). I suppose that the radical effects are largely suppressed due to the adhesion of contaminants, which would protect the sample surface (sec. 3.2).

7.4. Substrate charging effects

There has been an attempt to explain the electric property changes by irradiation-induced charging of the substrate (back-gate dielectric) just under the SWCNT (Marquardt et al., 2008. Vijayaraghavan et al., 2010). They observed a large conductivity decrease of on-substrate SWCNT devices by electron irradiation in a SEM. Furthermore, they reversibly and repeatedly recovered the electric properties by applying a high bias voltage (~10 V) to the SWCNT. The phenomena they observed seem to be essentially the same as ours. However, they ascribed the conductivity decrease to a local band gap opening caused by irradiation-induced charging of the dielectric SiO_2 layer just under the SWCNT. Actually, theoretical calculations predict that a uniform (Li et al., 2003) or inhomogeneous (Rotkin & Hess, 2004) electric field can open a gap in a metallic SWCNT of specific chiralities. Marquardt et al. (Marquardt et al., 2008) and Vijayaraghavan et al. (Vijayaraghavan et al., 2010) think that electron irradiation in a SEM causes such local and inhomogeneous charging of the dielectric. In their model, the recovery is explained by a release of trapped charges in the vicinity of the SWCNT caused by the high-bias voltage applied to the drain electrodes. In a conventional on-substrate device, a large electric field may be produced by irradiation-induced charging, considering that the field strength is inversely proportional to the square of the distance.

However, the substrate charging model does not at all explain the fact that the irradiation-induced conductivity decrease is as ever observed for suspended SWCNTs, as shown in Fig. 4. The theoretical calculations have predicted that an extraoridinarily high electric field of ~1 Vnm^{-1} barely opens a band gap of several ten milli-electron volts. It is very unlikely that such a high electric field is formed at SWCNTs suspended 300 nm above the substrate. In fact, a simulation has been performed under a condition where the gate voltage was applied from a metal tip located only 0.5 nm from a SWCNT (Rotkin & Hess, 2004). Similarly, this model cannot explain the degradation of Raman and PL spectra of suspended SWCNTs (Figs. 1 and 9). Moreover, electron (hole) injection-induced band gap opening has been observed for a metallic SWCNT lying on a metal substrate, which does not have charge trap sites (Yamada et al., 2009). This model does not explain the observed band gap value, either. The calculations show that the maximum value of the field-induced band gap is at most ~0.1 eV, which is not sufficient to explain the almost insulating properties observed at room temperature. In fact, an energy barrier of ~0.6 eV was observed for a SWCNT whose room-temperature electric properties were converted from metallic to semiconducting by irradiation (sec. 5). Finally, it should be noted that the irradiation-induced conductivity decrease has been observed in all measured SWCNTs, whereas, in the theoretical calculations, band gaps open only in SWCNTs having certain chiralities. Considering that the irradiation-induced physical property changes can recover at a moderate temperature (~300º C [Fig. 7]), the high bias-induced recovery observed in refs. 7 and 8 seems to be due to annealing by Joule heating. I do not deny an electric field-induced band gap opening in a metallic SWNT. However, I do not think that such a high or inhomogeneous electric field is produced by simple SEM observation or line scans.

7.5. Irradiation-induced heating effects

Low-energy electron and photon irradiation may increase the temperature of the irradiated SWCNTs. However, the heating effect itself does not explain the low-energy irradiation damage at all because less damage is observed at higher temperatures, as shown in Fig. 1. Originally, SWCNTs are thermally very stable materials. Thus, at least under usual conditions, irradiation-induced heating itself would not damage the SWCNTs in a vacuum, if ever. In practice, damage is often observed during Raman measurements in air when the excitation laser power is too large. However, this is not low-energy irradiation damage, but instead would be combustion, because this damage is not observed in a vacuum or an inert gas atmosphere.

CONCLUSION

I have shown that low-energy electron and photon irradiation solely damages SWCNTs. The low-energy irradiation damage extinguishes the characteristic optical and electric properties and reduces chemical tolerance. Thus, we have to pay attention to the damage when we use analytical tools that use low-energy electrons (SEM, LEEM etc.) and VUV light or soft X-rays (photoemission spectroscopy using bright light). The defects have some unique properties. The damage and recovery are reversible, indicating that the number of carbon atoms is preserved. The damage strongly depends on diameter. That is, thinner SWCNTs are more severely damaged. The damage has been observed in SWCNTs but not in MWCNTs, suggesting that it is characteristic of low-dimensional structures or nanostructures. The activation energy of the defect healing depends on the extent of the damage and was evaluated to be about 0.7 to 1.4 eV. Because of the relatively small activation energy, the defects can be healed even at room temperature or below, and less damage occurs at higher temperatures. I also showed that the irradiation-induced defects can convert the room temperature electric properties of a metallic SWCNT to semiconducting. The conversion can be explained by the local band gap opening caused by the defect and gate-voltage-induced band bending in the metallic SWCNT. Energetically, the low-energy is still sufficiently larger than the C-C bond energy and can therefore break the bonds. Future studies should address the detailed defect structure.

ACKNOWLEDGEMENTS

This work has been done through cooperation of many coworkers. I thank all of my coworkers for their cooperation and assistance in this work.

REFERENCES

1. K. H. An, J. G. Heo, K. G. Jeon, D. J. Bae, C. Jo, C. W. Yang, C. Y. Park, Y. H. Lee, Y. S. Lee, Y. S. Chung, 2002 X-ray photoemission spectroscopy study of fluorinated single-walled carbon nanotubes. Appl. Phys. Lett. 80 22 April 2002), 42354237 .

2. S. Arima, S. Lee, Y. Mera, S. Ogura, K. Fukutani, S. Sato, J. Tohji, K. Maeda, 2009 Electron-stimulated defect formation in single-walled carbon nanotubes studied by hydrogen thermal desorption spectroscopy. Appl. Surf. Sci. 256 4 June 2009), 11961199 .

3. A. Bachtold, M. Henny, C. Terrier, C. Strunk, C. Schonenberger, J. P. Salvetat, J. M. Bonard, L. Forro, 1998 Contacting carbon nanotubes selectively with low-ohmic contacts for four-probe electric measurements. Appl. Phys. Lett. 73 2 July 1998), 274276 .

4. M. Berthe, S. Yoshida, Y. Ebine, K. Kanazawa, A. Okada, A. Taninaka, O. Takeuchi, N. Fukui, H. Shinohara, S. Suzuki, K. Sumitomo, Y. Kobayashi, B. Grandidier, D. Stievenard, H. Shigekawa, 2007 Reversible defect engineering of single-walled carbon nanotubes using scanning tunneling microscopy. Nano Lett. 7 12 November 2007), 36233627 .

5. B. H. Chen, J. H. Wei, P. Y. Lo, Z. W. Pei, T. S. Chao, H. C. Lin, T. Y. Huang, 2006 Novel method of converting metallic-type carbon nanotubes to semiconducting-type carbon nanotube field-effect transistors. Jpn. J. Appl. Phys. 45 4B April 2006), 36803685 .

6. M. S. Dresselhaus, G. Dresselhaus, R. Saito, A. Jorio, 2005 Raman spectroscopy of carbon nanotubes. Phys. Rep., 409 (2005), 4799 .

7. A.Hashimoto, K. Suenaga, A. Gloter, K. Urlta, S. Iijima, 2004 Direct evidence for atomic defects in graphene layers. Nature 430 7002 August 2004), 870873 .

8. R. Hobara, S. Yoshimoto, T. Ikuno, M. Katayama, N. Yamauchi, W. Wongwiriyapan, S. Honda, I. Matsuda, S. Hasegawa, K. Oura, 2004 Electric transport in multiwalled carbon nanotubes contacted with patterned electrodes. Jpn. J. Appl. Phys. 43 8B July 2004), L1081L1084
    .

9. K. Kanzaki, S. Suzuki, H. Inokawa, Y. Ono, A. Vijayaraghavan, Y. Kobayashi, 2007 Mechanism of metal-semiconductor transition in electric properties of single-walled carbon nanotubes induced by low-

energy electron irradiation. J. Appl. Phys. 101 3 Feburuary 2007), 034317034311 -4.

10. A. V. Krasheninnikov, K. Nordlund, P. O. Lehtinen, A. Foster, S. Ayuela, R. M. Nieminen, 2004 Adsorption and migration of carbon adatoms on zigzag carbon nanotubes. Carbon 42 5-6 , (January 2004), 10211025 .

11. V. Krasheninnikov, K. Nordlund, 2010 Ion and electron irradiation-induced effects in nanostructured materials. J. Appl. Phys. 107 7 April 2010), 071301071301 -70.

12. Y. Li, V. R. Rotkin, U. Ravaioli, 2003 Electronic response and bandstructure modulation of carbon nanotubes in a transverse electric field. Nano Lett. 3 2 November 2003), 183187 .

13. W. Marquardt, S. Dehm, A. Vijayaraghavan, S. Blatt, F. Hennrich, R. Krupke, 2008 Reversible metal-insulator transitions in metallic single-walled carbon nanotubes. Nano Lett. 8 9 August 2008), 27672772 .

14. S. Lee, G. Kim, H. Kim, B. Choi, J. Lee, B. W. Jeong, J. Ihm, Y. Kuk, S. J. Kahng, 2005 Paired gap states in a semiconducting carbon nanotube: Deep and shallow levels. Phys. Rev. Lett. 95 16 October 2005), 166402166401 -4.

15. K. Matsumoto, S. Kinoshita, Y. Gotoh, K. Kurachi, T. Kamimura, M. Maeda, K. Sakamoto, M. Kuwahara, N. Atoda, Y. Awano, 2003 Jpn. J. Appl. Phys. 42 4B Janurary 2003), 24152418 .

16. Y. Mera, Y. Harada, S. Arima, K. Hata, S. Shin, K. Maeda, 2009 Defects generation in single-walled carbon nanotubes induced by soft x-ray illumination. Chem. Phys.Lett. 473 1-3 (March 2009), 138141 .

17. Y. Mera, T. Fujikawa, K. Ishizaki, R. Xiang, J. Shimomi, S. Maruyama, T. Kakiuchi, K. Mase, K. Maeda, 2010 Ion desorption from single-walled carbon nanotubes induced by soft x-ray illumination. Jpn. J. Appl. Phys. 49 10 October 2010), 105104105101 -5.

18. T. Mizutani, H. Ohnaka, Y. Okigawa, S. Kishimoto, Y. Ohno, 2009 A study of preferential growth of carbon nanotubes with semiconducting behavior grown by plasma-enhanced chemical vapor deposition. J. Appl. Phys. 106 7 October 2009), 073705073701 -5.

19. H. Ohnaka, Y. Kojima, S. Kishimoto, Y. Ohno, T. Mizutani, 2006 Fabrication of carbon nanotube field effect transistors using plasma-enhanced chemical vapor deposition grown nanotubes. Jpn. J. Appl. Phys. 45 6B June 2006), 54855489 .

20. Y. Ohno, 2010 Environmentaleffects on photoluminescence of single-walled carbon nanotubes. In: Carbon Nanotubes, InTech, 978-9-53307-054-4

21. S. Okada, 2007 Energetics and electronic structures of carbon nanotubes with adatom-vacancy defects. Chem. Phys. Lett. 447 4-6 (September 2007), 263267 .

22. J. Onoe, T. Nakayama, M. Aono, T. Hara, 2003 Structural and electrical properties of an electron-beam-irradiated C60 film. Appl. Phys. Lett. 82 4 January 2003), 595597 ..

23. V. R. Rotkin, K. Hess, 2004 Possibility of a metallic field effect transistor. Appl. Phys. Lett. 84 16 February 2004), 31393141 .

24. S. Suzuki, Y. Watanabe, T. Ogino, Y. Homma, D. Takagi, S. Heun, L. Gregoratti, A. Barinov, M. Kiskinova, 2004a Observation of single-walled carbon nanotubes by photoemission microscopy. Carbon, 42 3 (January 2004), 559L63 .

25. S. Suzuki, K. Kanzaki, Y. Homma, S. Fukuba, 2004b Low-acceleration-voltage electron irradiation damage in single-walled carbon nanotubes. Jpn. J. Appl. Phys., 43 8B (July 2004), L1118L1120 .

26. S. Suzuki, D. Takagi, Y. Homma, Y. Kobayashi, 2005a Selective removal of carbon nanotubes utilizing low-acceleration-voltage electron irradiation damage. Jpn. J. Appl. Phys. 44 4 January 2005), L133L135 .

27. S. Suzuki, S. Fukuba, K. Kanzaki, Y. Homma, Y. Kobayashi, 2005b Spatially selective removal of carbon nanotubes for fabricating nanotube circuits. Proceedings of 5th IEEE Conference on Nanotechnology. Nagoya, July 2005.

28. S. Suzuki, Y. Kobayashi, 2005 Conductivity decrease in carbon nanotubes casued by low-acceleration-voltage electron irradiation. Jpn. J. Appl. Phys. 44 49 November 2005), L1498L1501 .

29. S. Suzuki, F. Maeda, Y. Kobayashi, 2006 Photon-induced damage creation in carbon nanotubes. 30th Fullerenes Nanotube General Symposium. Nagoya, January 2006.

30. S. Suzuki, Y. Kobayashi, 2006a Diameter dependence of low-energy electron and photon irradiation damage in single-walled carbon nanotubes. Chem. Phys. Lett. 430 1-3 (September 2006), 370374 .

31. S. Suzuki, Y. Kobayashi, 2006b Processing and electric property control of carbon nanotubes by low-energy electron irradiation. NTT Technical. Review 4 11 November 2006), 2530 .

32. S. Suzuki, Y. Kobayashi, 2007a Healing of low-energy irradiation-induced defects in single-walled carbon nanotubes at room temperature. J. Phys. Chem. C 111 12 March 2007), 45244528 .

33. S. Suzuki, Y. Kobayashi, 2007b Low-energy irradiation damage in single-walled carbon nanotubes. Mater. Res. Soc. Symp. Proc. 994 San Francisco, (April 2007), F04F02 -1-12.

34. S. Suzuki, Y. Kobayashi, 2008 Threshold energy of low-energy irradiation damage in single-walled carbon nanotubes. Jpn. J. Appl.Phys. 47 4 April 2008), 20402043 .

35. S. Suzuki, J. Hashimoto, T. Ogino, Y. Kobayashi, 2008 Electric property control of carbon nanotubes by defects. Jpn. J. Appl.Phys. 47 4 April 2008), 32923295 .

36. S. Suzuki, K. Yamaya, Y. Homma, Y. Kobayashi, 2010 Activation energy of healing of low-energy irradiation-induced defects in single-wall carbon nanotubes. Carbon 48 11 May 2010), 32113217 .

37. S. Suzuki, 2011 Origin of the electric property change of a single-wall carbon nanotube caused by low-energy irradiation: defects or substrate charging? e-J. Surf. Sci. & Nanotechnol. 9 (March 2011), 103106 .

38. S. Suzuki, H. Hibino, 2011 Characterization of doped single-wall carbon nanotubes by Raman spectroscopy. Carbon 49 7 January 2011), 22642272 .

39. A.Vijayaraghavan, C. W. Marquardt, S. Dehm, F. Hennrich, R. Krupke, 2010 Imaging defects and junctions in single-walled carbon nanotubes by voltage-contrast scanning electron microscopy. Carbon 48 2 September 2009), 494500 .

40. K. Yamada, H. Sato, T. Komaguchi, Y. Mera, K. Maeda, 2009 Local opening of a large bandgap in metallic single-walled carbon nanotubes induced by tunnel injection of low-energy electrons. Appl. Phys. Lett. 94 25 June 2009), 253103253101 -3.

41. Y. H. Yan, M. B. Chan-Park, Q. Zhou, C. M. Li, C. Y. Yue, 2005 Functionalization of carbon nanotubes by argon plasma-assisted ultraviolet grafting. Appl. Phys. Lett. 87 21 November 2005), 213101213101 -3.

42. M. Yang, K. H. An, J. S. Park, K. A. Park, S. C. Lim, S. H. Cho, Y. S. Lee, W. Park, C. Y. Park, Y. H. Lee, 2006 Preferential etching of metallic single-

walled carbon nanotubes with small diameter by fluorine gas. Phys. Rev. B 73 7 February 2006), 075419075411 -7.

43. T. D. Yuzvinsky, A. M. Fennimore, W. Nickelson, C. Esquivias, A. Zettl, 2005 Precision cutting of nanotubes with a low-energy electron beam. Appl. Phys. Lett. 86 5 January 2005), 053109053101 -3.

44. G. Zhang, P. Qi, X. Wang, Y. Lu, D. Mann, X. Li, H. Dai, 2006a Hydrogenation and hydrocarbonation and etching of single-walled carbon nantoubes. J. Am. Chem. Soc. 128 18 April 2006), 60266027 .

45. G. Zhang, P. Qi, X. Wang, Y. Lu, X. Li, R. Tu, S. Bangsaruntip, D. Mann, Li. Zhang, H. Dai, 2006b Selective etching of metallic carbon nanotubes by gas-phase reaction. Science 314 5801 November 2006), 974977 .

46. Y. B. Zhao, D. M. Poirier, R. J. Pechman, J. H. Weaver, 1994 Electron stimulated polymerization of solid C60. Appl. Phys. Lett. 64 5 January 1994), 577579 .

47. S. Y. Zhou, C. O. Girit, A. Scholl, C. J. Jozwiak, D. A. Siegel, P. Yu, J. T. Robinson, F. Wang, A. Zettl, 2009 Instability of two-dimensional graphene: Breaking sp2 bonds with soft x rays. Phys. Rev. B 80 12 September 2009), 121409121401 -4.

Electrochemical Sensors Based on Carbon Nanotubes

J. Saleh Ahammad[1], Jae-Joon Lee[1,2,*], and Md. Aminur Rahman[2,*],

[1]Department of Advanced Technology Fusion, Konkuk University, Seoul 143-701, Korea

[2]Department of Applied Chemistry, Konkuk University, Chungju 380-701, Korea

ABSTRACT

This review focuses on recent contributions in the development of the electrochemical sensors based on carbon nanotubes (CNTs). CNTs have unique mechanical and electronic properties, combined with chemical stability, and behave electrically as a metal or semiconductor, depending on their structure. For sensing applications, CNTs have many advantages such as small size with larger surface area, excellent electron transfer promoting ability when used as electrodes modifier in electrochemical reactions, and easy protein immobilization with retention of its activity for potential biosensors. CNTs play an important role in the performance of electrochemical biosensors, immunosensors, and DNA biosensors. Various methods have been developed for the design of sensors using CNTs in recent years. Herein we summarize the applications of CNTs in the construction of electrochemical sensors and biosensors along with other nanomaterials and conducting polymers.

Keywords: Carbon Nanotubes; Modified Electrodes; Electrochemical Sensors; Biosensors; Immunosensors; DNA sensors

1. INTRODUCTION

A sensor is a device which detects a variable quantity, usually electronically, and converts the measurement into signals to be recorded elsewhere. The most important aspect of investigation of sensors is sensitivity, selectivity, and stability. Sensors can be classified, according to the type of energy transfer, as thermal, electromagnetic, mechanical, and electrochemical. Among them, the electrochemical sensors are very promising analytical methods because of their high degree of selectivity and sensitivity. They are more useful and easy to determine the concentrations of various analytes in samples such as fluids and dissolved solid materials. They are frequently used in clinical diagnostics, occupational safety, medical engineering, process measuring engineering, and environmental analysis.

Currently, much attention has been focused on developing nanomaterials, which are used for signal amplification in electrochemical sensors. Nanomaterials are usually used to take advantage of a larger surface area for biomolecules to be immobilized. This generally increases the number of binding sites available for the detection of a specific chemical analyte [1]. Various types of nanomaterials are used in electrochemical sensors. Carbon nanotubes (CNTs) [2] are one of the most exciting materials because of their unique electronic, chemical, and mechanical properties [3]. CNTs possessed sp^2 carbon units with several nanometers in diameter and many microns in length. Two groups of CNTs, multi-walled (MW) and single-walled (SW), can be synthesized by electrical arc discharge, laser vaporization, and chemical vapor deposition methods. CNTs behave as either metals or semiconductors, depending on the diameter and the degree of helicity [4]. They are suitable for the modification of various electrodes due to their high electronic conductivity for the electron transfer reactions and better electrochemical and chemical stabilities in both aqueous and non-aqueous solutions [5]. Furthermore, construction of efficient electrochemical sensors using the CNTs-modified electrodes is very promising in that they promote electron-transfer reactions in several small biologically important molecules and large biomolecules [6,7].

This review focuses on the use of CNTs for electrochemical sensors and biosensors. Thousands of paper have been published in this field during last decade and therefore the references that appeared before 2006 were not included in this manuscript. Electrochemical detection based on a voltammetric and chronopotentiometric techniques was mostly discussed, whereas some other detection techniques related with electrophoresis, chromatography, and lab-on-a-chip were not included.

2. ELECTROCHEMISTRY OF CARBON NANOTUBES

CNTs are electrochemically inert materials similar to other carbon-based materials used in electrochemistry, i.e. glassy carbon, graphite, and diamond. They possess distinct electrochemical properties because of their unique electronic structure. The carbon atoms of CNTs at the sidewall and the end of the tubes are not same and their behavior can be compared with the basal plane and edge plane of highly oriented pyrolytic graphite (HOPG), respectively [8]. Compton *et al.* studied the redox reaction of ferricyanide at the C_{60}- and nanotube modified electrodes and compared these results with basal and edge planes pyrolytic graphite electrodes. They observed similar electron transfer rate constants for CNTs-modified and the edge plane HOPG electrodes. They reported that the CNTs acted as an efficient electron transfer promoter [Figure 1] [9–11]. To illustrate the electron transfer kinetics, Holloway *et al.* studied the voltammetry of two standard redox processes of $Fe(CN)_6^{4-}$ and $Ru(NH_3)_6^{3+}$ using three types of MWCNTs having oxygenated edge-plane, edge-plane, and almost no edge-plane like defects [12]. The rate of electron transfer was determined to be faster in the case of edge-plane defect sites. This further indicates that the electroactive sites on MWCNTs are located at the tube ends. The electrochemical properties of ferricyanide redox couple at aligned and randomly dispersed SWNTs-modified electrodes was studied by Gooding *et al.* [13,14]. For an acid-treated aligned SWNTs-modified electrode, a peak separation of 59 mV with a half-wave potential ($E_{1/2}$) of 0.231 V (*vs.* Ag/AgCl) for the ferricyanide redox couple was observed. On the other hand, the peak separation was observed to be 99 mV when the acid-treated but randomly dispersed SWNTs modified electrode was used. The aligned SWNTs modified electrode shows the better electrochemical properties. It means that the electrochemistry of CNTs is dominated by the ends of the tubes.

The length of the aligned CNTs also has a significant effect on the electron transfer rate. The electron transfer rate constant varied inversely with the mean length of the CNTs. Gooding *et al.* investigated the effect of the length of CNTs on the apparent electron transfer rate constant of the surface attached ferrocenemethylamine on to the vertically aligned CNTs with cutting times of 2, 4, and 6 hrs during acid treatment [15]. The apparent rate constants were determined to be 98 ± 25, 187 ± 98, and 459 ± 132 s^{-1} for the mean lengths of 1175, 507, and 257 nm, respectively. When the nanotube dispersed randomly, the rate constant was found to be 12 ± 8 s^{-1} for a mean tube length of 257 nm, which was 40 times slower than that obtained at vertically aligned nanotubes modified electrode. The electron transfer rate at the CNTs modified electrode could be further increased by dialyzing the CNTs after purification and shortening treatment [16]. During purification and shortening of the SWNTs in concentrated nitric and sulfuric acids mixture, some residual acid moieties adsorbed on single-walled carbon nanotubes (SWNTs). Using TEM and HR-TEM techniques, Pumera *et al.* confirmed that CNTs contain residual metal impurities after acid wash with concentrated nitric acid at temperature of 80 °C [17]. These acid moieties can be reduced by dialyzing the purified and dialyzed CNTs against a solution of Triton° X-100. The electrochemical measurements using self-assembled ferrocene-functionalized nanotube monolayers on a gold electrode showed that the dialyzed nanotubes exhibited a faster rate of electron transfer compared to the nondialyzed nanotubes [16].

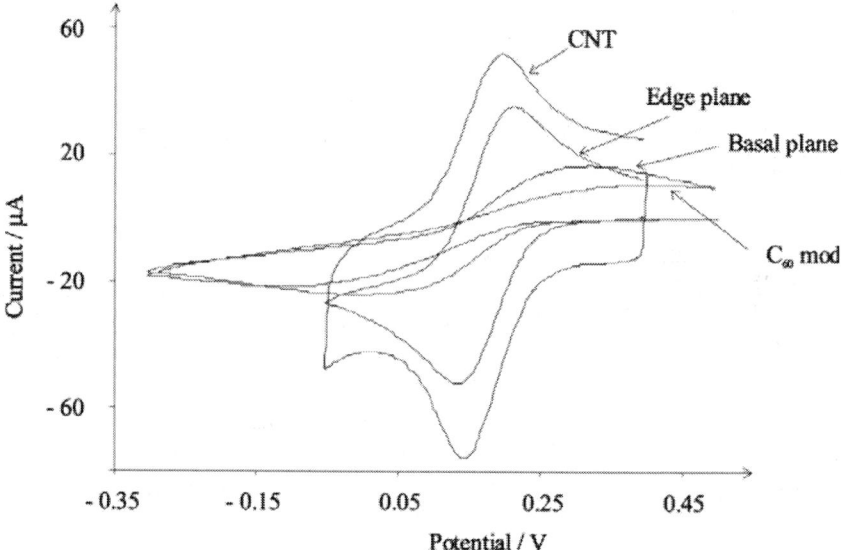

Figure 1. Cyclic voltammograms for the reduction of 1 mM ferricyanide for different electrodes at a scan rate of 100 mV s^{-1} [10]. Reproduced by permission of The Royal Society of Chemistry.

The adsorbed acid moieties during purification and acid-treatment processes can also decrease the electrocatalytic activity of CNTs in electroanalysis. Pumera and his colleague suggested the use of dc magnetic susceptibility and electron paramagnetic resonance for screening and quality control of CNTs before use them in electroanalysis [18]. Recently, Dai and co-workers reported that vertically aligned nitrogen-doped CNTs can act as a metal-free electrode with a much better electrocatalytic activity [19]. The electrocatalytic activity and the electroanalytical performance at CNTs modified electrodes are strongly depended on the mode of production of the CNTs, either by chemical vapor deposition (CVD) or the ARC discharge process [20]. CNTs produced by CVD appear to be more electrochemically reactive in their voltammetric study than those produced using the ARC methodology. The differences in the electrochemical reactivity are attributed to the smaller fraction of exposed edge planes at ARC-CNTs and higher density of edge plane defects at CVD-CNTs. The electrocatalytic activity of ARC-CNTs can be increased after pre-anodization. Wang's group illustrated the effect of the electrochemical pretreatment of ARC- and CVD-prepared multi-walled CNTs using nicotinamide adenine dinucleotide, ascorbic acid, hydrazine, and hydrogen peroxide model redox systems [21]. The fact that the ARC-CNTs display a marked improvement in their electrochemical reactivity, which indicates that the pre-anodization effectively breaks the end caps of ARC-CNTs to expose new edge plane-like sites.

3. CARBON NANOTUBE-BASED ELECTROCHEMICAL SENSORS

There have been many reports on CNTs-based electrochemical sensors. Various electrochemical techniques were used for sensing of biomolecules. Some common techniques are voltammetry, amperometry, potentiometry, impedemetry, and conductometry, which have been described in our previous review paper [22]. Voltammetry, measuring the current as a response to the applied potential, is one of the most useful and widespread technique among them. For enhanced current response, it is very important to develop a stable and highly target specific interface by various surface modification of conventional electrodes. The sensitivity and the selectivity are the crucial issues for the development of sensors for detecting biologically important molecules. Much effort has been made in the development of a highly sensitive and selective method for the detection of dopamine (DA), which is one of the important catecholamine neurotransmitters in the mammalian central nervous system. Conventional electrodes are not suitable for the determination of DA due to the interference from ascorbic acid (AA) and uric acid (UA), which are co-existed in a real sample at 100 times higher concentration than DA. These compounds can be easily oxidized at the similar potential of DA and thus always interfere with DA detection. The CNTs-modified electrodes have been widely used to resolve this problem. Wang *et al.* reported the fabrication of a poly (3-methylthiophene) modified glassy carbon electrode (GCE) coated with Nafion/SWCNTs for highly selective and sensitive determination of DA [23]. The modified electrode enhanced the voltammetric signal of DA and effectively suppressed the interferences of AA and UA as well. A lower detection limit of 5.0 nM was achieved for DA. It was also successfully applied for the determination of DA in healthy human blood serum. A CNT-polymer composite-modified electrode, with poly(styrenesulfonic acid) sodium salt and SWCNTs, were used for selective detection of DA [24]. The negatively charged poly (styrene sulfonic acid) sodium salt attracted positively charged DA in pH 7 PBS and selectively detected it from the interference of AA.

Polymer-MWCNTs composite can be used for the fabrication of DA sensor. Yin *et al.* developed a DA sensor using β-cyclodextrin-incorporated MWCNTs on a polyaniline (PANI) modified GCE [25]. A superior transducing property of PANI, a rapid electron transfer capability of MWCNTs, and the preconcentration by β-cyclodextrin showed the excellent sensitivity, selectivity, stability, and reproducibility in the determination of DA. Recently, Angeles *et al.* also developed an amperometric sensor for DA detection by using β-cyclodextrin and MWCNTs without polymer [26]. The enhancement of the oxidation current of DA was possible due to its diffusion through the β-cyclo-dextrin cavities and the facile contact with the dispersed MWCNTs. Moreover, Li *et al.* found that polypyrrole-SWCNTs composite film can detect DA, AA, and UA simultaneously, and it showed the electrocatalytic activity towards the oxidation of nitrite [27]. Zhou's group reported that the modified electrode with poly (acrylic acid) and MWCNTs can suppress the oxidation peak of AA but enhance the DA and UA signals [28]. The application of poly (neutral red) functionalized MWCNTs was also reported by Yogeswaran and Chen for the simultaneous detection of AA, UA, and DA [29]. The clear separation of peaks was attributed to the electrostatic and hydrophobic interaction between the three analytes and the fixed cationic sites on polymer backbone as well as the functionalized MWCNTs, which are negatively charged.

Another important catecholamine neurotransmitter is epinephrine (EP), which is involved in the message transfer of the mammalian central nervous system. In biological fluids such as blood and urine, EP coexists with AA, and UA, so AA and UA may interfere during the electrochemical detection of EP at an unmodified electrode. CNTs-modified electrodes have been successfully used for the determination of EP. Chen's group developed a method for simultaneous determination of AA, EP, and UA at physiologically relevant conditions by using the composite film composed of functionalized-MWCNTs and Nafion incorporating platinum and gold nanoparticles [30]. The oxidation peaks for AA, EP, and UA were separately observed and thus, the detections of these compounds did not interfere with each other. An EP sensor prepared by an *in-situ* electropolymerization of brilliant cresol blue (BCB) was reported by Yi *et al.* [31]. In this work, the GCE modified by the film of polymeric BCB and functionalized MWCNTs composite were used to detect EP. A low detection limit of 10 nM was obtained by using the BCB and functionalized MWCNTs nanocomposite. However, the authors did not discuss the issue of interference from other biological compounds such as AA. Valentini *et al.* used functionalized SWCNTs instead of MWCNTs for the selective detection of EP in the presence of AA [32]. They used the stainless steel microelectrodes modified by hydroxyl group functionalized SWCNTs, which were deposited electrophoretically. The presence of electron-donating -OH groups on SWCNTs repels AA and attracts the positively charged EP. It provided a relatively high electrochemical sensitivity for EP up to the detection limit of 2.0 nM.

Nicotinamide adenine dinucleotide (NADH) is a coenzyme involved in a wide range of enzymatic reactions. The direct oxidation of NADH at a bare electrode needs a high overpotential. CNT-modified electrode can be used for the stable low potential detection of NADH. Zhai *et al.* developed a multilayer film of MWCNTs and chitosan (CS) using the layer-by-layer (LBL) method by taking advantage of the interaction between a positively charged CS and the negatively charged MWCNTs [33]. They assembled nine-layers of CS/MWCNTs successfully, which showed a very rapid and stable response of NADH oxidation at about 400 mV with the detection limit (S/N = 3) of 0.3 μM. A layer-by-layer approach is an efficient way to increase the amount of catalyst or enzyme at the sensor surface. However, the thickness of the multilayer need to be optimized as the sensor response can be suppressed by the very thick multilayer. Wang *et al.* fabricated an electrode by mixing a room-temperature ionic liquid, 1-butyl-3-methylimidazolium tetrafluoroborate, and MWCNTs along with CS. This electrode was used for NADH sensing with the detection limit of *ca.* 0.06 μM [34]. The use of ionic liquid and MWCNTs at the sensor surface may increase the surface ionic and electrical conductivities, thus, may enhance the sensitivity of the sensor. Radoi *et al.* used the covalently linked variamine blue, as a redox mediator to the oxidized SWCNTs for the detection of NADH [35]. The NADH oxidation potential was found to be lowered, from the changes of formal redox potential of the mediator, and therefore the sensor efficiency was improved due to the electrocatalytic activity of the mediator. Another very common approach for NADH sensors is the use of polymer composites with MWCNTs such as poly (acrylic acid)-wrapped MWCNTs complex [36], MWCNTs-poly (1,2-diaminobenzene) nanoporous composite [37], quinone-amine polymer-MWCNTs nanocomposite [38], and poly-(3-methylthiophene)-MWCNTs hybrid composite [39]. The sensitivities of the polymer-MWCNTs nanocomposite-based NADH detections were found to be enhanced due to the excellent electrocatalytic activities of the nanocomposites.

Hydrogen peroxide (H_2O_2) is a product of several biological, enzyme-catalyzed reactions. The detection of H_2O_2 plays an important role in food industry, environmental protection, and in medical diagnostics. For the sensitive detection of H_2O_2, Tkac and Ruzgas have used an electrode modified with SWCNTs. The sensitivity was highly dependent on the dispersing agent in the organic solvents and charging status of polymers (e.g. Nafion and CS) [40]. They found that the dispersion of both polymers is highly stable but the SWCNTs in the CS dispersion showed higher sensitivity for H_2O_2 compared to that in Nafion. Sun et al. introduced the modification with the ferrocene-filled SWCNTs for a H_2O_2 sensor with a good stability and reproducibility [41]. Ferrocene/ferrocenium (Fc/Fc$^+$) was used as an electron-transfer mediator for the redox reaction of H_2O_2.

Electrochemical detections of metal ions have widely been studied using CNTs-modified electrode. For example, Yuan et al. developed a mercury-free electrode system by casting a dispersed solution of MWCNTs in Nafion on GCE [42]. It was applied for the detection of europium (III) by differential pulse adsorptive stripping voltammetry (DPASV) and a wide linear range from 40 nM to 10 mM with lower detection limit of 10 nM was obtained. Sun et al. used SWCNTs-Nafion film for the determination of Cd^{2+} in water [43]. Profumo et al. have prepared chemically modified MWCNTs electrode to detect the trace amount of As (III) and Bi (III) in a natural and high-salinity waters [44]. An oxidation of MWCNTs to introduce carboxyl groups and the subsequent chemical treatments were required to get a robust modification of the electrode surface for reliable measurement of saline water without desalting the sample (Figure 2).

Phosphate containing molecules such as phytic acid (PA), phosphomolybdic acid (PMA), dihexadecylphosphate (DHP), and dicetyl phosphate (DCP) have been widely used with CNTs for sensing applications. Jeon's group recently developed an electrochemical sensor based on the modification of platinum electrode with SWCNTs and PA for the selective detection of DA in the presence of AA and UA [45]. The PA-SWCNTs films promoted the electron transfer reaction of DA while the PA in the films act as a binder and a negatively charged linker as well. The PA-SWCNTs/Pt electrode can separate the oxidation peak of DA from the interferences of electrochemical responses of AA and UA. Li et al. used PMA with MWCNTs to make an amperometric bromate sensor [46]. Hu and co-workers described a method for the preparation of a MWCNTs-DHP composite film on the GCE surface for the determination of lincomycin in tablets [47]. They found a well-defined oxidation peak for lincomycin by using cyclic voltammetry (CV) and a linear response range from 0.45 μM to 0.15 mM with the detection limit of 0.2 μM. The similar approach was applied by Ming et al. recently for the determination of trace Sudan I contamination in chili powder [48] and the same electrode was also used to determine the acyclovir voltammetrically [49]. DHP was replaced by DCP and 4-aminobenzenesulfonic acid (4-ABSA) for the determination of 2-chlorophenol [50] and tyrosine [51], respectively.

The GCE modified with MWCNTs even without DHP was also used for the determination of procaine [52] and phenylephrine [53].

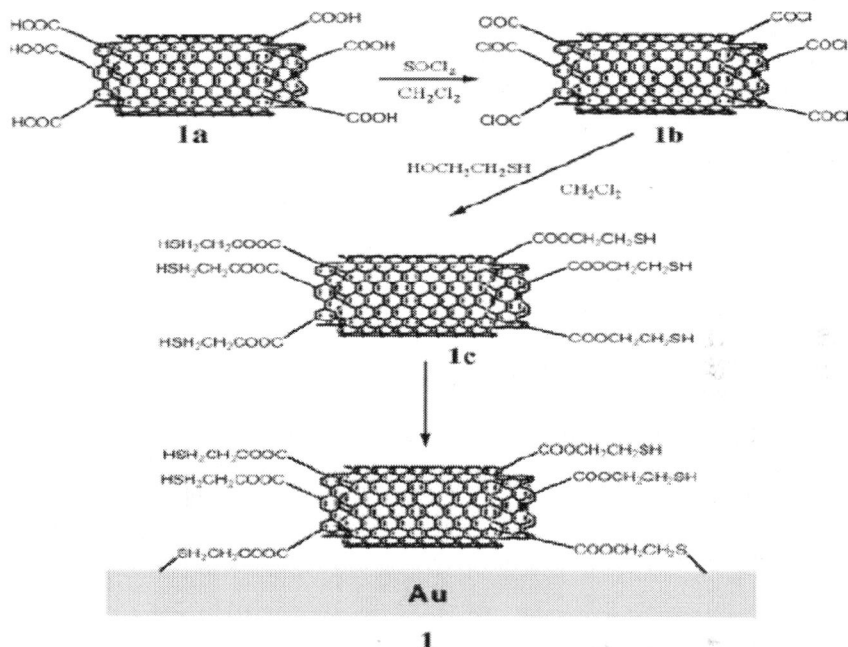

Figure 2. Scheme for the preparation of MWCNTs chemically modified electrode bearing SH groups. Reprinted with permission from [44]. Copyright (2006) American Chemical Society.

Rutin is a kind of flavonoid glycoside that has a wide range of physiological activities such as anti-inflammatory, antitumor, and antibacterial. CNT modified electrodes have been successfully using for the determination of rutin. A gold electrode modified with SWCNTs was fabricated, by Zeng *et al.*, to investigate the voltammetric behavior of rutin [54]. At this electrode rutin exhibited an adsorption-controlled electrode reaction with a pair of peaks at ~ 0.35V, implying two electron transfer process at the electrode surface. The linear range from 20 nM to 5.0 µM and the detection limit of 10 nM were determined. Yu's group developed a rutin sensor based on GCE modified with β-cyclodextrin incorporated MWCNTs to take advantage of the inclusion interaction of β-cyclodextrin and rutin [55]. GCE modified by SWCNTs/poly (Neutral Red) composite film also exhibited a good catalytic activity on the redox properties of rutin [56]. Rutin, a kind of mediator was also used with MWCNTs for the fabrication of a hydroxylamine sensor [57].

Metal nanoparticles have received considerable attention in recent years. Incorporation of nanoparticles to CNTs for the modification of electrodes has been demonstrated to enhance the electrocatalytic activity of many electrochemical processes and therefore be suitable for sensing applications. Hrapovic et al. focused on metal nanoparticles/CNTs nanocomposites for electrochemical detection of trinitrotoluene (TNT) and other nitroaromatics [58]. They found that Cu nanoparticles and SWCNTs solubilized in Nafion provided the highest sensitivity for TNT with the detection limit of 1 ppb for analysis of TNT in tap water, river water, and contaminated soil. The composites of SWCNTs, gold nanoparticle (GNP), and ionic liquid (i.e. 1-octyl-3-methylimidazolium hexafluorophosphate) were used to fabricate a modified GCE for the sensitive voltammetric detection of chloramphenicol [59]. The composition of the film and the operation conditions played an important role in the voltammetric response and the detection limit under optimum condition was 5.0 nM. Electrochemically deposited Pt nano-clusters onto MWCNTs-modified GCE showed a strong electrocatalytic activity toward the oxidation of estrogens involving estradiol, estrone, and estriol [60]. This electrode showed the linear responses between 0.5–15 µM, 2.0–50 µM, and 1.0–75 µM for estradiol, estrone, and estriol, respectively, in the square-wave voltammetry. Wei and co-workers used composite of nano-silver coated MWCNTs to determine a trace of thiocyanate in urine and saliva samples from smokers and nonsmokers and observed the decrease of anodic current upon addition of trace amount of thiocyanate in nM level [61]. Liu et al. designed a sensor by electropolymerization of thionine at the GCE modified with GNPs/MWCNTs composites for simultaneous determination of adenine and guanine in DNA [62]. The detection limits (S/N = 3) of 10 nM and 8.0 nM were obtained for guanine and adenine, respectively.

Carbon-paste electrodes (CPEs) have been widely used as a suitable matrix for preparation of modified electrodes due to their simple preparation, renewability, and compatibility with various types of modifiers. CNT-modified CPEs showed considerable improvements in electrochemical behavior of materials. Zhuang et al. reported the fabrication of CPE modified with MWCNTs for the determination of bergenin using CV and differential pulse voltammetry (DPV) [63]. They found that the oxidation peak current of bergenin increased significantly upon increase of the contents of MWCNT and the linear dynamic range of 0.6 ~ 10 µM was observed at an optimum experimental condition with a detection limit of 70 nM. Shahrokhian and Fotouhi added cobalt salophen (CoSal) for the determination of tryptophan [64]. The MWCNTs/CoSal CPE exhibited an electrocatalytic activity on the oxidation of tryptophan. Furthermore, this electrode could be used to detect tryptophan separately from the interferences of AA and cysteine. They also showed Nafion-incorporated MWCNTs/CoSal modified CPE can be used for the simultaneous detection of UA and AA. The detection limits were determined to be 60 and 100 nM, respectively [65]. When thionine-Nafion was incorporated into MWCNTs/CPE, it could be used to detect DA and AA simultaneously [66]. Recently, an electrode coated with SWCNTs paste was developed for sensitive voltammetric detection of methylparathion [67]. Fan *et al.* fabricated such electrode using ionic liquid (1-butyl-3-methyl-imidazolium hexafluophosphate) as a binder and they found a good electro-catalytic behavior to the electrochemical reduction of methylparathion. It was also found that the same electrode can be used to detect methylparathion and its hydrolysate (*p*-nitrophenol) simultaneously. The similar approach was applied for the voltammetric determination of xanthine [68] and UA [69]. A highly sensitive and fast responding electrochemical sensor was also prepared for piroxicam with MWCNTs paste electrodes by Abbaspour and Mirzajani [70]. It exhibited a high catalytic activity towards electrooxidation of piroxicam in a mild pH medium, showing the linear response range of 0.15–5 µg/mL, with the detection limit of 0.1 µg/mL. The same electrode was applied for the determinations of urapidil [71] and quercetin [72].

Composite film of CNTs with other materials such as conducting polymers, ceramics etc are very attractive combination of materials for the development of electrochemical sensors. Yang *et al.* fabricated a novel electrode using a poly(acid chrome blue K)/ MWCNTs composite film modified GCE for simultaneous detection of dihydroxybenzene isomers in real water samples by applying the first order linear sweep derivative voltammetry [73]. The detection limits for hydroquinone, catechol, and resorcinol were 100, 100, and 90 nM, respectively with this electrode. Wang *et al.* developed a novel sensor system by electrochemical oxidation of the mixture of L–cysteine and CNTs at GCE for the determination of terbinafine in human serum [74]. A composite film, found on GCE after oxidation, was responsible for the significant increase in the current response of terbinafine. It showed a wide range of linear response, 80 nM-50 µM, with the detection limit of 25 nM for terbinafine determination. Zhu *et al.* prepared CNTs ceramic composites electrode for electrochemical sensing by dispersing MWCNTs into the methyltrimethoxysilane derived sol–gel solution [75]. The electrode provided a better reversible behavior with a substantial decrease of overpotential, and a higher sensitivity compared to the unmodified one towards the electrochemical detection of ascorbic acid, uric acid, acetaminophenol, NADH, EP, DA, cysteine, and H_2O_2. Wang and coworkers have prepared a RuOx/CNTs-modified GCE by drop and dry process [76]. This electrode showed an electrocatalytic activity towards insulin and exhibited a wide range of linear response of 10 – 800 nM with a detection limit of 1 nM. Recently, Snider *et al.* developed a MWCNTs/dihydropyran composite film for the electrochemical detection of insulin secreted by islets in a microfluidic system [77]. Vega *et al.* developed a tetracycline antibiotic sensor based on MWCNTs modification of GCE [78]. A highly sensitive and reproducible electroanalytical response of tetracycline was attributed to the antifouling capability of the MWCNTs. Rezaei and Zare described a simple and highly sensitive method based on the modification of GCE by MWCNTs for the direct voltammetric determination of noscapine in pharmaceutical and clinical samples [79]. It significantly enhanced the noscapine signal and showed a wide linear range of 0.4 µM-10 mM, under the optimum condition, with the detection limit of 80 nM. They also applied the same electrode to study the determination of captopril with effective electrocatalysts, hexacyanoferrate (II) [80]. The system of MWCNTs and hexacyanoferrate strongly enhances the oxidation of captopril and a detection limit of 0.2 µM was obtained. Buratti *et al.* introduced cobalt oxide with the MWCNTs modified GCE for the detection of carbohydrates and thiols [81]. The modified electrode showed excellent electrocatalytic activity towards the oxidation of carbohydrates and thiols.

4. CARBON NANOTUBE-BASED ELECTROCHEMICAL BIOSENSORS

An electrochemical biosensor is an analytical tool for sensitive and selective detection of biomolecules. Increasing attention has been given to biosensors due to their potential applications in clinical chemistry, food industry, and environmental fields. Glucose oxidase (GOx)-based glucose biosensors are still considered as a primary model system in the development of new sensing materials and methods. They are the most extensively studied enzyme biosensors because of their high demand for blood glucose monitoring. A summary of recent progress in the field of glucose biosensors can be found in two excellent recent reviews by Wang and Heller [82,83]. CNTs are extremely attractive for fabricating electrochemical biosensors due to their outstanding properties, especially the excellent conductive, adsorptive and biocompatibility. Vertically aligned CNTs can be coupled with enzymes to provide a favorable surface orientation and act as an electrical connector between their redox center and the electrode surface [82]. Figure 3 showed the assembly of the CNT electrically contacted GOx electrode. Plugging enzymes into the CNTs by this way is an extremely efficient approach for the development of glucose biosensor.

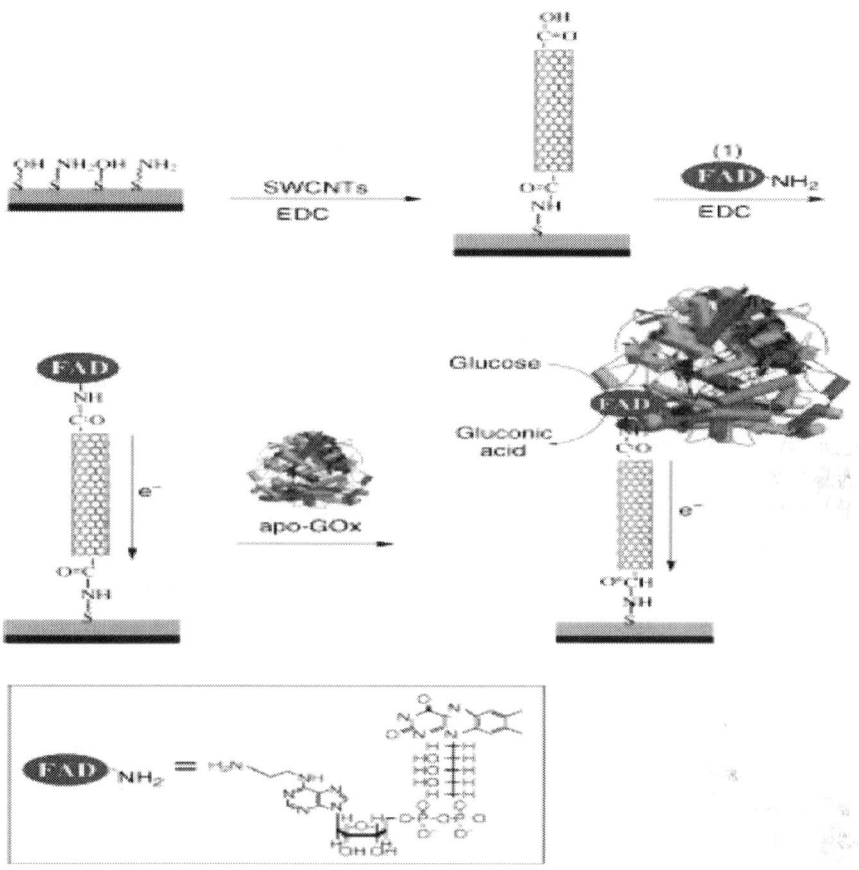

Figure 3. Assembly of the CNT electrically contacted glucose oxidase electrode. Reprinted with permission from [82]. Copyright (2008) American Chemical Society.

Different types of glucose biosensors have been developed in recent years. Tsai and co-workers developed a nanobiocomposite film by incorporating functionalized MWCNTs and GOx into polypyrrole (PPy) film for a highly sensitive glucose biosensor [84]. The amperometric response of the optimized biosensor displayed a sensitivity of 95 nA/mM, a linear range up to 4 mM, and a response time of about 8 sec. Huang *et al.* loaded MWCNTs and GOx on a graphite disk using a LBL assembly technique to construct a glucose biosensor [85]. The current response to glucose was highly dependent on the number of layers and the maximum response was obtained at 6 layers of MWCNTs/GOx with the detection limit of 90 µM. Liu and Lin also applied LBL assembly technique to construct a sandwich-like structure, PDDA/GOx/PDDA/CNTs, for a reproducible and stable glucose biosensor while Zhao and Ju added poly (sodium 4-styrenesulfonate) with PDDA to construct multilayer membranes [86,87]. They modified gold electrode with 3-mercapto-1-propanesulfonic-acid and then bilayers of the PDDA and PSS were formed on the modified Au surface. PDDA wrapped MWCNTs and GOx was then assembled through LBL technique. Wang et al. functionalized gold electrodes with the negatively charged 11-mercaptoundecanoic acid (MUA) and then apply the LBL assembly of a positively charged redox polymer, poly [(vinylpyridine)Os(bipyridyl)$_2$Cl$^{2+/3+}$], and the negatively charged GOx/SWCNTs for glucose sensor. Liu et al. developed an amperometric glucose biosensor based on electrostatic assembly of GNPs/MWCNTs/GOx [88]. Positively charged poly (dimethyldiallylammonium chloride) was used to connect them in a LBL pattern. The electrode showed an excellent electrocatalytic activity for glucose sensing at a relatively low potential (−0.2 V).

Xu et al. described an amperometric glucose biosensor based on an alternating electrostatic self-assembling GOx and dendrimer-encapsulated Pt nanoparticles (Pt-DENs) on MWCNTs [89]. The excellent electrocatalytic activity of CNTs and Pt-DENs toward H$_2$O$_2$ and special three-dimensional structure of the enzyme electrode resulted in a low detection limit with a wide linear response range, a high sensitivity with a good precision, and an enhanced operational stability. Shirsat et al. fabricated an amperometric glucose biosensor by applying a LBL assembly of SWCNTs and PPy multilayer film on a platinum coated with polyvinylidene fluoride (PVDF) membrane [90]. GOx was immobilized on the film by cross linking through glutaraldehyde (GA) (0.1%) and a linear response range from 1 mM to 50 mM of glucose concentration with the sensitivity of 7.06 uA/mM was achieved. A glucose sensor based on the LBL assembly of functionalized MWCNTs and poly (neutral red, PNR) multilayer film was also suggested [91]. This electrode showed a significant improvement of redox activity showing a synergic effect of excellent electron-transfer capability of CNTs and PNR. Another type of glucose biosensor was constructed by immobilizing GOx onto the electrode surface using GA. Yao and Shiu constructed a mediator type glucose sensor based on the immobilization of GOx at electropolymerized poly (toluidine blue O) film on CNTs modified GCE [92]. Poly (toluidine blue O) provided the polymer matrices to maintain the sensing activity and served as a redox mediator for enzymatic glucose oxidation. This biosensor showed enhanced current response at low potential (−0.1V) and therefore common interferences from AA, UA, and acetaminophen could be avoided. Zhu and co-workers suggested a Prussian Blue (PB) based amperometric glucose biosensor by assembling the PB nanoparticles on the surface of MWCNTs modified GCE followed by immobilization of GOx [93]. It showed good sensitivity, fast response with a detection limit of 12.7 mM. Similar PB based glucose biosensors were also prepared by immobilizing GOx in a film of LBL assembly of CS and MWCNTs [94] and on the nanocomposite film of PB nanoparticles/MWCNTs/poly (1,2-diamino-benzene) [95]. Manesh et al. fabricated a glucose biosensor based on the immobilization of GOx into an electrospun composite membrane consisting of polymethylmethacrylate (PMMA) dispersed with MWCNTs wrapped by a cationic PDDA polymer [96]. This nanofibrous electrode exhibited excellent electrocatalytic activity towards H_2O_2 with a pronounced oxidation current at +100 mV. Glucose was detected amperometrically with this nanofibrous electrode with a detection limit of 1 μM. A highly sensitive and selective glucose biosensor based on immobilization of GOx within mesoporous CNTs-titania-Nafion composite film coated on a platinized GCE, was also developed recently by Lee and co-workers [97]. It responded linearly to glucose in the wide range from 50 μM to 5.0 mM with sensitivity of 154 mA $M^{-1}cm^{-2}$.

A mediatorless glucose biosensor, based on the incorporation of GOx to the composite electrode of colloidal gold-CNTs-Teflon showed a remarkably higher sensitivity than that achieved with other GOx-CNTs bioelectrodes [98]. It could be used for ethanol biosensor by incorporating alcohol dehydrogenase (ADH) [99]. Chu et al. developed an amperometric glucose biosensor based on adsorption of GOx at the gold and platinum nanoparticles modified CNTs electrode where CNTs were covalently immobilized on cysteamine modified gold electrode [100]. The GOx/Au nano/Pt nano/CNTs/Au electrode was then covered with a thin layer of Nafion to avoid the loss of GOx and suppress the interfering signals from UA and AA. Recently, Zou et al. developed an amperometric glucose biosensor based on electrodeposition of platinum nanoparticles (PtNPs) onto MWCNTs and entrapping an enzyme in CS-SiO₂ sol-gel [101]. This electrode showed an excellent electrocatalytic activity and high stability as well due to the synergistic action of Pt and MWCNTs and the biocompatibility of CS-SiO₂sol-gel. A wide linear range from 1 μM to 23 mM and a low detection limit of 1 μM was achieved for glucose sensing. Zhao et al. recently investigated an amperometric glucose biosensor based on PtNPs combined aligned CNTs electrode [102]. The combination of PtNPs and CNTs in this glucose biosensor showed a highly sensitive detection of glucose. Kang et al. constructed another glucose biosensor based on the integration of CNTs with gold-platinum alloy nanoparticles (Au-PtNPs) [103]. In this sensor, GOx was immobilized in biocompatible CS through cross-linking with GA on the Au–PtNPs/CNTs/CS film. It showed a low potential (0.1 V) detection of glucose with high sensitivity, low detection limit, good reproducibility, long-term stability, fast response, and high specificity. This biosensor was applied in the determination of glucose in real blood and urine samples with satisfactory results. Yang et al. prepared MWCNTs composite using Pt–NP doped sol/gel solution as a binder and incorporated GOx for glucose biosensor [104]. The sensitivity enhanced 4 times when Pt nanoparticles were loaded. A glucose biosensor, developed by Rivas and co-workers was based on the electrocatalytic activity of copper and iridium microparticles incorporated within the CNTs paste electrode containing GOx [105]. This biosensor detected glucose at very low potentials (−0.1 V) with high sensitivity and selectivity. Yao and Shiu examined the electrochemical and electrocatalytic properties of different types of CNTs material and used them for fabricating glucose biosensors [106]. They found that the electrodes modified with SWCNTs usually had better electron-transfer and electrocatalytic properties than the corresponding MWCNTs-modified electrodes. Recently, Jia *et al.* reported the fabrication of needle-type glucose biosensor by packing a mixture of MWCNTs, graphite powder, and freeze-dried GOx powder into a glass capillary of 0.5 mm inner diameter [107]. It showed an improved sensitivity and stability when the experimental condition was optimized. Zhu and co-workers proposed a bienzymatic mediatorless glucose biosensor based on co-immobilization of GOx and horseradish peroxidase (HRP) in an electropolymerized PPy film on a SWCNTs modified electrode [108]. They took advantage of direct electron transfer characteristics of HRP with CNTs electrode and realized a lower operational potential for selective determination with a minimized interference.

The detection of hydrogen peroxide (H_2O_2) is very important because many enzymatic biosensors rely on the detection of H_2O_2 generated by an enzymatic reaction. Since the amount of generated H_2O_2 from an enzymatic reaction is very low, the fabrication of a highly sensitive H_2O_2 biosensor is needed. CNTs can be used in the fabrication of highly sensitive H_2O_2 biosensors. There have been many reports on CNTs based H_2O_2 biosensors. Chen and Lu reported the encapsulation of hemoglobin (Hb) in the composite film of carboxylic acid functionalized MWCNTs and polyelectrolyte-surfactant polymer to develop a H_2O_2 biosensor [109]. Faradic response of the Hb was observed and it exhibited excellent electrocatalytic activity to reduce H_2O_2. Chen *et al.* proposed an amperometric third-generation H_2O_2 biosensor based on the immobilization of Hb on the nanohybrid film of MWCNTs and gold colloidal nanoparticles [110]. A wide range of linear response from 0.21 µM to 3.0 mM with a detection limit of 80 nM was obtained. Tripathi *et al.* entrapped HRP in an ormosil composite doped with ferrocene monocarboxylic acid−bovine serum albumin conjugate and MWCNTs for a H_2O_2 biosensor [111]. MWCNTs improved the conductivity of the composite film and HRP provided a fast amperometric response to H_2O_2. A wide linear range between 20 µM and 4.0 mM with a detection limit of 5.0 µM (S/N = 3) was achieved. Luo *et al.* developed a H_2O_2 biosensor with an improved performance based on the immobilization of HRP onto electropolymerized PANI films doped with CNTs [112]. It was found that the existence of CNTs in the biosensor system could effectively increase the amount and stability of the immobilized HRP as well as the performance of the biosensor. A H_2O_2 biosensor based on the modification of graphite electrode with toluidine blue (Tb) modified MWCNTs was developed by H. Ju and co-workers [113]. HRP was immobilized with the aid of CS for sensing H_2O_2 with a good stability and reproducibility. Qian and Yang developed a mediator free amperometric biosensor for H_2O_2 based on composite film of MWCNTs/CS [114]. HRP was cross-linked with composite film using GA. Sanchez et al. reported the fabrication of MWCNTs/polysulfone biocomposite membrane, which allows the incorporation of HRP enzyme by phase inversion technique. This biocomposite membrane was used for the construction of a H_2O_2 biosensor [115]. Qu et al. took the combined advantages of CNTs and nano-Fe_3O_4 to prepare a magnetic CNTs/nano-Fe_3O_4 composite by co-precipitation. It exhibited higher electrocatalytic activity toward the redox processes of H_2O_2 [116]. They also introduced CS into the bulk of the composite by co-precipitation to immobilize GOx covalently to make an amperometric glucose sensor.

Choi et al. constructed a highly sensitive and stable amperometric ethanol biosensor based on the immobilization of ADH within a thin composite film of CNTs-titania-Nafion [117]. Due to the mesoporous nature of this composite film, the present ethanol biosensor exhibited remarkably fast response time and wide linear response range. Santos et al. constructed an amperometric ethanol biosensor based on co-immobilization of ADH and Methylene Blue (MB) on MWCNTs through the cross-linking with GA and agglutination with mineral oil [118]. The amperometric response of this biosensor showed an excellent operational stability and wide linear response range. Cai and co-workers fabricated a nanocomposite by the functionalization of SWCNTs with poly(Nile Blue A) for ethanol biosensor [119]. Immobilization of ADH onto the modified electrode showed electrocatalytic activity toward the oxidation of ethanol with a good stability, reproducibility, and higher biological affinity. Liu and Cai developed an ethanol biosensor based on the nanocomposites of positively charged PDDA wrapped SWCNTs [120]. The negatively charged ADH was immobilized on the nanocomposites via the charge interaction and this biosensor provided a good electrocatalytic activity toward the oxidation of ethanol with a good stability, reproducibility, and high biological affinity.

Yang et al. reported the modification of gold electrodes by self-assembling the positively charged Pt nanoparticle-MWCNTs-CS and negatively charged poly (sodium-p-styrenesulfonate) salt (PSS) for a sensitive cholesterol biosensor [121]. MWCNTs were dispersed in the Pt nanoparticle-doped CS solution to obtained Pt-CNTs-CHIT material. Cholesterol oxidase was immobilized onto the modified electrode surface using GA.

Tang et al. developed an amperometric glutamate biosensor based on self-assembly of glutamate dehydrogenase (GLDH) and poly (amidoamine) dendrimer-encapsulated platinum nanoparticles (Pt-PAMAM) onto carboxylic acid group-functionalized MWCNTs [122]. The modified electrode showed electrocatalytic activity toward the oxidation of glutamate with a good reproducibility and high sensitivity. Lin's group fabricated a simple and inexpensive choline biosensor based on the immobilization of choline oxidase (ChO), ChO and HRP bienzymes onto MWCNTs modified GCE using LBL assembly technique [123]. With this configuration, a wide linear response range from 50 µM to 5.0 mM with a detection limit of about 10 µM for choline was achieved. Song et al. fabricated an another choline biosensor by immobilizing ChO into a sol-gel silicate film on MWCNTs modified platinum electrode [124]. This biosensor was used to detect choline released from lecithin by phospholipase D (PLD) in serum samples with high sensitivity and the detection limit was 0.1 µM. A stable and sensitive acetylthiocholine sensor based on immobilization of acetylcholinesterase (AChE) on the CS-MWCNTs composite was developed by Du et al. [125]. GA was used as cross linker to covalently bond AChE and efficiently prevent leakage of the enzyme from the film.

Lee et al. developed an amperometric tyrosinase biosensor based on MWCNTs dispersed in mesoporous composite films of sol-gel-derived titania and Nafion [126]. Tyrosinase was immobilized within the composite film and phenolic compounds were determined by the direct reduction of biocatalytically liberated quinone species. This sensor exhibited remarkably fast response time less than 3 sec and a good performance in terms of the sensitivity (417 mA/M) and the detection limit (0.95 nM) due to the large pore size of the composite film. An amperometric biosensor was based on the immobilization of HRP on MB modified MWCNTs for phenolic compounds and it showed a very wide linear response with a good sensitivity for catechol [127]. Recently, Lopezet al. described a biosensor fabricated by the modification of GCE with a matrix based on MWCNTs, tetrahydrofuran (THF) mixed with poly(vinylchloride) (PVC), and with a GA solution (MWCNTs-TPG/GC) for NADH detection [128]. The modified electrode showed a relatively higher sensitivity, a promotion of electron transfer, and it facilitated the amperometric determination of NADH starting in a potential of +0.40 V. Male et al. developed a biosensor for arsenite by depositing molybdenum-containing arsenite oxidase galvanostatically onto the active surface of a MWCNTs on GCE [129]. The detection limit of 1 ppb was found for arsenite but there was a severe interference caused by common metal ions found in tap and river waters. Mita et al. constructed a bisphenol A biosensor using a various tyrosinase containing CPE and they optimized the experimental condition with the composition of 10% tyrosinase, 45% SWCNTs, and 45% mineral oil [130]. It showed a good reproducibility with a detection limit of 20 nM. Liu and Lin fabricated an amperometric biosensor based on LBL self-assembling AChE on CNTs-modified GCE and integrated it within a flow injection-detection system. It was highly sensitive for organophosphate pesticides and nerve agents and showed a good precision, reproducibility, and stability [131]. CNTs play a dual significant role in this structure. It provides a robust immobilization sites for a suitable microenvironment to retain the enzyme activity and as a transducer, which amplifies the electrochemical signal of the product of the enzymatic reaction. Rahman *et al.* fabricated an amperometric lactate biosensor based on MWCNTs and conducting polymer (CP) by covalently immobilizing the lactate dehydrogenase and NADH onto the MWCNTs/CP assembly [132]. The MWCNTs/CP nanocomposite assembly was obtained through the electrochemical polymerization of monomer containing MWCNTs. The analytical results such as sensitivity, selectivity, and stability were found to be improved significantly using MWCNTs/ CP nanocomposite assembly.

The detection of DNA is currently an area of tremendous interest in genetics, clinics, pathology, criminology, pharmacogenetics, food safety, and many other fields. Most of DNA biosensors are developed based on the immobilization of single-stranded DNA onto the electrode surface labeled with an electrochemical indicator to recognize its complementary target sequence. CNTs are promising materials for the development of electrochemical DNA hybridization biosensors. The unique properties of CNTs can be united with the specific molecular-recognition features of DNA by coupling SWNTs to peptide nucleic acid and hybridizing these macromolecular wires with complementary DNA [133]. Both covalent and non-covalent linkage of DNA with CNTs have been reported where the former provide the best stability, accessibility, and selectivity during competitive hybridization [134]. Figure 4 shows an overview of the covalent attachment process. By this attachment, it was found that DNA molecules are accessible to hybridization and strongly favor hybridization with molecules having complementary sequences compared with noncomplementary sequences. The integration of CNTs with other materials has been also used for the immobilization of DNA. Yang et al. described a sensitive DNA hybridization biosensor based on ZrO_2 nanoparticles and MWCNTs [135]. The MWCNTs/nano ZrO_2/CS-modified GCE was fabricated by dispersing ZrO_2 nanoparticles and MWCNTs in CS and oligonucleotides were immobilized on a modified GCE. The hybridization reaction on the electrode was monitored by DPV analysis where electroactive daunomycin was used as an indicator. Jiao and co-workers applied the same approach for DNA biosensor using ZnO nanoparticles instead of ZrO_2 nanoparticles [136]. Recently, Ma et al. fabricated an electrochemical DNA biosensor based on LBL self-assembly of MWCNTs and GNPs via covalent-bonding interaction [137]. Doxorubicin was used as an intercalator and the hybridization events were monitored electrochemically by DPV measurement. The biosensor showed an improved sensitivity with an excellent reproducibility due to the high catalytic activities of GNPs and the ability of CNTs to promote electron-transfer reactions. A wide linear response range from 0.5 to 0.01 nM with a detection limit of 7.5 pM for target DNA was achieved. Recently, Niu et al. used manganese complex of rutin as a redox intercalator with carboxylic acid group-functionalized MWCNTs and fabricated DNA biosensor for DNA hybridization detection [138]. The modified electrode dramatically increased the amount of DNA attachment and the sensitivity of the complementary ssDNA detection mostly due to the large surface area and good charge-transport characteristics of CNTs. Erdem et al. described a new DNA biosensor based on the enhancement of guanine signal at MWCNTs-modified pencil graphite electrode (PGE) using DPV [139]. PGE behaved as a microelectrode array coupled with its higher porosity and showed improved performance compared to GCE. Another new DNA biosensor based on electrochemical impedance was described by Fang's group [140]. They modified GCE using a composite material of PPy and carboxylic group-terminated MWCNTs. A probe with an amino group-termination was linked onto the PPy/MWCNTs-COOH/GCE by using EDAC and it was found that the hybridization reaction with its complementary decreased the electron transfer resistance. Zhang's group also used PPy and MWCNTs to develop a high sensitive and selective biosensor for DNA hybridization based on the immobilization of DNA probe within electropolymerized PPy on a MWCNTs paste electrode [141]. Ethidium bromide (EB) was used as an intercalator and the current change generated from it was monitored. Only the complementary DNA, compared to the five-point mismatched and non-complementary sequences, gave an obvious current flow and a detection limit of 0.85 pM was obtained. An electrochemical DNA biosensor based on palladium nanoparticles combined with MWCNTs was suggested by Chang et al. [142]. MB was used as an indicator and the hybridization was monitored by DPV measurement. A lower detection limit of 120 fM for the target DNA was achieved and the improved sensitivity was attributed to the ability of CNTs promoting electron-transfer process and the high catalytic activities of palladium nanoparticles for electrochemical reaction of MB.

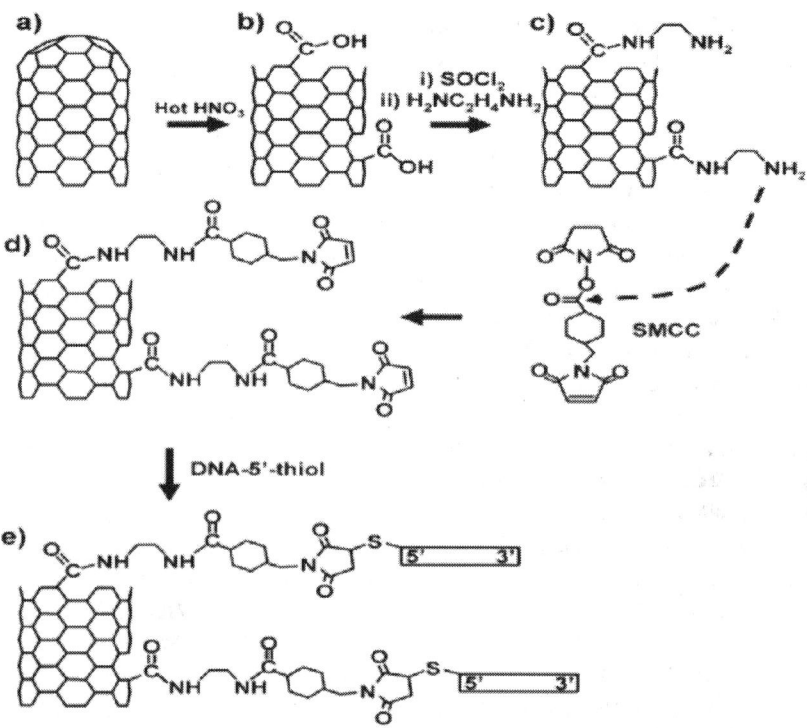

Figure 4. Overall scheme for fabrication of covalently linked DNA-nanotube adducts. Reprinted with permission from [134]. Copyright (2002) American Chemical Society.

Electrochemical immunosensors are the popular area of study due to their high specificity, sensitivity, and stability. Most of approaches are based on an enzyme-linked immunosorbent assay (ELISA) system built on electrode surfaces. The amount of enzyme-linked antigen bound to the immobilized antibody is determined by the relative concentration of the free and conjugated antigen and quantified by the rate of enzymic reaction. To avoid the additional steps for labeling, the label-free immunosensors are getting much interest. CNTs are potential materials for the fabrication of labeled or label-free immunosensors. The carboxylic acid groups-functionalized CNTs can be used as an antibody immobilizing platform. CNTs can also be used in the detection probe of an immunosensor. Yun *et al.* developed a label-free immunosensor based on CNTs array electrodes for direct electrochemical detection of antigen–antibody binding reaction [143]. Anti-mouse IgG was covalently immobilized on the carboxylic acid-terminated nanotube array and the binding was characterized by CV and electrochemical impedance spectroscopy (EIS). The detection limit was found to be 200 ng/mL. EIS was chosen here as the analytical approach because of its ability to analyze the electrochemical response of the electrode over a wide frequency range. Electrodes modified with SWCNTs array have also been used by Okuno *et al.* to fabricate a label-free electrochemical immunosensor for prostate-specific antigen [144]. Yu *et al.* described an electrochemical immunosensor using SWCNTs forest platforms with multi-label secondary antibody-nanotube bioconjugates for highly sensitive detection of a cancer biomarker in serum and tissue lysates [145]. A great amplification of the sensitivity was achieved by using bioconjugates featuring HRP labels and secondary antibodies (Ab2) linked to MWCNTs at high HRP/Ab2 ratio (Figure 5). A low detection limit of 4 pg/mL for prostate specific antigen was obtained in 10 µL undiluted calf serum with this strategy.

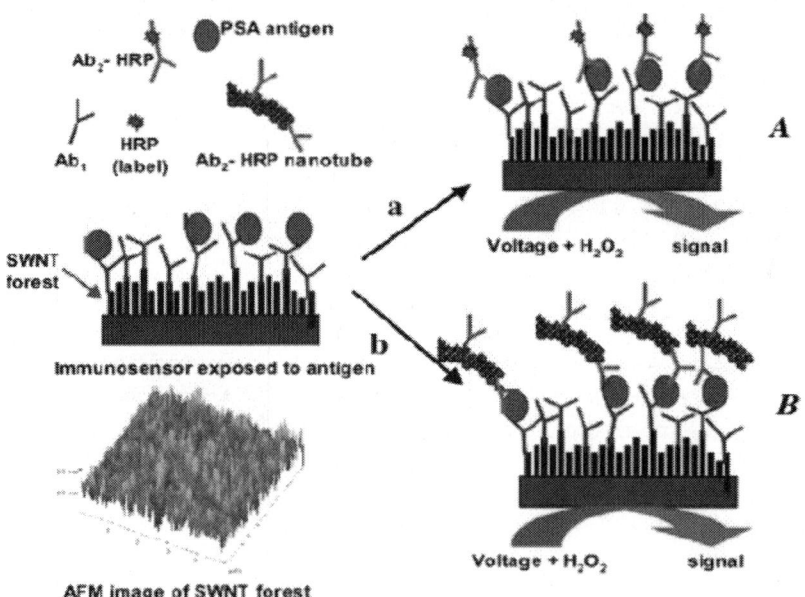

Figure 5. Configuration of the nanotube immunosensor. Reprinted with permission from [145]. Copyright (2006) American Chemical Society.

Polymer-CNTs composite films have been widely used for the fabrication of electrochemical immunosensors. Cataldo *et al.* evaluated an amperometric immunosensor based on the covalently bound anti-biotin antibodies (Ab) embedded into a polylysine (PLL)-SWCNTs composite layer [146]. An improved amperometric detection limit of 10 pM was obtained for biotin (Ag) labeled with HRP by incorporating SWCNTs into PLL-antibody assemblies. One advantage of this proposed immunosensor is the improved thermal stability of the anti-biotin embedded into the PLL matrix. Poly (3,4-ethylene-dioxythiophene)-coated MWCNTs have been used for the first time as a voltammetric sensor in an immunoassay for Cholera Toxin (CT) by Ho's group [147]. Ganglioside (GM1)-functionalized liposomes encapsulated with potassium ferrocyanide (an electroactive redox marker) were used for the detection of CT. A wide linear range from 10 fg/mL to 0.1 μg/mL with a detection limit of 0.1 fg/mL of CT was achieved. Sánchez *et al.* described an electrochemical immunosensor method based on polysulfone membrane encapsulating MWCNTs and immunoreagents layered on disposable screen-printed electrodes [148]. The immunocomposite acts both as reservoir of immunological material and transducer while offering high surface area, high toughness, and mechanical flexibility.

Chitosan (CS), a biological cationic macromolecule, provides a suitable environment for biomolecules to keep their activity. Their combination with highly conductive CNTs as an immunosensing platform has attracted considerable interest. Recently, Li *et al.* proposed a novel amperometric immunosensor for human chorionic gonadotropin (HCG) assay incorporating Tb and Hb on the MWCNTs–CS modified GCE, followed by electrostatic adsorption of a conducting GNPs film as sensing interface [149]. The MWCNTs-CS matrix provided a congenial microenvironment for the immobilization of biomolecules and promoted the electron transfer of the redox active species, thus, enhanced the sensitivity of the immunosensor. CS can be also used for the development of a highly sensitive and label-free amperometric immunosensor for carcinoembryonic antigen detection [150]. The fabrication of the immunosensor was based on LBL assembly of GNPs, MWCNTs-thionine, and CS on 3-mercaptopropanesulfonic acid (MPS)-modified gold electrode surface. The detection limit of carcinoembryonic antigen with this label-free amperometric immunosensor was determined to be 0.01 ng/ml.

The practical application of CNTs can be found in a very limited area only. It is partially attributable to the lack of simple modification skill of electrode surfaces with CNTs. To overcome this obstacle, Yang and coworkers developed two different methods for the modification of ITO electrode with CNTs. The ITO electrodes are advantageous to use in optical experiment and to make electrochemical (bio) sensors because it is optically transparent and display high conductivity with low background current. Initially, they fabricated thin film of MWCNTs by simply physisorbing of CNTs on the ITO electrode from the aqueous solution of carboxylated MWCNTs [151]. To make thin film of CNTs on ITO electrode, they covalently immobilized the carboxylated SWCNTs on an amine-functionalized ITO surface using dicyclohexylcarbodiimide as a coupling agent [152]. Both films showed low background currents and good electrocatalytic properties toward the oxidation of *p*-aminophenol (*p*-AP), which is an enzymatic product of enzyme substrate (*p*-aminophenyl phosphate). Both thin CNT films could be used to fabricate sandwich-type immunosensors because of the good electrocatlytic property with a low background current, and the ability of immobilization of biomolecules on the sidewall of the CNTs. To fabricate the sandwich type immunosensor, avidin was immobilized first on the hydrophobic sidewalls of CNTs to capture biotinylated anti-mouse IgG. Then, the target mouse IgG was sandwiched between the biotinylated anti-mouse IgG and alkaline phosphatase-conjugated antimouse IgG. The alkaline phosphatase catalyzed the conversion of electroactive *p*-AP from non electroactive *p*-aminophenyl phosphate. The enhancement of the generation of *p*-AP upon increase of the concentration of target mouse IgG resulted in the strength of the signal. The thin films of MWCNTs and SWCNTs gave the detection limits of 10 pg/mL and 100 pg/mL, respectively. They are comparable to those of sensors with a commonly employed enzyme-linked immunosorbent assay (pM range).

5. APPLICATION OF CNTS-BASED SENSORS TO REAL SAMPLE ANALYSIS

CNTs-based sensors can be applied in real sample analysis in different areas such as biomedical, food, agriculture, and fishing industries. There are many biomedical sensing applications where CNTs-based sensors perform better in real sample analysis. CNTs-based sensors can be used in commercial food samples to detect undesired chemical residues resulting from animal drugs, food additives, pesticides, and other environmental contaminants in raw and processed foods. Rahman *et al.* determined the lactate concentration in commercial milk using CNTs-based electrochemical biosensor and compared the result with that obtained from a biochemical analyzer [132]. The biosensor result was found to be in good agreement with the result from biochemical analyzer (0.18 ± 0.006 mM and 0.174 ± 0.01 mM, respectively). CNTs based electrochemical sensors are also widely used in real blood and urine samples analyses. Table 1 summarizes the applications of some CNTs-based electrochemical sensors to real sample analyses.

Table 1. Applications of CNTs-based electrochemical sensors in real samples.

Electrode	Analyte	Real sample	Detection limit	Reference
GCE/P3MT/SWNTs/Nafion	dopamine	serum	5.00 nM	[23]
GCE/ PANI /MWNTs /β-CD	dopamine	injection	12.0 nM	[25]
Au/SWNTs	rutin	tablet (drug)	10.0 nM	[54]
GCE/ MWNTs /β-CD	rutin	urine	0.20 μM	[55]
GCE/ MWNTs / PtNC	estrogen	serum	0.18 μM	[60]
CPE/ MWCNTs/CoSal	tryptophan	serum	0.10 μM	[64]
GCE /SWNTs /BMIMPF$_6$	methylparathion	lake water /apple	1.00 nM	[67]
GCE /SWNTs /BMIMPF$_6$	xanthine	serum / urine	2.00 nM	[68]
GCE/ MWNTs / poly-ACBK	dihydroxybenzene	water	0.10 μM	[73]

6. CONCLUSIONS

CNTs are now used extensively in the fabrication of novel nanostructured electrochemical sensors. CNTs-modified electrodes have many advantages over other forms of carbon electrodes due to their small size, high electrical and thermal conductivity, high chemical stability, high mechanical strength, and high specific surface area. Their small diameter and long length allow them to be plugged into proteins with better electro-activity compared to other carbon based electrodes. The promoted electron transfer and direct electrochemistry of proteins at CNTs-based electrochemical sensing films are now well documented. Due to its faster electron transfer over other carbon based materials, CNTs show excellent electrocatalytic activity in redox behavior of different compounds. Analytical sensing at CNTs-modified electrodes results in low detection limits, high sensitivities, reduction of overpotentials, and resistance to surface fouling. The aforementioned outstanding properties of CNTs make them an exciting alternative for the development of novel electrochemical sensors and biosensors. However, there are a number of challenges to be addressed to fulfill the application of CNTs for sensors. The commercial production of pure and defect-free CNTs is difficult and costly. Processing of CNTs is still not fully controlled. For example, aggregations of tubes are not prevented, lengths are not uniformly obtainable, and non-specific adsorptions of proteins to the walls of nanotubes are not prevented. CNTs can cause health risks due to their toxicity and harmful effects in the lung, where they can agglomerate leading to suffocation. The toxicological impact of nanotubes is an obstacle for the application of nanotubes in bioelectronics and any subject integrated with living biological systems. Another limitation is that CNTs are commonly insoluble in most solvents, which has greatly hindered their promising practical applications. Covalent and non-covalent functionalizations of CNTs are not very effective in overcoming this limitation. CNTs can be dispersed in Nafion, Teflon, CS, mineral oils, sol-gel silica, and in some polymer but these methods possibly impair the chemical properties of CNTs or decrease their conductivity. While there have some disadvantages of CNT-modified electrode for sensing application, the continuous growing research interest in this field is contributing to overcome them. It is believed that the merits of CNT-based sensors will bring dramatic changes to future sensor industry.

ACKNOWLEDGMENTS

This work was supported by the Regional Innovation Center Program of the Ministry of Knowledge Economy through the Bio-Food & Drug Research Center at Konkuk University, Korea.

REFERENCES

1. Padigi, S.K.; Reddy, R.K.K.; Prasad, S. Carbon nanotube based aliphatic hydrocarbon sensor. *Biosens. Bioelectron* **2007**, *22*, 829–837.

2. Iijima, S. Helical microtubules of graphitic carbon. *Nature* **1991**, *354*, 56–58.

3. Sun, G.; Liu, S.; Hua, K.; Lv, X.; Huang, L.; Wang, Y. Electrochemical chlorine sensor with multi-walled carbon nanotubes as electrocatalysts. *Electrochem. Commun* **2007**, *9*, 2436–2440.

4. Zhu, L.; Yang, R.; Zhai, J.; Tian, C. Bienzymatic glucose biosensor based on co-immobilization of peroxidase and glucose oxidase on a carbon nanotubes electrode. *Biosens. Bioelectron* **2007**, *23*, 528–535.

5. Santhosh, P.; Manesh, K.M.; Gopalan, A.; Lee, K.P. Novel amperometric carbon monoxide sensor based on multi-wall carbon nanotubes grafted with polydiphenylamine—fabrication and performance. *Sens. Actuat. B-Chem* **2007**, *125*, 92–99.

6. Du, P.; Liu, S.; Wu, P.; Cai, C. Preparation and characterization of room temperature ionic liquid/single-walled carbon nanotube

nanocomposites and their application to the direct electrochemistry of heme-containing proteins/enzymes. *Electrochim. Acta* **2007**, *52*, 6534–6547.

7. Sato, N.; Okuma, H. Development of single-wall carbon nanotubes modified screen-printed electrode using a ferrocene-modified cationic surfactant for amperometric glucose biosensor applications. *Sens. Actuat. B-Chem* **2008**, *129*, 188–194.

8. Banks, C.E.; Compton, R.G. New electrodes for old: from carbon nanotubes to edge plane pyrolytic graphite. *Analyst* **2006**, *131*, 15–21.

9. Moore, R.R.; Banks, C.E.; Compton, R.G. Basal plane pyrolytic graphite modified electrodes: comparison of carbon nanotubes and graphite powder as electrocatalysts. *Anal. Chem* **2004**, *76*, 2677–2682.

10. Banks, C.E.; Moore, R.R.; Davies, T.J.; Compton, R.G. Investigation of modified basal plane pyrolytic graphite electrodes:definitive evidence for the electrocatalytic properties of the ends of carbon nanotubes.*Chem. Commun* **2004**, 1804–1805.

11. Banks, C.E.; Davies, T.J.; Wildgoose, G.G.; Compton, R.G. Electrocatalysis at graphite and carbon nanotube modified electrodes: edge-plane sites and tube ends are the reactive sites. *Chem. Commun* **2005**, 829–841.

12. Holloway, A.F.; Wildgoose, G.G.; Compton, R.G.; Shao, L.; Green, M.L.H. The influence of edge-plane defects and oxygen-containing surface groups on the voltammetry of acid-treated, annealed and "super-annealed" multiwalled carbon nanotubes. *J. Solid State Electrochem.* **2008**, *12*, 1337–1348.

13. Chou, A.; Böcking, T.; Singh, N.K.; Gooding, J.J. Demonstration of the importance of oxygenated species at the ends of carbon nanotubes for their favourable electrochemical properties. *Chem. Commun* **2005**, 842–844.

14. Liu, J.; Chou, A.; Rahmat, W.; Paddon-Row, M.N.; Gooding, J.J. Achieving direct electrical connection to glucose oxidase using aligned single walled carbon nanotube arrays. *Electroanal* **2005**, *17*, 38–46.

15. Gooding, J.J.; Chou, A.; Liu, J.; Losic, D.; Shapter, J.G.; Hibbert, D.B. The effects of the lengths and orientations of single-walled carbon

nanotubes on the electrochemistry of nanotube-modified electrodes.*Electrochem. Commun* **2007**, *9*, 1677–1683.

16. Chou, A.; Böcking, T.; Liu, R.; Singh, N.K.; Moran, G.; Gooding, J.J. Effect of dialysis on the electrochemical properties of acid-oxidized single-walled carbon nanotubes. *J. Phys. Chem. C* **2008**, *112*, 14131–14138.

17. Pumera, M. Carbon nanotubes contain residual metal catalyst nanoparticles even after washing with nitric acid at elevated temperature because these metal nanoparticles are sheathed by several graphene sheets.*Langmuir* **2007**, *23*, 6453–6458.

18. Kolodiazhnyi, T.; Pumera, M. Towards an ultrasensitive method for the determination of metal impurities in carbon nanotubes. *Small* **2008**, *9*, 1476–1484.

19. Gong, K.; Du, F.; Xia, Z.; Durstock, M.; Dai, L. Nitrogen-doped carbon nanotube arrays with high electrocatalytic activity for oxygen reduction. *Science* **2009**, *323*, 760–764.

20. Lawrence, N.S.; Deo, R.P.; Wang, J. Comparison of the electrochemical reactivity of electrodes modified with carbon nanotubes from different sources. *Electroanalysis* **2005**, *17*, 65–72.

21. Musameh, M.; Lawrence, N.S.; Wang, J. Electrochemical activation of carbon nanotubes. *Electrochem. Commun* **2005**, *7*, 14–18.

22. Rahman, M.A.; Kumar, P.; Park, D.S.; Shim, Y.B. Electrochemical sensors based on organic conjugated polymers. *Sensors* **2008**, *8*, 118–141.

23. Wang, H.S.; Li, T.H.; Jia, W.L.; Xu, H.Y. Highly selective and sensitive determination of dopamine using a Nafion/carbon nanotubes coated poly (3-methylthiophene) modified electrode. *Biosens. Bioelectron* **2006**, *22*, 664–669.

24. Zhang, Y.; Cai, Y.; Su, S. Determination of dopamine in the presence of ascorbic acid by poly (styrene sulfonic acid) sodium salt/single-wall carbon nanotube modified glassy carbon electrode. *Anal. Biochem***2006**, *350*, 285–291.

25. Yin, T.; Wei, W.; Zeng, J. Selective detection of dopamine in the presence of ascorbic acid by use of glassy-carbon electrodes modified with both polyaniline film and multi-walled carbon nanotubes with incorporated β-cyclodextrin. *Anal. Bioanal. Chem* **2006**, *386*, 2087–2094.

26.   Angeles, G.A.; López, B.P.; Pardave, M.P.; Silva, M.T.R.; Alegret, S.; Merkoci, A. Enhanced host–guest electrochemical recognition of dopamine using cyclodextrin in the presence of carbon nanotubes. *Carbon* **2008**, *46*, 898–906.

27.   Li, Y.; Wang, P.; Wang, L.; Lin, X. Overoxidized polypyrrole film directed single-walled carbon nanotubes immobilization on glassy carbon electrode and its sensing applications. *Biosens. Bioelectron* **2007**, *22*, 3120–3125.

28.   Liu, A.; Honma, I.; Zhou, H. Simultaneous voltammetric detection of dopamine and uric acid at their physiological level in the presence of ascorbic acid using poly (acrylic acid)-multiwalled carbon-nanotube composite-covered glassy-carbon electrode. *Biosens. Bioelectron* **2007**, *23*, 74–80.

29.   Yogeswaran, U.; Chen, S.M. Separation and concentration effect of f-MWCNTs on electrocatalytic responses of ascorbic acid, dopamine and uric acid at f-MWCNTs incorporated with poly (neutral red) composite films. *Electrochim. Acta* **2007**, *52*, 5985–5996.

30.   Yogeswaran, U.; Thiagarajan, S.; Chen, S.M. Nanocomposite of functionalized multiwall carbon nanotubes with nafion, nano platinum, and nano gold biosensing film for simultaneous determination of ascorbic acid, epinephrine, and uric acid. *Anal. Biochem* **2007**, *365*, 122–131.

31.   Yi, H.; Zheng, D.; Hu, C.; Hu, S. Functionalized multiwalled carbon nanotubes through in situ electropolymerization of brilliant cresyl blue for determination of epinephrine. *Electroanalysis* **2008**, *20*, 1143–1146.

32.   Valentini, F.; Palleschi, G.; Morales, E.L.; Orlanducci, S.; Tamburri, E.; Terranova, M.L. Functionalized single-walled carbon nanotubes modified microsensors for the selective response of epinephrine in presence of ascorbic acid. *Electroanal* **2007**, *19*, 859–869.

33.   Zhai, X.; Wei, W.; Zeng, J.; Gong, S.; Yin, J. Layer-by-layer assembled film based on chitosan/carbon nanotubes, and its application to electrocatalytic oxidation of NADH. *Microchim. Acta* **2006**, *154*, 315–320.

34.   Wang, Q.; Tang, H.; Xie, Q.; Tan, L.; Zhang, Y.; Li, B.; Yao, S. Room-temperature ionic liquids/multi-walled carbon nanotubes/chitosan

composite   electrode   for   electrochemical   analysis   of NADH. *Electrochim. Acta* **2007**, *52*, 6630–6637.

35.   Radoi, A.; Compagnone, D.; Valcarcel, M.A.; Placidi, P.; Materazzi, S.; Moscone, D.; Palleschi, G. Detection of NADH via electrocatalytic oxidation at single-walled carbon nanotubes modified with Variamine blue.*Electrochim. Acta* **2008**, *53*, 2161–2169.

36.   Liu, A.; Watanabe, T.; Honma, I.; Wang, J.; Zhou, H. Effect of solution pH and ionic strength on the stability of poly(acrylic acid)-encapsulated multiwalled carbon nanotubes aqueous dispersion and its application for NADH sensor. *Biosens. Bioelectron* **2006**, *22*, 694–699.

37.   Zeng, J.; Gao, X.; Wei, W.; Zhai, X.; Yin, J.; Wu, L.; Liu, X.; Liu, K.; Gong,   S.   Fabrication   of   carbon   nanotubes/poly(1,2-diaminobenzene)   nanoporous   composite   via   multipulse chronoamperometric electropolymerization process and its electrocatalytic property toward oxidation of NADH. *Sens. Actuat. B-Chem* **2007**, *120*, 595–602.

38.   Tu, X.; Xie, Q.; Huang, Z.; Yang, Q.; Yao, S. Synthesis and characterization   of   novel   quinone-amine   polymer/carbon nanotubes composite for sensitive electrocatalytic detection of NADH. *Electroanalysis* **2007**,*19*, 1815–1821.

39.   Agüí, L.; Farfal, C.P.; Sedeñno, P.Y.; Pingarrón, J.M. Poly-(3-methylthiophene)/carbon nanotubes hybrid composite-modified electrodes. *Electrochim. Acta* **2007**, *52*, 7946–7952.

40.   Tkac, J.; Ruzgas, T. Dispersion of single walled carbon nanotubes. Comparison of different dispersing strategies for preparation of modified   electrodes   toward   hydrogen   peroxide detection. *Electrochem. Commun* **2006**, *8*, 899–903.

41.   Sun, N.; Guan, L.; Shi, Z.; Li, N.; Gu, Z.; Zhu, Z.; Li, M.; Shao, Y. Ferrocene   peapod   modified   electrodes: preparation, characterization, and mediation of H_2O_2. *Anal. Chem* **2006**, *78*, 6050–6057.

42.   Yuan, S.; He, Q.; Yao, S.; Hu, S. Mercury-free detection of europium (III) at a glassy carbon electrode modified with carbon nanotubes by adsorptive stripping voltammetry. *Anal. Lett* **2006**, *39*, 373–385.

43. Suna, D.; Xie, X.; Cai, Y.; Zhang, H.; Wu, K. Voltammetric determination of Cd^{2+} based on the bifunctionality of single-walled carbon nanotubes-Nafion film. *Anal. Chim. Acta* **2007**, *581*, 27–31.

44. Profumo, A.; Fagnoni, M.; Merli, D.; Quartarone, E.; Protti, S.; Dondi, D.; Albini, A. Multiwalled carbon nanotube chemically modified gold electrode for inorganic as speciation and Bi(III) determination. *Anal. Chem* **2006**, *78*, 4194–4199.

45. Jo, S.; Jeong, H.; Bae, S.R.; Jeon, S. Modified platinum electrode with phytic acid and single-walled carbon nanotube: Application to the selective determination of dopamine in the presence of ascorbic and uric acids. *Microchem. J* **2008**, *88*, 1–6.

46. Li, Z.; Chen, J.; Pan, D.; Tao, W.; Nie, L.; Yao, S. A sensitive amperometric bromate sensor based on multi-walled carbon nanotubes/phosphomolybdic acid composite film. *Electrochim. Acta* **2006**, *51*, 4255–4261.

47. Wu, Y.; Ye, S.; Hu, S. Electrochemical study of lincomycin on a multi-wall carbon nanotubes modified glassy carbon electrode and its determination in tablets. *J. Pharm. Biomed. Anal* **2006**, *41*, 820–824.

48. Ming, L.; Xi, X.; Chen, T.; Liu, J. Electrochemical determination of trace sudan I contamination in chili powder at carbon nanotube modified electrodes. *Sensors* **2008**, *8*, 1890–1900.

49. Wang, F.; Chen, L.; Chen, X.; Hu, S. Studies on electrochemical behaviors of acyclovir and its voltammetric determination with nano-structured film electrode. *Anal. Chim. Acta* **2006**, *576*, 17–22.

50. Li, C. Voltammetric determination of 2-chlorophenol using a glassy carbon electrode coated with multi-wall carbon nanotube-dicetyl phosphate film. *Microchim. Acta* **2007**, *157*, 21–26.

51. Huang, K.J.; Luo, D.F.; Xie, W.Z.; Yu, Y.S. Sensitive voltammetric determination of tyrosine using multi-walled carbon nanotubes/4-aminobenzeresulfonic acid film-coated glassy carbon electrode. *Colloid Surf. B* **2008**, *61*, 176–181.

52. Wu, K.; Wang, H.; Chen, F.; Hu, S. Electrochemistry and voltammetry of procaine using a carbon nanotube film coated electrode. *Bioelectrochemistry* **2006**, *68*, 144–149.

53. Zhua, Y.; Zhang, Z.; Zhao, W.; Pang, D. Voltammetric behavior and determination of phenylephrine at a glassy carbon electrode

modified with multi-wall carbon nanotubes. *Sens. Actuat. B-Chem* **2006**, *119*, 308–314.

54.   Zeng, B.; Wei, S.; Xiao, F.; Zhao, F. Voltammetric behavior and determination of rutin at a single-walled carbon nanotubes modified gold electrode. *Sens. Actuat. B-Chem* **2006**, *115*, 240–246.

55.   He, J.L.; Yang, Y.; Yang, X.; Liu, Y.L.; Liu, Z.H.; Shen, G.L.; Yu, R.Q. β-Cyclodextrin incorporated carbon nanotube-modified electrode as an electrochemical sensor for rutin. *Sens. Actuat. B-Chem* **2006**, *114*, 94–100.

56.   Wang, G.; Hu, N.; Wang, W.; Li, P.; Gu, H.; Fang, B. Preparation of carbon nanotubes/neutral red composite film modified electrode and its catalysis on rutin. *Electroanalysis* **2007**, *19*, 2329–2334.

57.   Zare, H.R.; Sobhani, Z.; Ardakani, M.M. Electrocatalytic oxidation of hydroxylamine at a rutin multi-wall carbon nanotubes modified glassy carbon electrode: Improvement of the catalytic activity. *Sens. Actuat. B-Chem* **2007**, *126*, 641–647.

58.   Hrapovic, S.; Majid, E.; Liu, Y.; Male, K.; Luong, J.H.T. Metallic nanoparticle-carbon nanotube composites for electrochemical determination of explosive nitroaromatic compounds. *Anal. Chem* **2006**, *78*, 5504–5512.

59.   Xiao, F.; Zhao, F.; Li, J.; Yan, R.; Yu, J.; Zeng, B. Sensitive voltammetric determination of chloramphenicol by using single-wall carbon nanotube–gold nanoparticle–ionic liquid composite film modified glassy carbon electrodes. *Anal. Chim. Acta* **2007**, *596*, 79–85.

60.   Lin, X.; Li, Y. A sensitive determination of estrogens with a Pt nano-clusters/multi-walled carbon nanotubes modified glassy carbon electrode. *Biosens. Bioelectron* **2006**, *22*, 253–259.

61.   Yang, P.; Wei, W.; Tao, C. Determination of trace thiocyanate with nano-silver coated multi-walled carbon nanotubes modified glassy carbon electrode. *Anal. Chim. Acta* **2007**, *585*, 331–336.

62.   Liu, H.; Wang, G.; Chen, D.; Zhang, W.; Li, C.; Fang, B. Fabrication of polythionine/NPAu/MWNTs modified electrode for simultaneous determination of adenine and guanine in DNA. *Sens. Actuat. B-Chem* **2008**, *128*, 414–421.

63.   Zhuang, Q.; Chen, J.; Chen, J.; Lin, X. Electrocatalytical properties of bergenin on a multi-wall carbon nanotubes modified carbon paste

electrode and its determination in tablets. *Sens. Actuat. B-Chem* **2008**, *128*, 500–506.

64. Shahrokhian, S.; Fotouhi, L. Carbon paste electrode incorporating multi-walled carbon nanotube/cobalt salophen for sensitive voltammetric determination of tryptophan. *Sens. Actuat. B-Chem* **2007**, *123*, 942–949.

65. Shahrokhian, S.; Mehrjardi, H.R.Z. Simultaneous voltammetric determination of uric acid and ascorbic acid using a carbon-paste electrode modified with multi-walled carbon nanotubes/nafion and cobalt (II)-nitrosalophen. *Electroanalysis* **2007**, *19*, 2234–2242.

66. Shahrokhian, S.; Mehrjardi, H.R.Z. Application of thionine-nafion supported on multi-walled carbon nanotube for preparation of a modified electrode in simultaneous voltammetric detection of dopamine and ascorbic acid. *Electrochim. Acta* **2007**, *52*, 6310–6317.

67. Fan, S.; Xiao, F.; Liu, L.; Zhao, F.; Zeng, B. Sensitive voltammetric response of methylparathion on single-walled carbon nanotube paste coated electrodes using ionic liquid as binder. *Sens. Actuat. B-Chem* **2008**, *132*, 34–39.

68. Xiao, F.; Ruan, C.; Li, J.; Liu, L.; Zhao, F.; Zeng, B. Voltammetric determination of xanthine with a single-walled carbon nanotube-ionic liquid paste modified glassy carbon electrode. *Electroanalysis* **2008**, *20*, 361–366.

69. Yan, Q.; Zhao, F.; Li, G.; Zeng, B. Voltammetric determination of uric acid with a glassy carbon electrode coated by paste of multiwalled carbon nanotubes and ionic liquid. *Electroanalysis* **2006**, *18*, 1075–1080.

70. Abbaspour, A.; Mirzajani, R. Electrochemical monitoring of piroxicam in different pharmaceutical forms with multi-walled carbon nanotubes paste electrode. *J. Pharm. Biomed. Anal* **2007**, *44*, 41–48.

71. Li, Z.; Junfeng, S. Voltammetric behavior of urapidil and its determination at multi-wall carbon nanotube paste electrode. *Talanta* **2007**, *73*, 943–947.

72. Xiao, P.; Zhao, F.; Zeng, B. Voltammetric determination of quercetin at a multi-walled carbon nanotubes paste electrode. *Microchem. J* **2007**, *85*, 244–249.

73. Yang, P.; Wei1, W.; Yang, L. Simultaneous voltammetric determination of dihydroxybenzene isomers using a poly (acid chrome blue K)/carbon nanotube composite electrode. *Microchim. Acta* **2007**, *157*, 229–235.

74. Wang, C.; Mao, Y.; Wang, D.; Yang, G.; Qu, Q.; Hu, X. Voltammetric determination of terbinafine in biological fluid at glassy carbon electrode modified by cysteic acid/carbon nanotubes composite film.*Bioelectrochemistry* **2008**, *72*, 107–115.

75. Zhu, L.; Tian, C.; Zhai, J.; Yang, R. Sol–gel derived carbon nanotubes ceramic composite electrodes for electrochemical sensing. *Sens. Actuat. B-Chem* **2007**, *125*, 254–261.

76. Wang, J.; Tangkuaram, T.; Loyprasert, S.; Alvarez, T.V.; Veerasai, W.; Kanatharana, P.; Thavarungkul, P. Electrocatalytic detection of insulin at RuOx/carbon nanotube-modified carbon electrodes. *Anal. Chim. Acta***2007**, *581*, 1–6.

77. Snider, R. M.; Ciobanu, M.; Rue, A.E.; Cliffel, D.E. A multiwalled carbon nanotube/dihydropyran composite film electrode for insulin detection in a microphysiometer chamber. *Anal. Chim. Acta* **2008**, *609*, 44–52.

78. Vega, D.; Agüí, L.; Cortés, A.G.; Sedeño, P.Y.; Pingarrón, J. M. Voltammetry and amperometric detection of tetracyclines at multi-wall carbon nanotube modified electrodes. *Anal. Bioanal. Chem* **2007**, *389*, 951–958.

79. Rezaei, B.; Zare, S.Z.M. Modified glassy carbon electrode with multiwall carbon nanotubes as a voltammetric sensor for determination of noscapine in biological and pharmaceutical samples. *Sens. Actuat. B-Chem* **2008**, *134*, 292–299.

80. Rezaei, B.; Damiri, S. Voltammetric behavior of multi-walled carbon nanotubes modified electrode-hexacyanoferrate (II) electrocatalyst system as a sensor for determination of captopril. *Sens. Actuat. B-Chem***2008**, *134*, 324–331.

81. Buratti, S.; Brunetti, B.; Mannino, S. Amperometric detection of carbohydrates and thiols by using a glassy carbon electrode coated with Co oxide/multi-wall carbon nanotubes catalytic system. *Talanta* **2008**, *76*, 454–457.

82. Wang, J. Electrochemical glucose biosensors. *Chem. Rev* **2008**, *108*, 814–825.

83. Heller, A.; Feldman, B. Electrochemical glucose sensors and their applications in diabetes management.*Chem. Rev* **2008**, *108*, 2482–2505.

84. Tsai, Y.C.; Li, S.C.; Liao, S.W. Electrodeposition of polypyrrole–multiwalled carbon nanotube–glucose oxidase nanobiocomposite film for the detection of glucose. *Biosens. Bioelectron* **2006**, *22*, 495–500.

85. Huang, J.; Yang, Y.; Shi, H.; Song, Z.; Zhao, Z.; Anzai, J.; Osa, T.; Chen, Q. Multi-walled carbon nanotubes-based glucose biosensor prepared by a layer-by-layer technique. *Mater. Sci. Eng. C* **2006**, *26*, 113–117.

86. Liu, G.; Lin, Y. Amperometric glucose biosensor based on self-assembling glucose oxidase on carbon nanotubes. *Electrochem. Commun* **2006**, *8*, 251–256.

87. Zhao, H.; Ju, H. Multilayer membranes for glucose biosensing via layer-by-layer assembly of multiwall carbon nanotubes and glucose oxidase. *Anal. Biochem* **2006**, *350*, 138–144.

88. Liu, Y.; Wu, S.; Ju, H.; Xu, Li. Amperometric glucose biosensing of gold nanoparticles and carbon nanotube multilayer membranes. *Electroanalysis* **2007**, *19*, 986–992.

89. Xu, L.; Zhu, Y.; Tang, L.; Yang, X.; Li, C. Biosensor based on self-assembling glucose oxidase and dendrimer-encapsulated Pt nanoparticles on carbon nanotubes for glucose detection. *Electroanalysis* **2007**,*19*, 717–722.

90. Shirsat, M.D.; Too, C.O.; Wallace, G.G. Amperometric glucose biosensor on layer by layer assembled carbon nanotube and polypyrrole multilayer film. *Electroanalysis* **2008**, *20*, 150–156.

91. Qu, F.; Yang, M.; Chen, J.; Shen, G.; Yu, R. Amperometric biosensors for glucose based on layer-by-layer assembled functionalized carbon nanotube and poly (neutral red) multilayer film. *Anal. Lett* **2006**, *39*, 1785–1799.

92. Yao, Y.L.; Shiu, K.K. Low potential detection of glucose at carbon nanotube modified glassy carbon electrode with electropolymerized poly(toluidine blue O) film. *Anal. Chim. Acta* **2007**, *53*, 278–284.

93. Zhu, L.; Zhai, J.; Guo, Y.; Tian, C.; Yang, R. Amperometric glucose biosensors based on integration of glucose oxidase onto prussian

blue/carbon nanotubes nanocomposite electrodes. *Electroanalysis* **2006**, *18*, 1842–1846.

94.   Zoua, Y.; Xiang, C.; Suna, L.; Xu, F. Amperometric glucose biosensor prepared with biocompatible material and carbon nanotube by layer-by-layer self-assembly technique. *Anal. Chim. Acta* **2008**, *53*, 4089–4095.

95.   Zeng, J.; Wei, W.; Liu, X.; Wang, Y.; Luo, G. A simple method to fabricate a Prussian Blue nanoparticles/carbon nanotubes/poly (1,2-diaminobenzene) based glucose biosensor. *Microchim. Acta* **2008**,*160*, 261–267.

96.   Manesh, K.M.; Kim, H.T.; Santhosh, P.; Gopalan, A.I.; Lee, K. A novel glucose biosensor based on immobilization of glucose oxidase into multiwall carbon nanotubes–polyelectrolyte-loaded electrospun nanofibrous membrane. *Biosens. Bioelectron* **2008**, *23*, 771–779.

97.   Choi, H.N.; Han, J.H.; Park, J.A.; Lee, J.M.; Lee, W.Y. Amperometric glucose biosensor based on glucose oxidase encapsulated in carbon nanotube-titania-Nafion composite film on platinized glassy carbon electrode. *Electroanalysis* **2007**, *19*, 1757–1763.

98.   Manso, J.; Mena, M.L.; Sedeño, P.Y.; Pingarrón, J. Electrochemical biosensors based on colloidal gold–carbon nanotubes composite electrodes. *J. Electroanal. Chem* **2007**, *603*, 1–7.

99.   Manso, J.; Mena, M.L.; Sedeño, P.Y.; Pingarrón, J.M. Alcohol dehydrogenase amperometric biosensor based on a colloidal gold–carbon nanotubes composite electrode. *Electrochim. Acta* **2008**, *53*, 4007–4012.

100. Chu, X.; Duan, D.; Shen, G.; Yu, R. Amperometric glucose biosensor based on electrodeposition of platinum nanoparticles onto covalently immobilized carbon nanotube electrode. *Talanta* **2007**, *71*, 2040–2047.

101. Zou, Y.; Xiang, C.; Suna, L.X.; Xu, F. Glucose biosensor based on electrodeposition of platinum nanoparticles onto carbon nanotubes and immobilizing enzyme with chitosan-SiO2 sol–gel. *Biosens. Bioelectron* **2008**, *23*, 1010–1016.

102. Zhao, K.; Zhuang, S.; Chang, Z.; Songm, H.; Dai, L.; He, P.; Fang, Y. Amperometric glucose biosensor based on platinum nanoparticles combined aligned carbon nanotubes electrode. *Electroanalysis* **2007**, *19*, 1069–1074.

103. Kang, X.; Mai, Z.; Zou, X.; Cai, P.; Mo, J. A novel glucose biosensor based on immobilization of glucose oxidase in chitosan on a glassy carbon electrode modified with gold–platinum alloy nanoparticles/multiwall carbon nanotubes. *Anal. Biochem* **2007**, *369*, 71–79.

104. Yang, M.; Yang, Y.; Liu, Y.; Shen, G. Yu, R. Platinum nanoparticles-doped sol–gel/carbon nanotubes composite electrochemical sensors and biosensors. *Biosens. Bioelectron* **2006**, *21*, 1125–1131.

105. Luque, G.L.; Ferreyra, N.F.; Rivas, G.A. Glucose biosensor based on the use of a carbon nanotube paste electrode modified with metallic particles. *Microchim. Acta* **2006**, *152*, 277–283.

106. Yao, Y.; Shiu, K.K. Electron-transfer properties of different carbon nanotube materials, and their use in glucose biosensors. *Anal. Bioanal. Chem* **2007**, *387*, 303–309.

107. Jia, J.; Guan, W.; Sim, M.; Li, Y.; Li, H. Carbon nanotubes based glucose needle-type Biosensor. *Sensors***2008**, *8*, 1712–1718.

108. Zhu, L.; Yang, R.; Zhai, J.; Tian, C. Bienzymatic glucose biosensor based on co-immobilization of peroxidase and glucose oxidase on a carbon nanotubes electrode. *Biosens. Bioelectron* **2007**, *23*, 528–535.

109. Chen, L.; Lu, G. Novel amperometric biosensor based on composite film assembled by polyelectrolyte-surfactant polymer, carbon nanotubes and hemoglobin. *Biosens. Bioelectron* **2007**, *121*, 423–429.

110. Chen, S.; Yuan, R.; Chai, Y.; Zhang, L.; Wang, N.; Li, X. Amperometric third-generation hydrogen peroxide biosensor based on the immobilization of hemoglobin on multiwall carbon nanotubes and gold colloidal nanoparticles. *Biosens. Bioelectron* **2007**, *22*, 1268–1274.

111. Tripathi, V.S.; Kandimalla, V.B.; Ju, H. Amperometric biosensor for hydrogen peroxide based on ferrocene-bovine serum albumin and multiwall carbon nanotube modified ormosil composite. *Biosens. Bioelectron***2006**, *21*, 1529–1535.

112. Luo, X.; Killard, A.J.; Morrin, A.; Smyth, M.R. Enhancement of a conducting polymer-based biosensor using carbon nanotube-doped polyaniline. *Anal. Chim. Acta* **2006**, *575*, 39–44.

113. Liu, Y.; Lei, J.; Ju, H. Amperometric sensor for hydrogen peroxide based on electric wire composed of horseradish peroxidase and toluidine blue-multiwalled carbon nanotubes nanocomposite. *Talanta* **2008**, *74*, 965–970.

114. Qian, L.; Yang, X. Composite film of carbon nanotubes and chitosan for preparation of amperometric hydrogen peroxide biosensor. *Talanta* **2006**, *68*, 721–727.

115. Sánchez, S.; Pumera, M.; Cabruja, E.; Fàbregas, E. Carbon nanotube/polysulfone composite screen-printed electrochemical enzyme biosensors. *Analyst* **2007**, *132*, 142–147.

116. Qu, S.; Wang, J.; Kong, J.; Yang, P.; Chen, G. Magnetic loading of carbon nanotube/nano-Fe3O4 composite for electrochemical sensing. *Talanta* **2007**, *71*, 1096–1102.

117. Choi, H. N.; Lyu, Y.K.; Han, J. H.; Lee, W.Y. Amperometric ethanol biosensor based on carbon nanotubes dispersed in sol – gel-derived titania – nafion composite film. *Electroanalysis* **2007**, *19*, 1524–1530.

118. Santos, A.S.; Pereira, A.C.; Durán, N.; Kubota, L.T. Amperometric biosensor for ethanol based on co-immobilization of alcohol dehydrogenase and Meldola's Blue on multi-wall carbon nanotube. *Electrochim. Acta* **2006**, *52*, 215–220.

119. Du, P.; Liu, S.; Wu, P.; Cai, C. Single-walled carbon nanotubes functionalized with poly(nile blue A) and their application to dehydrogenase-based biosensors. *Electrochim. Acta* **2007**, *53*, 1811–1823.

120. Liu, S.; Cai, C. Immobilization and characterization of alcohol dehydrogenase on single-walled carbon nanotubes and its application in sensing ethanol. *J. Electroanal. Chem* **2007**, *602*, 103–114.

121. Yang, M.; Yang, Y.; Yang, H.; Shen, G.; Yu, R. Layer-by-layer self-assembled multilayer films of carbon nanotubes and platinum nanoparticles withpolyelectrolyte for the fabrication of biosensors. *Biomaterials* **2006**, *27*, 246–255.

122. Tang, L.; Zhu, Y.; Xu, L.; Yang, X.; Li, C. Amperometric glutamate biosensor based on self-assembling glutamate dehydrogenase and dendrimer-encapsulated platinum nanoparticles onto carbon nanotubes. *Talanta* **2007**, *73*, 438–443.

123. Wang, J.; Liu, G.; Lin, Y. Amperometric choline biosensor fabricated through electrostatic assembly of bienzyme/polyelectrolyte hybrid layers on carbon nanotubes. *Analyst* **2006**, *131*, 477–483.

124. Song, Z.; Huang, J.D.; Wu, B.Y.; Shi, H.B.; Anzai, J.I.; Chen, Q. Amperometric aqueous sol–gel biosensor for low-potential stable choline detection at multi-wall carbon nanotube modified platinum electrode. *Sens. Actuat. B-Chem* **2006**, *115*, 626–633.

125. Du, D.; Huang, X.; Cai, J.; Zhang, A.; Ding, J.; Chen, S. An amperometric acetylthiocholine sensor based on immobilization of acetylcholinesterase on a multiwall carbon nanotube–cross-linked chitosan composite.*Anal. Bioanal. Chem* **2007**, *387*, 1059–1065.

126. Lee, Y.J.; Lyu, Y.K.; Choi, H.N.; Lee, W.Y. Amperometric tyrosinase biosensor based on carbon nanotube – titania – nafion composite film. *Electroanalysis* **2007**, *19*, 1048–1054.

127. Santos, A.S.; Pereira, A.C.; Sotomayor, M.D.P.T.; Tarley, C.R.T.; Durán, N.; Kubota, L.T. Determination of phenolic compounds based on co-immobilization of methylene blue and HRP on multi-wall carbon nanotubes. *Electroanalysis* **2007**, *19*, 549–554.

128. López, P.B.; Sola, J.; Alegret, S.; MerkoÅi, A. A carbon nanotube PVC based matrix modified with glutaraldehyde suitable for biosensor applications. *Electroanalysis* **2008**, *20*, 603–610.

129. Male, K.B.; Hrapovic, S.; Santini, J.M.; Luong, J.H.T. Biosensor for arsenite using arsenite oxidase and multiwalled carbon nanotube modified electrodes. *Anal. Chem* **2007**, *79*, 7831–7837.

130. Mita, D.G.; Attanasio, A.; Arduini, F.; Diano, N.; Grano, V.; Bencivenga, U.; Rossi, S.; Aminee, A.; Moscone, D. Enzymatic determination of BPA by means of tyrosinase immobilized on different carbon carriers.*Biosens. Bioelectron* **2007**, *23*, 60–65.

131. Liu, G.; Lin, Y. Biosensor based on self-assembling acetylcholinesterase on carbon nanotubes for flow injection/amperometric detection of organophosphate pesticides and nerve Agents. *Anal. Chem* **2006**, *78*, 835–843.

132. Rahman, M.M.; Shiddiky, M.J.A.; Rahman, M.A.; Shim, Y.B. A lactate biosensor based on lactate dehydrogenase/nictotinamide adenine dinucleotide (oxidized form) immobilized on a conducting polymer/multiwall carbon nanotube composite film. *Anal. Biochem* **2009**, *384*, 159–165.

133.   Williams, K.A.; Veenhuizen, P.T.M.; Torre, B.G.; Eritja, R.; Dekker, C. Carbon nanotubes with DNA recognition. *Nature* **2002**, *420*, 761.

134.   Baker, S.E.; Cai, W.; Lasseter, T.L.; Weidkamp, K.P.; Hamers, R.J. Covalently bonded adducts of deoxyribonucleic acid (DNA) oligonucleotides with single-wall carbon nanotubes: synthesis and hybridization. *Nano Lett* **2002**, *2*, 1413–1417.

135.   Yang, Y.; Wang, Z.; Yang, M.; Li, J.; Zheng, F.; Shen, G.; Yu, R. Electrical detection of deoxyribonucleic acid hybridization based on carbon-nanotubes/nano zirconium dioxide/chitosan-modified electrodes. *Anal. Chim. Acta* **2007**, *584*, 268–274.

136.   Zhang, W.; Yang, T.; Huang, D. M.; Jiao, K. Electrochemical sensing of DNA immobilization and hybridization based on carbon nanotubes/nano zinc oxide/chitosan composite film. *Chin. Chem. Lett* **2008**,*19*, 589–591.

137.   Ma, H.; Zhang, L.; Pan, Y.; Zhang, K.; Zhang, Y. A Novel electrochemical DNA biosensor fabricated with layer-by-layer covalent attachment of multiwalled carbon nanotubes and gold nanoparticles. *Electroanalysis***2008**, *20*, 1220–1226.

138.   Niu, S.; Zhao, M.; Hu, L.; Zhang, S. Carbon nanotube-enhanced DNA biosensor for DNA hybridization detection using rutin-Mn as electrochemical indicator. *Sens. Actuat. B-Chem* **2008**, *135*, 200–205.

139.   Erdem, A.; Papakonstantinou, P.; Murphy, H. Direct DNA hybridization at disposable graphite electrodes modified with carbon nanotubes. *Anal. Chem* **2006**, *78*, 6656–6659.

140.   Xu, Y.; Ye, X.; Yang, L.; He, P.; Fang, Y. Impedance DNA Biosensor using electropolymerized polypyrrole/multiwalled carbon nanotubes modified electrode. *Electroanalysis* **2006**, *18*, 1471–1478.

141.   Qi, H.; Li, X.; Chen; Pei Zhang, C. Electrochemical detection of DNA hybridization based on polypyrrole/ss-DNA/multi-wall carbon nanotubes paste electrode. *Talanta* **2007**, *72*, 1030–1035.

142.   Chang, Z.; Fan, H.; Zhao, K.; Chen, M.; He, P.; Fang, Y. Electrochemical DNA biosensors based on palladium nanoparticles combined with carbon nanotubes. *Electroanalysis* **2008**, *20*, 131–136.

143. Yun, Y.H.; Bange, A.; Heineman, W.R.; Halsall, H.B.; Shanov, V.N.; Dong, Z.; Pixley, S.; Behbehani, M.; Jazieh, A.; Tu, Y.; Wong, D.K.Y.; Bhattacharya, A.; Schulz, M.J. A nanotube array immunosensor for direct electrochemical detection of antigen–antibody binding. *Sens. Actuat. B-Chem* **2007**, *123*, 177–182.

144. Okuno, J.; Maehashi, K.; Kerman, K.; Takamura, Y.; Matsumoto, K.; Tamiya, E. Label-free immunosensor for prostate-specific antigen based on single-walled carbon nanotube array-modified microelectrodes.*Biosens. Bioelectron* **2007**, *22*, 2377–2381.

145. Yu, X.; Munge, B.; Patel, V.; Jensen, G.; Bhirde, A.; Gong, J.D.; Kim, S.N.; Gillespie, J.; Gutkind, J.S.; Papadimitrakopoulos, F.; Rusling, J.F. Carbon nanotube amplification strategies for highly sensitive immunodetection of cancer Biomarkers. *J. Am. Chem. Soc* **2006**, *128*, 11199–11205.

146. Cataldo, V.; Vaze, A.; Rusling, J.F. Improved detection limit and stability of amperometric carbon nanotube-based immunosensors by crosslinking antibodies with polylysine. *Electroanalysis* **2008**, *20*, 115–122.

147. Viswanathan, S.; Wu, L.; Huang, M.R.; Ho, J.A. Electrochemical immunosensor for cholera toxin using liposomes and poly (3,4-ethylenedioxythiophene)-coated carbon nanotubes. *Anal. Chem* **2006**, *78*, 1115–1121.

148. Sánchez, S.; Pumera, M.; Fàbregas, E. Carbon nanotube/polysulfone screen-printed electrochemical immunosensor. *Biosens. Bioelectron* **2007**, *23*, 332–340.

149. Li, N.; Yuan, R.; Chai, Y.; Chen, S.; An, H. Sensitive immunoassay of human chorionic gonadotrophin based on multi-walled carbon nanotube–chitosan matrix. *Bioprocess Biosyst. Eng* **2008**, *31*, 551–558.

150. Ou, C.; Yuan, R.; Chai, Y.; Tang, M.; Chai, R.; He, X. A novel amperometric immunosensor based on layer-by-layer assembly of gold nanoparticles–multi-walled carbon nanotubes-thionine multilayer films on polyelectrolyte surface. *Anal. Chim. Acta* **2007**, *603*, 205–213.

151. Aziz, M.A.; Park, S.; Jon, S.; Yang, H. Amperometric immunosensing using an indium tin oxide electrode modified with multi-walled

carbon nanotube and poly (ethylene glycol)–silane copolymer. *Chem. Commun***2007**, *25*, 2610–2612.

152.   Aziz, M.A.; Yang, H. Electrochemical immunosensor using the modification of an amine-functionalized indium tin oxide electrode with carboxylated single-walled carbon nanotubes. *Bull. Kor. Chem. Soc* **2007**, *28*, 1171–1174.

Films of Carbon Nanomaterials for Transparent Conductors

Xinning Ho and Jun Wei

ABSTRACT

The demand for transparent conductors is expected to grow rapidly as electronic devices, such as touch screens, displays, solid state lighting and photovoltaics become ubiquitous in our lives. Doped metal oxides, especially indium tin oxide, are the commonly used materials for transparent conductors. As there are some drawbacks to this class of materials, exploration of alternative materials has been conducted. There is an interest in films of carbon nanomaterials such as, carbon nanotubes and graphene as they exhibit outstanding properties. This article reviews the synthesis and assembly of these films and their post-treatment. These processes determine the film performance and understanding of this platform will be useful for future work to improve the film performance.

Keywords: transparent conductor; graphene; carbon nanotube; hybrid film; percolation; transfer printing; doping; grain boundaries; contact resistance

1. INTRODUCTION

As electronic devices, such as touch screens, displays, solid state lighting and photovoltaic devices become more prevalent in our lives, the demand for transparent conductors increases. Doped metal oxides, especially indium tin oxide (ITO) [1,2] are often used as transparent conductors. They exhibit low electrical sheet resistance (<10 Ω/sq) and high optical transparency (>80%).

The production of ITO films can also be easily scaled up. However, the material suffers several drawbacks. As ITO is very brittle and can fracture at low strains [3], it cannot be integrated into flexible devices, which is an area many semiconductor giants are looking to develop. Furthermore, due to the increasing scarcity of indium, a component material of ITO, the production of ITO may become very expensive. Hence, there is an interest in the exploration of alternative materials. These materials include conducting polymers [4,5,6], metal nanowires [7,8,9,10], thin metal films [11,12] and carbon nanomaterials [13,14,15,16,17,18,19,20,21,22,23].

Conducting polymers have exhibited good electrical, optical and mechanical properties [4]. However, they suffer from electrical instability. Exposure to environmental elements like humidity, high temperature or UV light deteriorates the electrical conductivity [5,6]. Alternative materials based on metals also face similar challenges. Although metal nanowires and thin metal films display superb electrical conductivity intrinsically, they oxidize easily and their electrical conductivity degrades accordingly [7,8,9,10,11,12]. Films based on carbon nanomaterials, such as carbon nanotube (CNT) and graphene have been of particular interest due to their good electrical, optical and mechanical properties, as well as good chemical stability. They have been found to exhibit low sheet resistance, high optical transparency, good flexibility and stability over time. Hence, they are appealing in novel applications requiring flexibility. This article reviews the synthesis, assembly and post-treatment of carbon nanotube (CNT), graphene and carbon based hybrid films and their impact on the film performance which can provide insights on opportunities for future work to improve the film performance.

2. CARBON NANOTUBE (CNT) FILMS

Carbon nanotube (CNT), a one-dimensional material, is extensively studied by various groups due to its attractive properties. CNT exhibits low electrical resistivity [24,25] and high current carrying capability [26], making it an ideal material for electrical conductors. A nanometer-sized CNT is also extremely small, which renders it transparent. Hence, a thin film of CNTs has emerged as a promising material for transparent conductors [13,14,15,16,17,18]. Besides good electrical conductivity and transparency, CNTs demonstrate exceptional mechanical properties. The fracture strain is up to 30% [27], which makes it ideal for use in flexible electronics. Despite the impressive properties exhibited by individual CNT, the use of an individual CNT is not practical for real world applications, which often require higher current output. The obvious solution to which is the use of a CNT film. However, such films suffer from poorer electrical conductivity and transparency than individual CNTs. In this section, we examine the factors, which degrade the

CNT film performance to provide insights for improving the film properties as a transparent conductor.

2.1. Properties of Carbon Nanotube (CNT) Thin Films

Various techniques have been developed to synthesize CNTs. Some common techniques include arc discharge [28], laser ablation [29] and chemical vapor deposition [30]. CNTs synthesized via arc discharge and laser ablation are often dispersed in solution to form CNT ink, which can be deposited on various substrates by different methods to form CNT network films. Solution processed CNT films are very attractive for large area and low cost commercial applications.

Carbon nanotubes (CNTs) can be dispersed in solution via three main ways: dispersion in organic solvents [31], dispersion in aqueous media via dispersing agents like surfactants [32,33,34,35] and dispersion in solution by functionalizing the CNTs [36,37]. Direct dispersion of CNTs in organic solvent is a simple and straightforward method. However, CNTs can only be dispersed in low concentration in organic solvents, which is not practical for commercial applications. Use of dispersing agents like amphiphilic surfactants, assists the dispersion of hydrophobic CNTs in aqueous media like water. The hydrophobic interaction between the tails of the amphiphilic surfactants and the CNTs is very strong. A monolayer of surfactants is adsorbed on the CNT walls and a stable carbon nanotube-surfactant monolayer micelle is formed [32]. This allows CNTs to be dispersed in higher concentration in aqueous media. The presence of insulating surfactants in deposited CNT films formed from such solution decreases the electrical conductivity of the film. Hence, removal of surfactants after deposition is essential [38]. The last dispersion technique of functionalizing CNTs facilitates the dispersion of CNTs in solution by increasing the attraction between the CNTs and the solvent. Acid treatment of CNTs introduces carboxylic acid group (COOH) to the CNTs, which assists the dispersion [36]. However, excessive functionalization also introduces defects on the CNT structure, which degrades the electrical conductivity [37].

Carbon nanotubes (CNTs) are often dispersed in solution by ultrasonication. This step is essential as it breaks down the CNT aggregates to disperse them in the solvent. As the process is very harsh, defects are introduced, which can degrade the electrical conductivity of the CNT [39]. Besides introducing defects, sonication also shortens the tube length [40], which has adverse effects on the electrical conductivity of a CNT film, which will be discussed in detail in a subsequent section. The presence of impurities from the synthesis process, such as catalyst particles and amorphous carbon also degrades the electrical properties of the CNT film formed from solution. Parts a and b of Figure 1 show scanning electron microscope (SEM) images of "clean" and "dirty" network CNT films formed from solution respectively. The "clean" film has a relatively lower concentration of impurities on the film while huge particles are visibly present on the "dirty" film over large areas.

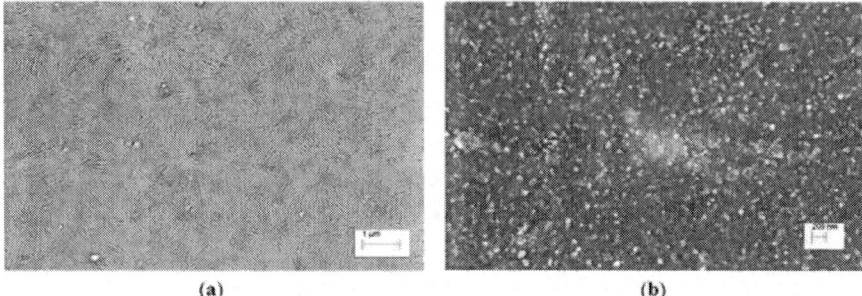

(a)   (b)

Figure 1. Scanning electron microscope (SEM) images of (**a**) "clean"; and (**b**) "dirty" carbon nanotube (CNT) network films formed from solution. Many impurity particles which are present in the "dirty" CNT network films are absent in the "clean" CNT network films.

Carbon nanotube (CNT) films formed directly via chemical vapor deposition (CVD) [41,42] generally exhibit higher structural perfection, longer tube lengths and higher purity than solution processed CNT films. Hence, CVD grown films tend to be of higher quality. However, solution processed films can potentially be processed by low cost and large scale production so more efforts have been focused on solution processed films for commercial applications.

Another important factor that determines the carbon nanotube (CNT) film property is the characteristics of individual CNTs. Synthesized CNTs are not uniform in conductivity type, length and diameter. Longer and larger diameter metallic CNTs are more desirable for transparent conductor applications.

The lower resistivity of metallic carbon nanotubes (CNTs) relative to semiconducting CNTs is attributed to the longer mean free path in metallic CNTs. Ballistic conduction in metallic CNTs can span up to micrometer range but semiconducting CNTs have a series of barriers to conduction along their lengths [43,44]. Resistivity of metallic CNTs have been found experimentally to be between 6 and 30 kΩ/µm [45,46,47] while semiconducting CNTs have higher resistivity, which is dependent on the gate voltage when used as the semiconducting channel in a transistor [45]. In fact, films based on solely metallic CNTs have demonstrated lower sheet resistance at a fixed transparency than films based on a mixture of metallic and semiconducting

CNTs (Figure 2) [48]. It is also evident from Figure 2 that a CNT film with lower sheet resistance has lower transparency. The transparency of CNT films is dominated by the absorption of CNTs in the film. Hence, when a thicker CNT film (with higher density of CNTs) is prepared, there is more absorption, which results in a lower transparency. However, the higher density of CNTs in the film also results in an improvement in the electrical conductivity by increasing the number of electrical pathways. Therefore, control of the density of CNTs in the film is essential for optimizing the sheet resistance and transparency of the films.

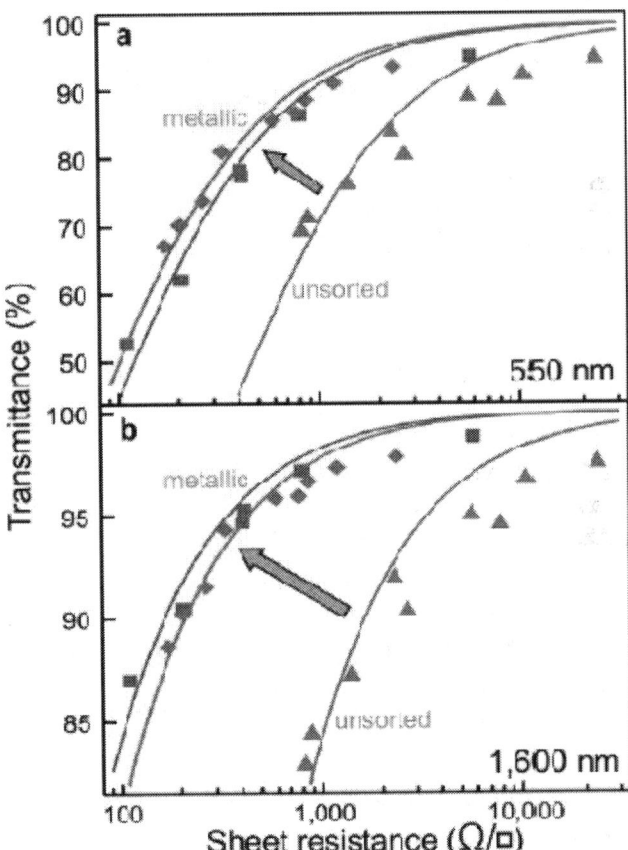

Figure 2. Transmittance versus sheet resistance of transparent conductive films generated from CNTs at (**a**) 550 nm; and (**b**) 1600 nm wavelengths. The materials used were metallic CNTs with principal diameters of 0.9 nm (red diamond symbols), 1.0 nm (blue square symbols) and a mixture of metallic and semiconducting CNTs (gray triangle symbols). Reprinted with permission from [48]. Copyright (2008) by the American Chemical Society.

Besides the lower sheet resistivity of metallic CNTs, the improvement in electrical conductivity of a film based on metallic CNTs only is attributed to the lower junction resistance between two metallic CNTs compared to junction resistance between one metallic CNT and one semiconducting CNT [49]. The junction resistance between two metallic CNTs or two semiconducting CNTs is low as there is a finite density of states at the junction for tunneling on either side of the junction. On the other hand, the junction resistance between a metallic CNT and semiconducting CNT is high as a Schottky barrier forms due to charge transfer from the metallic CNT to the semiconducting CNT [49].

Hence, sorting of carbon nanotubes (CNTs) or selective synthesis of metallic CNTs is very crucial. Various groups have explored different means of sorting or selective synthesis, such as selective chemical functionalization [50,51], selective electrical breakdown [52], density differentiation [53,54] and dielectrophoresis [55] with varying degrees of success. However, these methods are tedious and the yield is low, making it unsuitable for low cost commercial applications.

Larger diameter CNTs are preferred for transparent conductors because they carry more current than smaller diameter CNTs [56,57]. In fact, the peak mobility of CNT is found to scale with the square of the CNT diameter and the maximum conductance scales linearly with the CNT diameter [56]. Various groups have explored means to grow CNTs with selective CNT diameters [48,53,58,59,60,61]. Hersam group has been successful in achieving monodispersed CNTs via density differentiation [48]. However, such CNT films are colored and non-ideal as transparent conductors because CNTs with a monodisperse bandgap have a narrow range of absorption peaks.

Another important factor that influences the conductivity of the CNT film is the tube length of the CNTs [62,63]. The conductivity of CNT networks, σ, varies as $\sigma \sim L_{av}^{1.46}$ [62] (Figure 3). This is expected as the junction resistance between CNTs is higher than the intrinsic tube resistance of the CNTs. The resistance between two metallic CNTs or two semiconducting CNTs is between 200 and 400 kΩ and about 100 times smaller than the junction resistance between a metallic CNT and a semiconducting CNT. On the other hand, the tube resistivity of CNTs is between 6 and 30 kΩ/μm. As the tube length in the network approaches 20 μm, the intrinsic tube resistance is comparable to the junction resistance. Hence, beyond 20 μm, there is limited improvement in conductivity of the network with an increase in tube length.

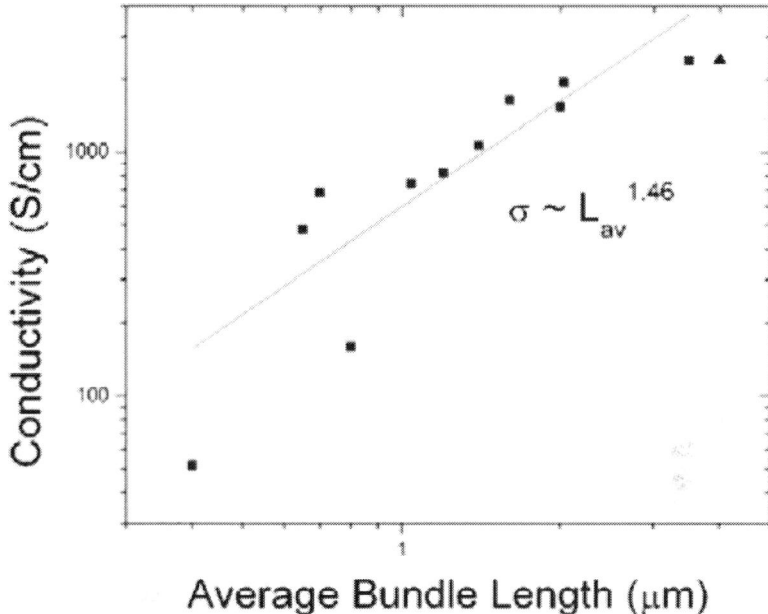

Figure 3. Variation of the CNT network electrical conductivity with the average bundle length of the CNTs in the network. The relationship between the conductivity of the CNT network and the average bundle length is $\sigma \sim L_{av}^{1.46}$. Reprinted with permission from [62], Copyright (2006) by the American Institute of Physics.

Sheet resistance (R_s) and transmittance (T) of carbon nanotube network are related by

$$T(\lambda) = (1 + \frac{188.5}{R_s} \frac{\sigma_{Op}(\lambda)}{\sigma_{DC}})^{-2} \tag{1}$$

where σ_{DC} and σ_{Op} are the electrical and optical conductivities respectively [16]. Hence, high σ_{DC}/σ_{Op} is desired for a low sheet resistance and high transparency network. The electrical conductivity, σ_{DC}, is found to correlate with the network morphology. The mean diameter of the CNT bundles (bundled due to agglomeration of solution processed CNTs) has an inverse relationship with σ_{DC}. This is because the smaller the bundle, the higher the junction density and number of electrical pathways in the network. Hence, debundling of CNTs is desirable [16].

The properties of CNTs in the CNT network film are examined in this section. Long and large diameter metallic carbon nanotubes, which are debundled are most desirable for transparent conductor applications.

2.2. Carbon Nanotube (CNT) Film Assembly

Besides the properties of CNTs in the CNT network film, the way that the CNT film is assembled is also very crucial to the electrical and optical properties of the film. There are various methods to deposit solution based CNTs onto a receiving substrate, such as vacuum filtration followed by transfer printing [15,64], spray coating [65,66] and controlled flocculation [67,68]. Transfer printing is often used to transfer chemical vapor deposition (CVD) grown CNT films from the growth substrate to another receiving substrate [69].

The vacuum filtration method is one of the most commonly used methods to assemble solution deposited CNT films. The process is simple and straightforward, and yields a uniform film, which is desired for many transparent conductor applications such as in photovoltaic cells and light emitting diodes. A uniform film with high surface roughness can potentially cause short circuit in thin film devices. However, the vacuum filtration method is limited in scale as it is determined by the size of the membrane filter.

Spray coating is another commonly used method. It is very versatile and can be used to coat surfaces with various shapes and curvatures. However, care must be taken to prevent re-agglomeration of the CNTs as they are deposited on the heated surface [66]. Re-agglomeration results in bundled CNTs, which decreases the electrical conductivity of the deposited film (as discussed in the earlier section). Careful selection of the surfactant used in the CNT solution and the parameters of spraying can mitigate this problem [66].

Controlled flocculation, another method to assemble solution based CNTs, deposits CNTs by adding liquids that are miscible with the suspending solvent and interact well with the surfactant. This drives the CNTs out and deposits them on the desired substrate. Methanol is often used [67,68]. In this case, re-agglomeration, which results in bundling is also an issue so the process must occur close to the surface of the receiving substrate to minimize agglomeration before deposition [67].

Finally, transfer printing of chemical vapor deposition (CVD) grown CNTs has proven to be an effective method to transfer CVD grown CNTs. The method is deterministic and the CNTs can be completely transferred from a donor substrate to a receiving substrate [69]. It is evident from this section that the assembly method of CNTs is very important and can impact the electrical conductivity of the film obtained. Hence, careful optimization of the assembly method is essential to yield high quality films.

2.3. Post Treatment

After the assembly of the CNT film, post deposition treatment is often carried out to improve the electrical conductivity by improving the junction resistance between two CNTs [70,71], removal of the insulating surfactants [72,73] and doping of the CNTs [74,75,76,77,78,79].

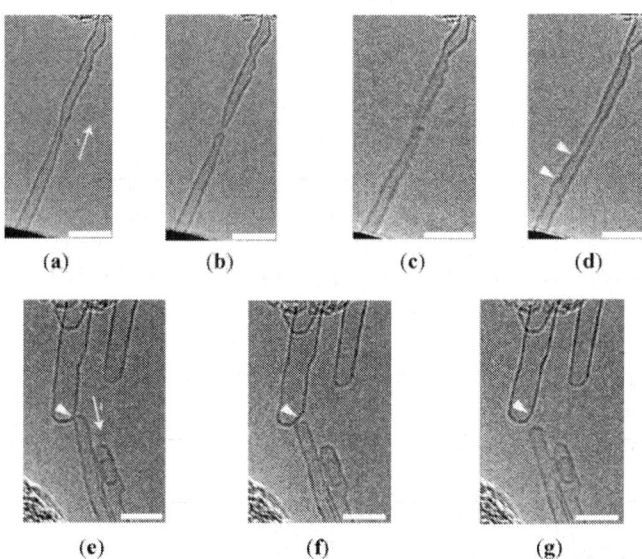

Figure 4. Joining of two CNTs with similar diameter works but fails for two CNTs with different diameters. (**a–d**) Successful joining of a CNT split into two separate CNTs by electrical breakdown (**a,b**). After contacting the two CNTs and applying a large voltage (**c**), the two CNTs bond and a new CNT forms (**d**); (**e–g**) Unsuccessful joining of two CNTs with different diameters. Scale bars are 5 nm in all parts of the Figure. Reprinted with permission from [70]. Copyright (2008) by the Nature Publishing Group.

The junction resistance between two CNTs is expected to decrease when they are joined. Two CNTs can be joined by Joule heating [70]. By contacting two CNTs and applying a large voltage across them, current flows across the two CNTs and electromigration occurs. Joining is only possible between CNTs of similar diameter as shown in Figure 4. Hence, this method is not feasible for CNT networks with a large range of diameters. Another way to join CNTs is via electron irradiation in a scanning electron microscope (SEM) [71]. By focusing the electron beam at the junction, carbon contamination is deposited selectively at the junction. The graphitic material, which connects two CNTs is expected to be electrically conductive. However, the two methods mentioned above are not scalable and will not be useful for large scale commercial use.

The electrical conductivity of CNT network can also be improved by removing surfactants often used to disperse CNTs in solution. Some surfactant residues are deposited together with CNTs during CNT film assembly. As the surfactant is insulating, the electrical conductivity of the CNT film is compromised. Hence, after CNT film assembly, removal of the surfactant improves the electrical conductivity. Surfactant can be removed by washing with water, followed by acetone [72] or by acid treatment, such as immersion in HNO_3[73].

Another commonly used method to improve the electrical conductivity of CNT network is by doping. Various types of dopants have been studied and found to improve the electrical conductivity by different extents. Doping of CNTs by vapor phase reactions with bromine (electron acceptor) or potassium (electron donor) has yielded decreased resistivity by a factor of 30 [74]. Doping by NO_2 shifts the Fermi level closer to the valence band and conductivity improves [75]. P doping by dopants, such as HNO_3, $SOCl_2$ or I_2 also prove effective [76,77]. Another commonly used p-type dopant is tetracyanoquinodimethane (TCNQ) [78,79]. However, all these methods suffer from a stability problem. The dopants desorb and electrical conductivity degrades with time. More work remains to find an alternative dopant. The successful commercialization of CNT transparent conductors is dependent upon further improvement of the electrical and optical properties, scalability, reproducibility and cost effectiveness of the production.

3. Graphene Films

Graphene, the two-dimensional allotrope of carbon, has generated much interest because of its high mobility [80], transparency [81] and flexibility [21,82]. Besides near ballistic transport in suspended graphene [80], the transmittance through a single layer of graphene is extremely high (97.7%) [81]. Graphene has also demonstrated extreme flexibility [21]. These outstanding properties of graphene can lead to its potential application as a flexible transparent conductor. Nonetheless, challenges remain for the material before its successful application in the real-world. Large scale production of low sheet resistance and high optical transparency graphene films that are electrically stable over time has yet to be established. In this section, we review the factors influencing the graphene film performance, in particular the synthesis, assembly and post-treatment of graphene films.

3.1. Synthesis and Assembly of Graphene Films

3.1.1. Solution Processed Films

Solution processed graphene films have been studied extensively because their production via roll-to-roll processing can potentially be scaled up for commercial applications. There are two main forms of precursors to graphene in solution: graphitic precursors [83,84,85,86] or graphite oxide precursors [82,87,88,89,90,91,92,93,94,95,96].

As graphite is hydrophobic, surfactants are often used to assist the graphite to disperse in organic solvents. Lotya *et al.* [85] demonstrated that the surfactant, sodium dodecylbenzene sulfonate (SDBS), aids the dispersion of graphite in water. The ionic surfactant adsorbs onto the graphite flakes and prevents the re-aggregation of graphite flakes suspended in water via Coulomb repulsion. Hence, a large percentage of flakes have less than five layers while ~3% of the flakes are monolayer. The graphite flakes are also found to be largely free of defects or oxides, which improves the electrical conductivity of the flakes. However, the flake dimensions (*i.e.*, the width and length) tend to be <400 nm, which will increase the number of inter-flake interfaces when a large surface area film is formed from such dispersions. The large number of inter-flake interfaces present will decrease the electrical conductivity of the film formed. Thin films can be formed from graphite dispersions via various techniques like vacuum filtration [83,85], spray coating [83,85] or Langmuir-Blodgett assembly [84].

Another group [84] reported dispersing graphite in dimethylformamide (DMF) with the aid of the surfactant, 1,2-distearoyl-sn-glycero-3-phosphoethanolamine-N-[methoxy(polyethyleneglycol)-5000] (DSPE-mPEG). Thin films formed from this dispersion via the Langmuir-Blodgett (LB) assembly method demonstrate a sheet resistance down to 8 kΩ/sq and transparency up to 93%, depending on the number of LB films deposited, as shown in Figure 5. Many challenges still remain as most applications for transparent conductors, such as touch screens require a sheet resistance <500 Ω/sq and transparency >85% [96].

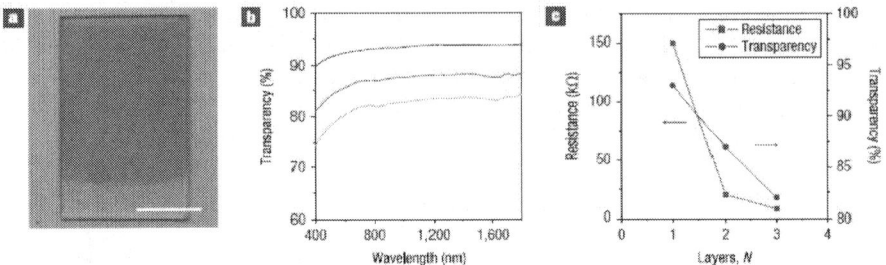

Figure 5. Langmuir-Blodgett (LB) films of graphene sheet. (**a**) A photograph of a film of graphene sheet deposited on the top portion of a quartz substrate. The scale bar is 10 mm; (**b**) Transmission spectra of the single layer (black curve), double layer (red curve) and triple layer (green curve) LB films; (**c**) Sheet resistance (red curve) and transparency at a wavelength of 1000 nm (blue curve) of the LB films with different number of layers. Reprinted with permission from [84]. Copyright (2008) by the Macmillan Publishers Limited.

Another form of precursor to graphene in solution is graphite oxide precursors. By oxidizing graphite via the Hummers method [97] or a modified Hummers method [98,99], the graphite oxide formed can now be dispersed in water. The addition of surfactant is no longer necessary. Graphite oxide films can be formed from the graphite oxide dispersion via similar methods to the graphite dispersions. Other methods of film assembly include spin coating [88] and dip coating [94]. However, upon film formation, the graphite oxide film has to be reduced to form graphene or graphite film. The graphite oxide film can be reduced via thermal annealing or chemical methods.

Becerril *et al.* compared the sheet resistance and transparency of films formed via various reduction treatments [88]. Three treatment methods are examined: reduction by hydrazine vapor, reduction by hydrazine vapor and annealing at 400 °C under argon flow and annealing at 1100 °C in vacuum. Figure 6 shows the outcome of his study. Thermal annealing at 1100 °C proves most effective while reduction via hydrazine vapor is the least effective. Characterization of the films after reduction using X-ray photoelectron spectroscopy (XPS) explains the phenomenon, as shown in Figure 7. Hydrazine treatment incorporates nitrogen into the samples by partially reducing the carbonyl functionalities to hydrazone groups. This decreases the relative content of carbon unbounded to oxygen or nitrogen and decreases the film conductivity of the sample. Annealing at 400 °C can desorb some of the nitrogen, leading to an improvement in film conductivity. Hydrazone groups are absent in samples that are annealed at

1100 °C and these samples show the best electrical conductivity. Complete reduction of graphene oxide films is very crucial as it has been found that lattice vacancies that cannot be healed during reduction can result in a three orders of magnitude decrease in electrical conductivity for such films relative to graphene film [95].

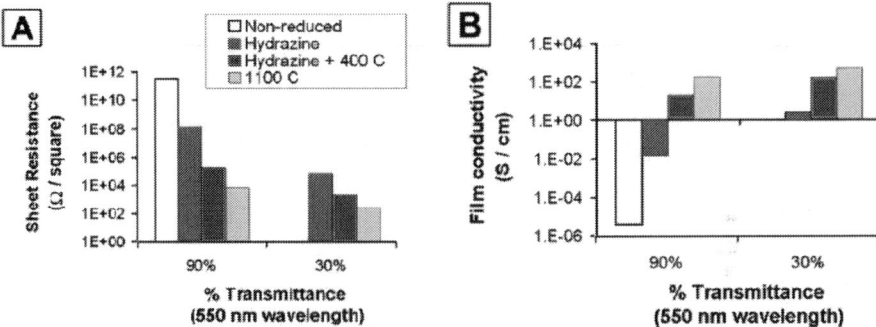

Figure 6. Electrical properties of reduced graphene oxide films that have been treated with different reduction methods. (**A**) Sheet resistance of the films with either 90% or 30% transmittance at a wavelength of 550 nm; (**B**) Film conductivity of the films with either 90% or 30% transmittance at a wavelength of 550 nm. Reprinted with permission from [88]. Copyright (2008) by the American Chemical Society.

Mattevi *et al.* also reported that incomplete reduction has detrimental effects on the electrical properties of reduced graphene films [100]. The residual oxygen in the graphene oxide film forms sp^3 bonds with the carbon atoms in the basal plane. The sp^3 bonds disrupt the transport of charge carriers delocalized in a sp^2 network. Hence, the electrical conductivity is lower in a partially reduced film relative to a completely reduced film.

Besides complete reduction of graphene oxide films, large graphene sheets are desired as they decrease the inter-flake interfaces. Large-scale graphene sheets with an area up to 20 × 40 µm have been demonstrated by Tung *et al.* [87] as shown in Figure 8. By dispersing graphene oxide in pure hydrazine, hydrazinium graphene dispersions are formed. A film can be formed from the dispersion via spin-coating.

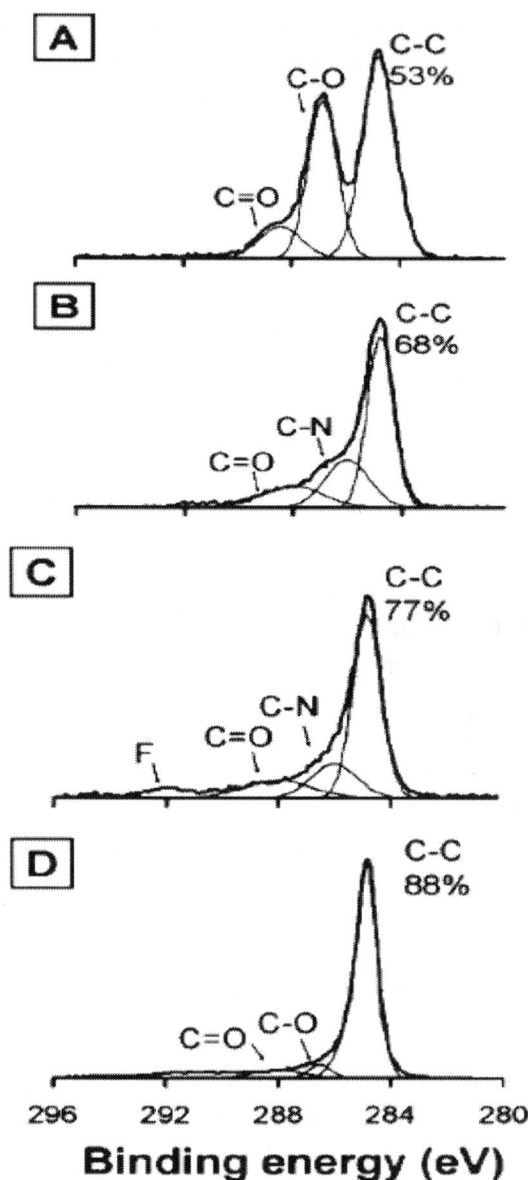

Figure 7. XPS analysis of the reduced graphene oxide films after different reduction treatments. **(A)** Non-reduced film; **(B)** Hydrazine-reduced film; **(C)**Film reduced by hydrazine and annealing at 400 °C; **(D)** Film reduced by annealing at 1100 °C. Reprinted with permission from [88]. Copyright (2008) by the American Chemical Society.

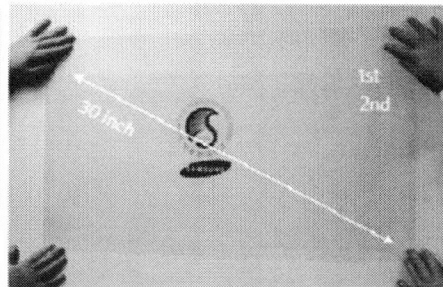

Figure 8. Atomic force microscope (AFM) image of a large scale graphene. Reprinted with permission from [87]. Copyright (2009) by the Macmillan Publishers Limited.

The main advantage of solution processed films is their compatibility with large scale and low cost roll-to-roll processing. The key disadvantages are the incomplete reduction of graphene oxide and the small dimensions of the graphene sheets formed. Development of methods to overcome these issues will result in films with better electrical conductivity, which is an important figure of merit for transparent conductors.

3.1.2. Chemical Vapor Deposition (CVD) Growth Films

Graphene films can also be formed by chemical vapor deposition (CVD) growth. The process usually involves breaking down a gaseous carbon feedstock (e.g., methane) in hydrogen gas at high temperature on a metal catalyst to form a graphene film. Graphene film growth via CVD is very appealing because large scale growth is possible and the electrical conductivity of CVD grown graphene films is generally better than that of solution processed graphene films.

High quality graphene films with dimensions up to 30 in have been demonstrated by Bae *et al.* [23]. This is achieved by growing the graphene film on a thin copper foil, which is wrapped around a large quartz tube to be placed in a furnace for CVD growth, as shown in Figure 9. The graphene film grown on the copper foil can be transferred to another arbitrary substrate via transfer printing. The different variants of transfer printing and the impact on the electrical performance of the transferred graphene film will be discussed in detail in the followingsection (3.1.3). The graphene film grown by Bae *et al.* has also demonstrated very outstanding sheet resistance of 30 Ω/sq and 90% transmittance.

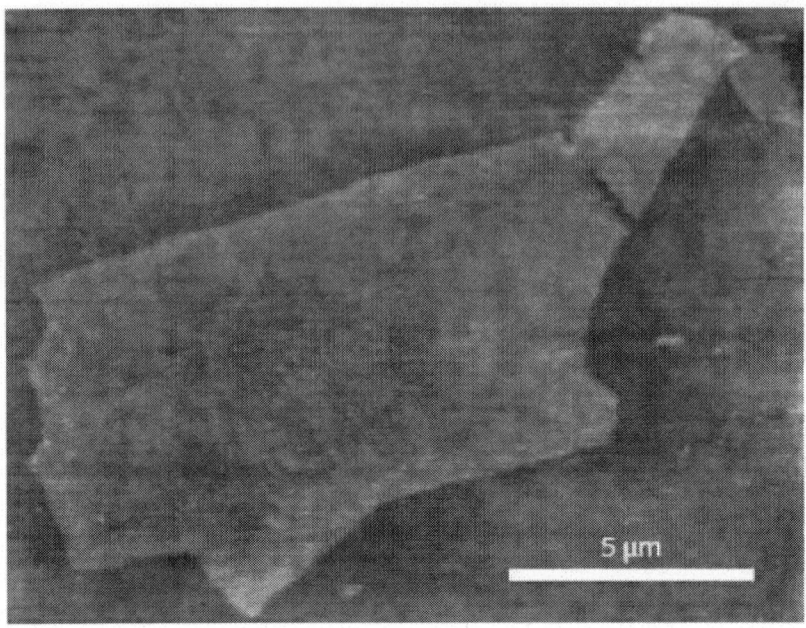

Figure 9. Image of a copper foil wrapped around a quartz tube to be placed in a furnace for chemical vapor deposition is shown on the left; A large area transparent graphene film transferred on a transparent substrate is shown on the right. Reprinted with permission from [23]. Copyright (2010) by the Macmillan Publishers Limited.

Despite the impressive progress in the field, some challenges remain to be overcome. CVD grown graphene films that demonstrate very low sheet resistance are often doped chemically. However, chemical doping is unstable and the sheet resistance of graphene films increases with time. This will be discussed in greater detail in a subsequent section (3.2). Undoped graphene films grown via CVD do not display superb electrical properties like an exfoliated graphene film due to the presence of grain boundaries [101,102,103,104,105,106,107,108] and wrinkles [109,110,111,112].

Grain boundaries in graphene films have been found to decrease the local work function, leading to potential barriers that scatter charge carriers by both backscattering and intervalley carrier scattering [102]. The carrier mobility decreases and impedes electrical transport. Hence, various groups have examined methods to optimize the CVD growth in order to decrease the density of grain boundaries and increase the grain size.

Yu *et al*. reported an approach that achieves grain size up to tens of micrometers by pre-patterning seed crystals to control the graphene nucleation [101]. In the absence of pre-patterned seed crystals, the nucleation sites form randomly and many grain boundaries can be expected.

Other groups have also adopted a similar strategy of decreasing the nucleation sites to increase the grain size and decrease the grain boundaries by various clever methods. Li *et al*. found that high temperature, low partial pressure and methane (carbon feedstock for graphene growth) flow rate during CVD growth result in fewer nucleation sites. Hence, they proposed a two-step CVD process [104]. The first step involves a high temperature and low methane flow rate and partial pressure process to generate a low density of graphene nuclei. Methane flow rate and partial pressure are increased in a subsequent step to increase the size of the graphene domain. The graphene films with larger domains that result from the two-step CVD process are shown to have high mobility due to reduced scattering at the inter-domain interfaces.

Millimeter-sized grains have been demonstrated by various groups [105,106,107,108] (Figure 10). This is achieved by a similar strategy that was mentioned earlier: initiating with a low nucleation rate and driving the nucleation sites to grow bigger in a subsequent step. In one instance, the low nucleation rate was achieved by low hydrogen flow rate and use of a polished copper catalyst substrate [105]. Polishing and pre-annealing the copper substrate reduces the defects on the substrate (e.g., impurities and surface irregularities), which can serve as nucleation centers. Higher temperature in a subsequent step drives the growth of the graphene domains to form large grains.

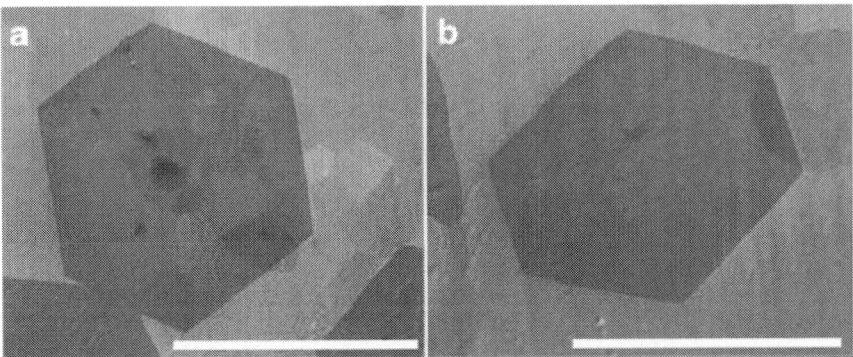

Figure 10. Scanning electron microscope (SEM) images of graphene grains on a Pt foil. The scale bars in are (**a**) 1mm; and (**b**) 0.5 mm respectively. Reprinted with permission from [108]. Copyright (2012) by the Macmillan Publishers Limited.

Besides grain boundaries, wrinkles in graphene films can impede electrical transport across the folds [110]. Wrinkles form due to thermal-induced stress during the CVD process [109,111]. High temperatures during CVD growth expands the metal catalyst film, which upon cooling, contracts. The thermal induced strain set up is accommodated by wrinkles formed in graphene films grown [111]. These wrinkles are often formed around step edges and defect lines of the substrate [109]. As mentioned earlier, polishing may be able to reduce the defect lines, which may suppress the wrinkle formation. If thin epitaxial metal films with small thermal expansion coefficients were used, the thermal induced strain and wrinkle formation are expected to decrease. The wrinkle formation is also found to be reversible by heating, which releases the strain. Hence, it may be possible to release the strain and eliminate the wrinkles during the transfer process when the graphene film is not fixed to a support [111]. The methods suggested to decrease wrinkles can potentially improve the electrical conductivity of the graphene films.

3.1.3. Transfer Printing of Graphene Films

A graphene film grown via chemical vapor deposition (CVD) is often grown on a metal foil or metal film on substrate. Graphene film, which is used for transparent conductor applications has to be transfer printed from the growth substrate to a transparent substrate. Polymeric substrates are often used as they are flexible and transparent. It is not feasible to grow graphene film directly on a polymeric substrate via CVD because the polymeric substrate degrades under the high temperature employed during CVD. In this section, we discuss the various techniques developed to transfer print graphene films and relate the transfer printing process to the quality of the film transferred.

The transfer printing process [21,23,113,114,115, 116,117,118, 119,120] generally involves attaching the graphene film onto a support substrate before etching the metal catalyst off. The graphene film on the support substrate is then transfer printed onto the desired receiving substrate from the support substrate. This is achieved by dissolving or peeling off the support substrate after attaching the graphene film—support substrate sandwich to the receiving substrate.

Various types of supporting substrates have been investigated. They include poly(methyl methacrylate) PMMA [114,115], polydimethylsiloxane (PDMS) [21,113], polyimide (PI) [116,117] and thermal release tape [23], of which, PMMA is one of the most commonly used supporting substrates. Suk *et al.* reported dry and wet transfer techniques using PMMA, as illustrated in Figure 11 [115].

In the dry transfer process, PMMA is spin-coated onto graphene film grown on copper foil. A PDMS frame was used to support the PMMA/graphene film sandwich while etching the copper foil using ammonium persulfate. This copper etchant is preferred over iron(III) nitrate, because it does not leave behind contamination residues like iron oxide. After the copper foil is completely etched, the graphene film is transferred to a receiving substrate. Heat treatment above the glass transition temperature of PMMA is performed after the graphene transfer to improve the contact between the graphene film and the receiving substrate. This is achieved by softening the PMMA so that the gap between the graphene film and the receiving substrate is reduced. The improved adhesion of the graphene film to the receiving substrate prevents cracks and tears from forming when the PMMA is removed. The wet transfer process is similar, as illustrated in Figure 11. Heat treatment is also performed to improve the quality of transferred graphene film. This method has been found to yield graphene films with lower sheet resistance.

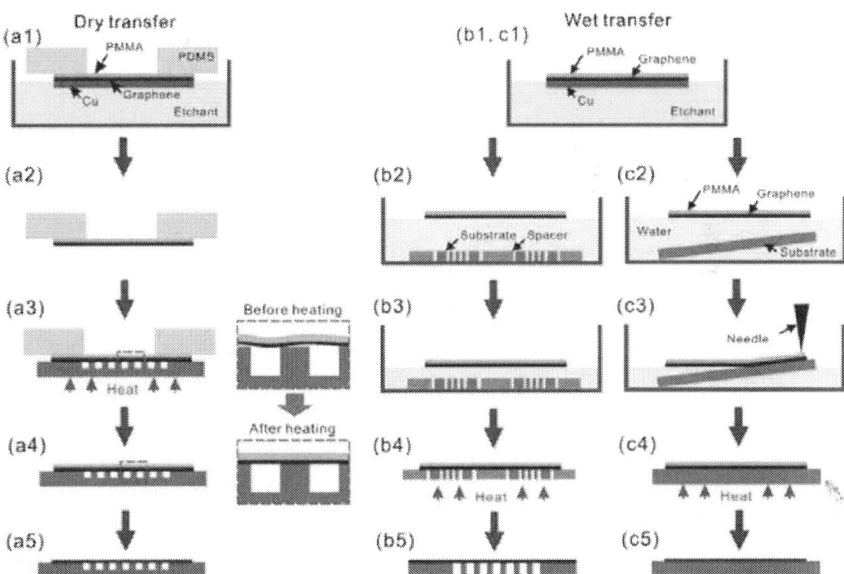

Figure 11. Schematic illustration of dry and wet transfer processes. (**a**) Dry transfer process onto perforated surface; (**b**) Wet transfer processes onto perforated surface and (**c**) non-perforated surface. Magnified views of (a3) and (a4) are provided. Reprinted with permission from [115]. Copyright (2011) by the American Chemical Society.

Liang *et al.* [114] also reported that the PMMA transfer process can be improved by performing a RCA clean, which removes any Cu or Fe residues from the copper etchant used to etch copper catalyst in graphene growth. Cracks formed in the graphene film during transfer are minimized when the adhesion of the graphene film to receiving substrate is improved. This can be achieved by increasing the hydrophilicity of the receiving substrate and baking.

The importance of adhesion between the transferred graphene film and receiving substrate is emphasized in [23]. When transfer is performed using thermal release tape, the first layer of transferred graphene film has relatively high sheet resistance (*i.e.*, ~275 Ω/sq), but subsequent transfers quickly decrease the sheet resistance. Hence, it was postulated that the adhesion between the first layer of film transferred directly onto the receiving substrate and the receiving substrate is poor, which results in mechanical damage of the film when the thermal release tape is removed. Hence, the sheet resistance is poor. Subsequent transferred layers do not interact directly with the substrate, so the sheet resistance is lower.

It is evident that the transfer printing process of graphene films to an arbitrary receiving substrate strongly influences the quality of the transferred film. This is because the transfer process can introduce residual contaminants and mechanical damage to the transferred film (Figure 12). Hence, an optimized process to minimize the contamination and damage will be expected to enhance the quality of the transferred film and improve the electrical properties.

3.2. Post Treatment

Typical sheet resistance of a transferred (via PMMA) graphene film is ~125 Ω/sq. This sheet resistance is low enough for certain applications, such as touch screens. However, when integrated into a solar cell or large area display, it will not be suitable as the series resistance is too high. Wet doping agents can be used to enhance the electrical conductivity of the graphene films [23,121,122,123], as shown in Figure 13. Some typical wet chemical p-dopants include $AuCl_3$ in nitromethane, HNO_3 in nitromethane and HCl. These strong oxidizing agents withdraw electrons from the graphene film and increase the doping density, leading to a decrease in sheet resistance [23]. Although more than 80% decrease in sheet resistance has been observed in films doped by $AuCl_3$ in nitromethane, the doping effect is transient. The sheet resistance increases by ~100% after 80 days in ambient condition at room temperature [121]. Another dopant studied, MoO_x, also suffers sheet resistance degradation with time [124]. A self-assembled monolayer of fluoroalkyltrichlorosilane (FTS) is another dopant investigated

[125]. However, it is not very effective in decreasing the sheet resistance (*i.e.*, only 8% decrease in sheet resistance).

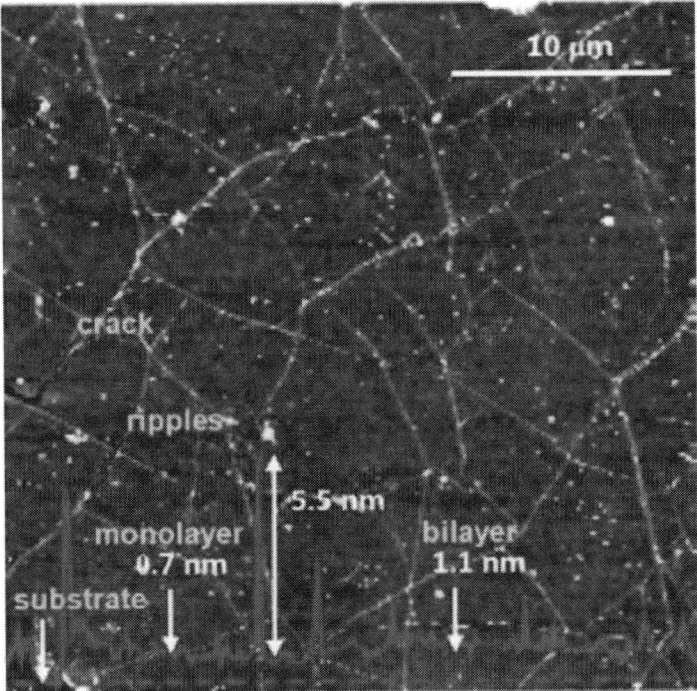

Figure 12. Atomic force microscope (AFM) image of a graphene film transferred onto a polyethylene terephthalate (PET) film using the thermal release tape. The solid red line is the height profile measured along the dashed red line. Reprinted with permission from [23]. Copyright (2010) by the Macmillan Publishers Limited.

Ni *et al.* reported a novel method of using nonvolatile ferroelectric polymer poly(vinylidenefluoride-co-trifluoroethylene) (P(VDF-TrFE)) gating to decrease the sheet resistance of the graphene film [126]. The advantage of the method is that electrostatic doping by the ferroelectric polymer is non-volatile so the sheet resistance stays low with time. Nonetheless, more work remains before this technology can be implemented in large-scale application. An opaque metal gate is used to apply a large electric field to polarize the ferroelectric polymer. Removal of the metal gate or an alternative way to polarize the ferroelectric polymer is essential to render transparency to the graphene-ferroelectric hybrid.

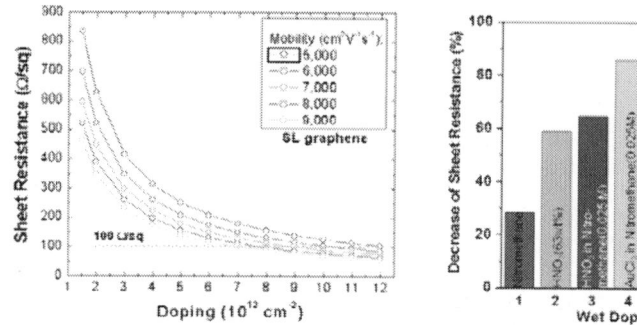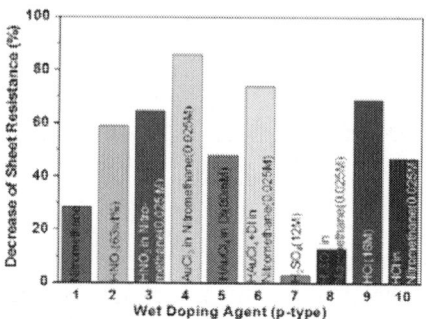

Figure 13. Calculated sheet resistance of a graphene film as a function of doping density with different charge carrier mobility is shown on the left. The decrease in sheet resistance using various wet doping agents is shown on the right. Reprinted with permission from [23]. Copyright (2010) by the Macmillan Publishers Limited.

Although single layer of chemical vapor deposition (CVD) grown graphene film has relatively low sheet resistance <500 Ω/sq, the search for a scalable, non-volatile and highly effective method to dope graphene continues so that graphene films can be used in more commercial applications [127]. The scalability, reproducibility and cost effectiveness of integrating graphene transparent conductors into practical devices also have to be carefully evaluated before successful commercialization.

4. HYBRID FILMS

Various groups have also explored hybrid films [128,129,130,131,132,133,134]. A hybrid film is desirable because it possesses the positive attributes of all its component materials. There are many variations of hybrid films, such as carbon nanotube—graphene hybrid [128,129], graphene—nanowire hybrid [130,131], graphene—metal grid hybrid [132,133] and carbon nanotube—(PEDOT) hybrid [134]. The electrical and optical properties can be improved by integrating different materials because one dimensional materials can often be used to bridge the gaps between non-uniform two dimensional materials. The selection of component materials and mixture of the component materials are critical parameters, which determine the electrical and optical properties of the hybrid films.

5. CONCLUSIONS

As demand for transparent conductors in flexible devices is expected to be high in future, alternative materials to substitute metal oxides are sought. Films of carbon nanomaterials have been identified as one potential class of substitute materials. More work remains as the sheet resistance of carbon nanomaterial films is still unable to reach <10 Ω/sq at 90% transparency (refer to Table 1 below), which deems them unsuitable for certain applications, such as transparent conductors in photovoltaic cells and large area displays. Better electrical and optical properties can possibly be achieved via a few means. Improved techniques to synthesize high quality carbon nanomaterials, whose properties are near those of a pristine carbon nanomaterial must be developed. Advanced methods of carbon nanomaterial film assembly or deposition, with good control of film properties and low introduction of impurities or contaminants, is essential. Better post-treatment methods to dope the films are desirable.

Table 1. Sheet resistance and optical transmittance of different carbon nanomaterials and their hybrids.

Material	Sheet resistance (Ω/sq)	Transmittance (%)	Reference
CNT			
CVD CNT	265	80	41
CVD CNT	265	70	42
Solution CNT	1000	80	13
Solution CNT	400	80	48
Solution CNT	100	70	63
Graphene			
CVD graphene	280	80	21
CVD graphene	30	90	23
CVD graphene	980	97.6	115
Reduced graphene oxide	8000	93	84
Reduced graphene oxide	1000	80	88
Reduced graphene oxide	1800	70	94
Hybrid			
CNT-graphene	240	85	128
Nanowire-graphene	64	94	130
Metal grid-graphene	20	90	132
CNT-PEDOT	80	75	134

Nonetheless, the films, which have been demonstrated may be suitable for less demanding applications, such as touch panels. Further exploration in their stability in devices over time is necessary. The scalability, reproducibility and cost effectiveness of integrating them into practical devices also have to be carefully evaluated. Finally, the success of practical application of carbon nanomaterial films in transparent conductors is also dependent on the development of competing alternative materials, such as thin metal films, metal nanowire films, conducting polymers and various other forms of hybrid films.

ACKNOWLEDGMENTS

This work was supported by the Agency of Science, Technology and Research (A*STAR), Singapore.

REFERENCES

1. Mizuhashi, M. Electrical properties of vacuum-deposited indium oxide and indium tin oxide films. *Thin Solid Films* **1980**, *70*, 91–100.

2. Kim, H.; Gilmore, C.M.; Pique, A.; Horwitz, J.S.; Mattoussi, H.; Murata, H.; Kafafi, Z.H.; Chrisy, D.B. Electrical, optical, and structural properties of indium-tin-oxide thin films for organic light-emitting devices. *J. Appl. Phys.* **1999**, *86*, 6451–6461.

3. Cairns, D.R.; Witte, R.P.; Sparacin, D.K.; Sachsman, S.M.; Paine, D.C.; Crawford, G.P.; Newton, R.R. Strain-dependent electrical resistance of tin-doped indium oxide on polymer substrates. *Appl. Phys. Lett.* **2000**, *76*, 1425.

4. Kirchmeyer, S.; Reuter, K. Scientific importance, properties and growing applications of Poly(3,4-ethylenedioxythiophene). *J. Mater. Chem.* **2005**, *15*, 2077–2088.

5. Nardes, A.M.; Kemerink, M.; de Kok, M.M.; Vinken, E.; Maturova, K.; Janssen, R.A.J. Conductivity, work function and environmental stability of pedot:pss thin films treated with sorbitol. *Org. Electron.* **2008**, *9*, 727–734.

6. Vitoratos, E.; Sakkopoulos, S.; Dalas, E.; Paliatsas, N.; Karageorgopoulos, D.; Petraki, F.; Kennou, S.; Choulis, S.A. Thermal degradation mechanisms of PEDOT:PSS. *Org. Electron.* **2009**, *10*, 61–66.

7. Lee, J.-Y.; Connor, S.T.; Cui, Y.; Peumans, P. Solution-processed metal nanowire mesh transparent electrodes. *Nano Lett.* **2008**, *8*, 689–692.

8. Wu, H.; Hu, L.; Rowell, M.W.; Kong, D.; Cha, J.J.; McDonough, J.R.; Zhu, J.; Yang, Y.; McGehee, M.D.; Cui, Y. Electrospun metal nanofiber webs as high-performance transparent electrode. *Nano Lett.* **2010**, *10*, 4242–4248.

9. Hu, L.; Kim, H.S.; Lee, J.-Y.; Peumans, P.; Cui, Y. Scalable coating and properties of transparent, flexible, silver nanowire electrodes. *ACS Nano* **2010**, *4*, 2955–2963.

10. De, S.; Higgins, T.M.; Lyons, P.E.; Doherty, E.M.; Mirmalraj, P.N.; Blau, W.J.; Boland, J.J.; Coleman, J.N. Silver nanowire networks as flexible, transparent, conducting films: Extremely high DC to optical conductivity ratios. *ACS Nano* **2009**, *3*, 1767–1774.

11. Ghosh, D.S.; Martinez, L.; Giurgola, S.; Vergani, P.; Pruneri, V. Widely transparent electrodes based on ultrathin metals. *Opt. Lett.* **2009**, *34*, 325–327.

12. Doriot-Werle, M.; Banakh, O.; Gay, P.A.; Matthey, J.; Steinmann, P.A. Tarnishing resistance of silver-palladium thin films. *Surf. Coat. Technol.* **2006**, *200*, 6696–6701.

13. Hu, L.; Hecht, D.S.; Gruner, G. Percolation in transparent and conducting carbon nanotube networks. *Nano Lett.* **2004**, *4*, 2513–2517.

14. Kaempgen, M.; Duesberg, G.S.; Roth, S. Transparent carbon nanotube coatings. *Appl. Surf. Sci.* **2005**, *252*, 425–429.

15. Wu, Z.C.; Chen, Z.H.; Du, X.; Logan, J.M.; Sippel, J.; Nikolou, M.; Kamaras, K.; Reynolds, J.R.; Tanner, D.B.; Hebard, A.F.; *et al.* Transparent, conductive carbon nanotube films. *Science* **2004**, *305*, 1273–1276.

16. Nirmalraj, P.N.; Lyons, P.E.; De, S.; Coleman, J.N.; Boland, J.J. Electrical connectivity in single-walled carbon nanotube networks. *Nano Lett.* **2009**, *9*, 3890–3895.

17. Li, Z.R.; Kandel, H.R.; Dervishi, E.; Saini, V.; Biris, A.S.; Biris, A.R.; Lupu, D. Does the wall number of carbon nanotubes matter as conductive transparent material? *Appl. Phys. Lett.* **2007**, *91*.

18. Hecht, D.S.; Heintz, A.M.; Lee, R.S.; Hu, L.; Moore, B.; Cucksey, C.; Risser, S. High conductivity transparent carbon nanotube films deposited from superacid. *Nanotechnology* **2011**, *22*.

19. Li, X.; Zhu, Y.; Cai, W.; Borysiak, M.; Han, B.; Chen, D.; Piner, R.D.; Colombo, L.; Ruoff, R.S. Transfer of large-area graphene films for high-performance transparent conductive electrodes. *Nano Lett.* **2009**, *9*, 4359–4363.

20. Lee, W.H.; Park, J.; Sim, S.H.; Jo, S.B.; Kim, K.S.; Hong, B.H.; Cho, K. Transparent flexible organic transistors based on monolayer graphene electrodes on plastic. *Adv. Mater.* **2011**, *23*, 1752–1756.

21. Kim, K.S.; Zhao, Y.; Jang, H.; Lee, S.Y.; Kim, J.M.; Kim, K.S.; Ahn, J.-H.; Kim, P.; Choi, J.-Y.; Hong, B.H. Large-scale pattern growth of graphene films for stretchable transparent electrodes. *Nature* **2009**, *457*, 706–710.

22. Kim, R.-H.; Bae, M.-H.; Kim, D.G.; Cheng, H.; Kim, B.H.; Kim, D.-H.; Li, M.; Wu, J.; Du, F.; Kim, H.-S.; *et al.* Stretchable transparent graphene interconnects for arrays of microscale inorganic light emitting diodes on rubber substrates. *Nano Lett.* **2011**, *11*, 3881–3886.

23. Bae, S.; Kim, H.; Lee, Y.; Xu, X.; Park, J.-S.; Zheng, Y.; Balakrishmam, J.; Lei, T.; Kim, H.R.; Song, Y.; *et al.* Roll-To-Roll Production of 30-inch graphene films for transparent electrodes. *Nat. Nanotechnol.* **2010**, *5*, 574–578.

24. White, C.T.; Todorov, T.N. Carbon nanotubes as long ballastic conductors. *Nature* **1998**, *393*, 240–242.

25. Quinn, B.M.; Lemay, S.G. Single-walled carbon nanotubes as templates and interconnects for nanoelectrodes. *Adv. Mater.* **2006**, *18*, 855–859.

26. Yao, Z.; Kane, C.L.; Dekker, C. High-field electrical transport in single-wall carbon nanotubes. *Phys. Rev. Lett.* **2000**, *84*, 2941–2944.

27. Bozovic, D.; Bockrath, M.; Hafner, J.H.; Lieber, C.M.; Park, H.; Tinkham, M. Plastic deformations in mechanically strained single-walled carbon nanotubes. *Phys. Rev. B* **2003**, *67*.

28. Ebbesen, T.W.; Ajayan, P.M. Large-scale synthesis of carbon nanotubes. *Nature* **1992**, *358*, 220–222.

29. Sadana, A.K.; Liang, F.; Brinson, B.; Arepalli, S.; Farhat, S.; Hauge, R.H.; Smalley, R.E.; Billups, W.E. Functionalization and extraction of

large fullerenes and carbon-coated metal formed during the synthesis of single wall carbon nanotubes by laser oven, direct current arc, and high-pressure carbon monoxide production methods. *J. Phys. Chem. B* **2005**, *109*, 4416–4418.

30. José-Yacamán, M.; Miki-Yoshida, M.; Rendón, L.; Santiesteban, J.G. Catalytic growth of carbon microtubules with fullerene structure. *Appl. Phys. Lett.* **1993**, *62*, 657–659.

31. Cheng, Q.; Debnath, S.; O'Neill, L.; Hedderman, T.G.; Gregan, E.; Byrne, H.J. Systematic study of the dispersion of SWNTs in organic solvents. *J. Phys. Chem. C* **2010**, *114*, 4857–4863.

32. Matarredona, O.; Rhoads, H.; Li, Z.; Harwell, J.H.; Balzano, L.; Resasco, D.E. Dispersion of single-walled carbon nanotubes in aqueous solutions of the anionic surfactant NaDDBS. *J. Phys. Chem. B* **2003**, *107*, 13357–13367.]

33. Moore, V.C.; Strano, M.S.; Haroz, E.H.; Hauge, R.H.; Smalley, R.E. Individually suspended single-walled carbon nanotubes in various surfactants. *Nano Lett.* **2003**, *3*, 1379–1382.

34. Lay, M.D.; Novak, J.P.; Snow, E.S. Simple route to large-scale ordered arrays of liquid-deposited carbon nanotubes. *Nano Lett.* **2004**, *4*, 603–606.

35. Saran, N.; Parikh, K.; Suh, D.-S.; Munoz, E.; Kolla, H.; Manohar, S.K. Fabrication and characterization of thin films of single-walled carbon nanotube bundles on flexible plastic substrates. *J. Am. Chem. Soc.* **2004**, *126*, 4462–4463.

36. Chen, J.; Rao, A.M.; Lyuksyutov, S.; Itkis, M.E.; Hamon, M.A.; Hu, H.; Cohn, R.W.; Eklund, P.C.; Colbert, D.T.; Smalley, R.E.; *et al.* Dissolution of full-length single-walled carbon nanotubes. *J. Phys. Chem. B* **2001**, *105*, 2525–2528.

37. Kaempgen, M.; Lebert, M.; Haluska, M.; Nicoloso, N.; Roth, S. Sonochemical optimization of the conductivity of single-wall carbon nanotube networks. *Adv. Mater.* **2008**, *20*, 616–620.

38. Wang, J.; Sun, J.; Gao, L.; Wang, Y.; Zhang, J.; Kajiura, H.; Li, Y.M.; Noda, K. Removal of the residual surfactants in transparent and conductive single-walled carbon nanotube films. *J. Phys. Chem. C* **2009**, *113*, 17685–17690.

39. Prisbrey, L.; Roundy, D.; Blank, K.; Fifield, L.S.; Minot, E.D. Electrical characteristics of carbon nanotube devices prepared with single oxidative point defects. *J. Phys. Chem. C.* **2012**, *116*, 1961–1965.

40. Huang, W.; Lin, Y.; Taylor, S.; Gaillard, J.; Rao, A.M.; Sun, Y.-P. Sonication-assisted functionalization and solubilization of carbon nanotubes. *Nano Lett.* **2002**, *2*, 231–234.

41. Cao, Q.; Hur, S.-H.; Zhu, Z.-T.; Sun, Y.; Wang, C.; Meitl, M.A.; Shim, M.; Rogers, J.A. Highly bendable, transparent thin-film transistors that use carbon-nanotube-based conductors and semiconductors with elastomeric dielectrics. *Adv. Mater.* **2006**, *18*, 304–309.

42. Cao, Q.; Zhu, Z.-T.; Lemaitre, M.G.; Xia, M.-G.; Shim, M.; Rogers, J.A. Transparent flexible organic thin-film transistors that use printed single-walled carbon nanotube electrodes. *Appl. Phys. Lett.* **2006**, *88*.

43. Bachtold, A.; Fuhrer, M.S.; Plyasunov, S.; Forero, M.; Anderson, E.H.; Zettl, A.; McEuen, P.L. Scanned probe microscopy of electronic transport in carbon nanotubes. *Phys. Rev. Lett.* **2000**, *84*, 6082–6085.

44. McEuen, P.L.; Bockrath, M.; Cobden, D.H.; Yoon, Y.-G.; Louie, S.G. Disorder, pseudospins, and backscattering in carbon nanotubes. *Phys. Rev. Lett.* **1999**, *83*, 5098–5101.

45. Ho, X.; Ye, L.; Rotkin, S.V.; Cao, Q.; Unarunotai, S.; Salamat, S.; Alam, M.A.; Rogers, J.A. Scaling properties in transistors that use aligned arrays of single-walled carbon nanotubes. *Nano Lett.* **2010**, *10*, 499–503.

46. Park, J.-Y.; Rosenblatt, S.; Yaish, Y.; Sazonova, V.; Ustunel, H.; Braig, S.; Arias, T.A.; Brouwer, P.W.; McEuen, P.L. Electron-phonon scattering in metallic single-walled carbon nanotubes. *Nano Lett.* **2004**, *4*, 517–520.

47. Javey, A.; Guo, J.; Paulsson, M.; Wang, Q.; Mann, D.; Lundstrom, M.; Dai, H. High-field quasiballistic transport in short carbon nanotubes. *Phys. Rev. Lett.* **2004**, *92*, 106804:1–106804:4.

48. Green, A.A.; Hersam, M.C. Colored semitransparent conductive coatings consisting of monodisperse metallic single-walled carbon nanotubes. *Nano Lett.* **2008**, *8*, 1417–1422.

49. Fuhrer, M.S.; Nygard, J.; Shih, L.; Forero, M.; Yoon, Y.-G.; Mazzoni, M.S.C.; Choi, H.J.; Ihm, J.; Louie, S.G.; Zettl, A.; et al. Crossed nanotube junctions. *Science* **2000**, *288*, 494–497.

50. Wang, C.; Cao, Q.; Ozel, T.; Gaur, A.; Rogers, J.A.; Shim, M. Electronically selective chemical functionalization of carbon nanotubes: Correlation between raman spectral and electrical responses. *J. Am. Chem. Soc.* **2005**, *127*, 11460–11468.

51. Strano, M.S.; Dyke, C.A.; Usrey, M.L.; Barone, P.W.; Allen, M.J.; Shan, H.; Kittrell, C.; Hauge, R.H.; Tour, J.M.; Smalley, R.E. Electronic structure control of single-walled carbon nanotube functionalization. *Science* **2003**, *301*, 1519–1522.

52. Collins, P.C.; Arnold, M.S.; Avouris, P. Engineering carbon nanotubes and nanotube circuits using electrical breakdown. *Science* **2001**, *292*, 706–709.

53. Arnold, M.S.; Green, A.A.; Hulvat, J.F.; Stupp, S.I.; Hersam, M.C. Sorting carbon nanotubes by electronic structure using density differentiation. *Nat. Nanotechnol.* **2006**, *1*, 60–65.

54. Hersam, M.C. Progress towards monodisperse single-walled carbon nanotubes. *Nat. Nanotechnol.* **2008**, *3*, 387–394.

55. Krupke, R.; Linden, S.; Rapp, M.; Hennrich, F. Thin films of metallic carbon nanotubes prepared by dielectrophoresis. *Adv. Mater.* **2006**, *18*, 1468–1470.

56. Zhou, X.; Park, J.-Y.; Huang, S.; Liu, J.; McEuen, P.L. Band structure, phonon scattering, and the performance limit of single-walled carbon nanotube transistors. *Phys. Rev. Lett.* **2005**, *95*.

57. Islam, A.E.; Du, F.; Ho, X.; Jin, S.H.; Dunham, S.; Rogers, J.A. Effect of variations in diameter and density on the statistics of aligned array carbon-nanotube field effect transistors. *J. Appl. Phy.* **2012**, *111*.

58. Li, Y.; Kim, W.; Zhang, Y.; Rolandi, M.; Wang, D.; Dai, H. Growth of single-walled carbon nanotubes from discrete catalytic nanoparticles of various sizes. *J. Phys. Chem. B* **2001**, *105*, 11424–11431.

59. Cheung, C.L.; Kurtz, A.; Park, H.; Lieber, C.M. Diameter-controlled synthesis of carbon nanotubes. *J. Phys. Chem. B* **2002**, *106*, 2429–2433.

60. Bachilo, S.M.; Balzano, L.; Herrera, J.E.; Pompeo, F.; Resasco, D.E.; Weisman, R.B. Narrow (n,m)-distribution of single-walled carbon

nanotubes grown using a solid supported catalyst. *J. Am. Chem. Soc.* **2003**, *125*, 11186–11187.

61. Ryu, K.; Badmaev, A.; Gomez, L.; Ishikawa, F.; Lei, B.; Zhou, C. Synthesis of aligned single-walled nanotubes using catalysts defined by nanosphere lithography. *J. Am. Chem. Soc.* **2007**, *129*, 10104–10105.

62. Hecht, D.; Hu, L.; Gruner, G. Conductivity scaling with bundle length and diameter in single walled carbon nanotube networks. *Appl. Phys. Lett.* **2006**, *89*.

63. De, S.; King, P.J.; Lyons, P.E.; Khan, U.; Coleman, J.N. Size effects and the problem with percolation in nanostructured transparent conductors. *ACS Nano* **2010**, *4*, 7064–7072.

64. Lim, C.; Min, D.-H.; Leea, S.-B. Direct patterning of carbon nanotube network devices by selective vacuum filtration. *Appl. Phys. Lett.* **2007**, *91*.

65. Liu, Q.; Fujigaya, T.; Cheng, H.-M.; Nakashima, N. Free-standing highly conductive transparent ultrathin single-walled carbon nanotube films. *J. Am. Chem. Soc.* **2010**, *132*, 16581–16586.

66. Tenet, R.C.; Barnes, T.M.; Bergeson, J.D.; Ferguson, A.J.; To, B.; Gedvilas, L.M.; Heben, M.J.; Blackburn, J.L. Ultrasmooth, large-area, high-uniformity, conductive transparent single-walled-carbon-nanotube films for photovoltaics produced by ultrasonic spraying. *Adv. Mater.* **2009**, *21*, 3210–3216.

67. Meitl, M.A.; Zhou, Y.; Gaur, A.; Jeon, S.; Usrey, M.L.; Strano, M.S.; Rogers, J.A. Solution casting and transfer printing single-walled carbon nanotube films. *Nano Lett.* **2004**, *4*, 1643–1647.

68. Park, J.-U.; Meitl, M.A.; Hur, S.-H.; Usrey, M.L.; Strano, M.S.; Kenis, P.J.A.; Rogers, J.A. *In situ* deposition and patterning of single-walled carbon nanotubes by laminar flow and controlled flocculation in microfluidic channels. *Angew. Chem. Int. Ed.* **2006**, *45*, 581–585.

69. Zhou, Y.; Hu, L.; Grüner, G. A method of printing carbon nanotube thin films. *Appl. Phys. Lett.* **2006**, *88*.

70. Jin, C.; Suenaga, K.; Iijima, S. Plumbing carbon nanotubes. *Nat. Nanotechnol.* **2008**, *3*, 17–21.

71. Banhart, F. The formation of a connection between carbon nanotubes in an electron beam. *Nano Lett.* **2001**, *1*, 329–332.

72. Dyke, C.A.; Tour, J.M. Unbundled and highly functionalized carbon nanotubes from aqueous reactions.*Nano Lett.* **2003**, *3*, 1215–1218.

73. Geng, H.-Z.; Kim, K.K.; So, K.P.; Lee, Y.S.; Chang, Y.; Lee, Y.H. Effect of acid treatment on carbon nanotube-based flexible transparent conducting films. *J. Am. Chem. Soc.* **2007**, *129*, 7758–7759.

74. Lee, R.S.; Kim, H.J.; Fischer, J.E.; Thess, A.; Smalley, R.E. Conductivity enhancement in single-walled carbon nanotube bundles doped with K and Br. *Nature* **1997**, *388*, 255–257.

75. Kong, J.; Franklin, N.R.; Zhou, C.; Chapline, M.G.; Peng, S.; Cho, K.; Dai, H. Nanotube molecular wires as chemical sensors. *Science* **2000**, *287*, 622–625.

76. Jackson, R.; Domercq, B.; Jain, R.; Kippelen, B.; Graham, S. Stability of doped transparent carbon nanotube electrodes. *Adv. Funct. Mater.* **2008**, *18*, 2548–2554.

77. Skakalova, V.; Kaiser, A.B.; Dettlaff-Weglikowska, U.; Hrncarikova, K.; Roth, S. Effect of chemical treatment on electrical conductivity, infrared absorption, and Raman spectra of single-walled carbon nanotubes. *J. Phys. Chem. B* **2005**, *109*, 7174–7181.

78. Takenobu, T.; Kanbara, T.; Akima, N.; Takahashi, T.; Shiraishi, M.; Tsukagoshi, K.; Kataura, H.; Aoyagi, Y.; Iwasa, Y. Control of carrier density by a solution method in carbon-nanotube devices. *Adv. Mater.* **2005**, *17*, 2430–2434.

79. Abdula, D.; Shim, M. Performance and photovoltaic response of polymer-doped carbon nanotube p-ndiodes. *ACS Nano* **2008**, *2*, 2154–2159.

80. Du, X.; Skachko, I.; Barker, A.; Andrei, E.Y. Approaching ballistic transport in suspended graphene. *Nat. Nanotechnol.* **2008**, *3*, 491–495.

81. Nair, R.R.; Blake, P.; Grigorenko, A.N.; Novoselov, K.S.; Booth, T.J.; Stauber, T.; Peres, N.M.R.; Geim, A.K. Fine structure constant defines visual transparency of graphene. *Science* **2008**, *320*.

82. Eda, G.; Fanchini, G.; Chhowalla, M. Large-area ultrathin films of reduced graphene oxide as a transparent and flexible electronic material. *Nat. Nanotechnol.* **2008**, *3*, 270–274.

83. Hernandez, Y.; Nicolosi, V.; Lotya, M.; Blighe, F.M.; Sun, Z.; De, S.; McGovern, I.T.; Holland, B.; Byrne, M.; Gun'ko, Y.K.; *et al.* High-yield

production of graphene by liquid-phase exfoliation of graphite. *Nat. Nanotechnol.* **2008**, *3*, 563–568.

84. Li, X.; Zhang, G.; Bai, X.; Sun, X.; Wang, X.; Wang, E.; Dai, H. Highly conducting graphene sheets and langmuir-blodgett films. *Nat. Nanotechnol.* **2008**, *3*, 538–542.

85. Lotya, M.; Hernandez, Y.; King, P.J.; Smith, R.J.; Nicolosi, V.; Karlsson, L.S.; Blighe, F.M.; De, S.; Wang, Z.; McGovern, I.T.; *et al.* Liquid phase production of graphene by exfoliation of graphite in surfactant / water solutions. *J. Am. Chem. Soc.* **2009**, *131*, 3611–3620.

86. De, S.; King, P.J.; Lotya, M.; O'Neill, A.; Doherty, E.M.; Hernandez, Y.; Duesberg, G.S.; Coleman, J.N. Flexible, transparent, conducting films of randomly stacked graphene from surfactant-stabilized, oxide-free graphene dispersions. *Small* **2010**, *6*, 458–464.

87. Tung, V.C.; Allen, M.J.; Yang, Y.; Kaner, R.B. High-throughput solution processing of large-scale graphene. *Nat. Nanotechnol.* **2009**, *4*, 25–29.

88. Becerril, H.A.; Mao, J.; Liu, Z.; Stoltenberg, R.M.; Bao, Z.; Chen, Y. Evaluation of solution-processed reduced graphene oxide films as transparent conductors. *ACS Nano* **2008**, *2*, 463–470.

89. Park, S.; An, J.; Jung, I.; Piner, R.D.; An, S.J.; Li, X.; Velamakanni, A.; Ruoff, R.S. Colloidal suspensions of highly reduced graphene oxide in a wide variety of organic solvents. *Nano Lett.* **2009**, *9*, 1593–1597.

90. Cote, L.J.; Kim, F.; Huang, J. Langmuir-blodgett assembly of graphite oxide single layers. *J. Am. Chem. Soc.* **2009**, *131*, 1043–1049.

91. Dikin, D.A.; Stankovich, S.; Zimney, E.J.; Piner, R.D.; Dommett, G.H.B.; Evmenenko, G.; Nguyen, S.T.; Ruoff, R.S. Preparation and characterization of graphene oxide paper. *Nature* **2007**, *448*, 457–460.

92. Gilje, S.; Han, S.; Wang, M.S.; Wang, K.L.; Kaner, R.B. A Chemical route to graphene for device applications. *Nano Lett.* **2007**, *7*, 3394–3398.

93. Li, D.; Muller, M.B.; Gilje, S.; Kaner, R.B.; Wallace, G.G. Processable aqueous dispersions of graphene nanosheets. *Nat. Nanotechnol.* **2008**, *3*, 101–105.

94. Wang, X.; Zhi, L.J.; Mullen, K. Transparent, conductive graphene electrodes for dye-sensitized solar cells.*Nano Lett.* **2008**, *8*, 323–327.

95. Gomez-Navarro, C.; Weitz, R.T.; Bittner, A.M.; Scolari, M.; Mews, A.; Burghard, M.; Kern, K. Electronic transport properties of individual chemically reduced graphene oxide sheets. *Nano Lett.* **2007**, *7*, 3499–3503.

96. Hecht, D.S.; Hu, L.; Irvin, G. Emerging transparent electrodes based on thin films of carbon nanotubes, graphene and metallic nanostructures. *Adv. Mater.* **2011**, *23*, 1482–1513.

97. Hummers, W.S.; Offeman, R.E. Preparation of graphitic oxide. *J. Am. Chem. Soc.* **1958**, *80*, 1339.

98. Park, S.; An, J.; Piner, R.D.; Jung, I.; Yang, D.; Velamakanni, A.; Nguyen, S.T.; Ruoff, R.S. Aqueous suspension and characterization of chemically modified graphene sheets. *Chem. Mater.* **2008**, *20*, 6592–6594.

99. Donner, S.; Li, H.W.; Yeung, E.S.; Porter, M.D. Fabrication of optically transparent carbon electrodes by the pyrolysis of photoresist films: approach to single-molecule spectroelectrochemistry. *Anal. Chem.* **2006**, *78*, 2816–2822.

100. Mattevi, C.; Eda, G.; Agnoli, S.; Miller, S.; Mkhoyan, K.A.; Celik, O.; Mastrogiovanni, D.; Granozzi, G.; Garfunkel, E.; Chhowalla, M. Evolution of electrical, chemical, and structural properties of transparent and conducting chemically derived graphene thin films. *Adv. Funct. Mater.* **2009**, *19*, 2577–2583.

101. Yu, Q.; Jauregui, L.A.; Wu, W.; Colby, R.; Tian, J.; Su, Z.; Cao, H.; Liu, Z.; Pandey, D.; Wei, D.; *et al.* Control and characterization of individual grains and grain boundaries in graphene grown by chemical vapor deposition. *Nat. Mater.* **2011**, *10*, 443–449.

102. Koepke, J.C.; Wood, J.D.; Estrada, D.; Ong, Z.-Y.; He, K.T.; Pop, E.; Lyding, J.W. Atomic-scale evidence for potential barriers and strong carrier scattering at graphene grain boundaries: A scanning tunneling microscopy study. *ACS Nano* **2013**, *7*, 75–86.

103. Vlassiouk, I.; Regmi, M.; Fulvio, P.; Dai, S.; Datskos, P.; Eres, G.; Smirnov, S. Role of hydrogen in chemical vapor deposition growth of large single-crystal graphene. *ACS Nano* **2011**, *5*, 6069–6076.

104. Li, X.; Magnuson, C.W.; Venugopal, A.; An, J.; Suk, J.W.; Han, B.; Borysiak, M.; Cai, W.; Velamakanni, A.; Zhu, Y.; *et al.* Graphene films with large domain size by a two-step chemical vapor deposition process. *Nano Lett.* **2010**, *10*, 4328–4334.

105. Wu, T.; Ding, G.; Shen, H.; Wang, H.; Sun, L.; Jiang, D.; Xie, X.; Jiang, M. Triggering the continuous growth of graphene toward millimeter-sized grains. *Adv. Funct. Mater.* **2013**, *23*, 198–203.

106. </i>Wang, H.; Wang, G.; Bao, P.; Yang, S.; Zhu, W.; Xie, X.; Zhang, W.-J. Controllable synthesis of submillimeter single-crystal monolayer graphene domains on copper foils by suppressing nucleation. *J. Am. Chem. Soc.* **2012**, *134*, 3627–3630.

107. Li, X.; Magnuson, C.W.; Venugopal, A.; Tromp, R.M.; Hannon, J.B.; Vogel, E.M.; Colombo, L.; Ruoff, R.S. Large-area graphene single crystals grown by low-pressure chemical vapor deposition of methane on copper. *J. Am. Chem. Soc.* **2011**, *133*, 2816–2819.

108. Gao, L.; Ren, W.; Xu, H.; Jin, L.; Wang, Z.; Ma, T.; Ma, L.-P.; Zhang, Z.; Fu, Q.; Peng, L.-M.; *et al.* Repeated growth and bubbling transfer of graphene with millimetre-size single-crystal grains using platinum. *Nat. Commun.* **2012**, *3*.

109. Chae, S.J.; Gunes, F.; Kim, K.K.; Kim, E.S.; Han, G.H.; Kim, S.M.; Shin, H.-J.; Yoon, S.-M.; Choi, J.-Y.; Park, M.H.; *et al.* Synthesis of large-area graphene layers on poly-nickel substrate by chemical vapor deposition: wrinkle formation. *Adv. Mater.* **2009**, *21*, 2328–2333.

110. Zhu, W.; Low, T.; Perebeinos, V.; Bol, A.A.; Zhu, Y.; Yan, H.; Tersoff, J.; Avouris, P. Structure and electronic transport in graphene wrinkles. *Nano Lett.* **2012**, *12*, 3431–3436.

111. Hattab, H.; N'Diaye, A.T.; Wall, D.; Klein, C.; Jnawali, G.; Coraux, J.; Busse, C.; van Gastel, R.; Poelsema, B.; Michely, T.; *et al.* Interplay of wrinkles, strain, and lattice parameter in graphene on iridium. *Nano Lett.* **2012**, *12*, 678–682.

112. Zhang, Y.; Gao, T.; Gao, Y.; Xie, S.; Ji, Q.; Yan, K.; Peng, H.; Liu, Z. Defect-like structures of graphene on copper foils for strain relief investigated by high-resolution scanning tunneling microscopy. *ACS Nano* **2011**, *5*, 4014–4022.

113. Lee, Y.; Bae, S.; Jang, H.; Jang, S.; Zhu, S.-E.; Sim, S.H.; Song, Y., Il; Hong, B.H.; Ahn, J.-H. Wafer-scale synthesis and transfer of graphene films. *Nano Lett.* **2010**, *10*, 490–493.

114. Liang, X.; Sperling, B.A.; Calizo, I.; Cheng, G.; Hacker, C.A.; Zhang, Q.; Obeng, Y.; Yan, K.; Peng, H.; Li, Q.;*et al.* Toward clean and crackless transfer of graphene. *ACS Nano* **2011**, *5*, 9144–9153.

115. Suk, J.W.; Kitt, A.; Magnuson, C.W.; Hao, Y.; Ahmed, S.; An, J.; Swan, A.K.; Goldberg, B.B.; Ruoff, R.S. Transfer of CVD-grown monolayer graphene onto arbitrary substrates. *ACS Nano* **2011**, *5*, 6916–6924.

116. Unarunotai, S.; Murata, Y.; Chialvo, C.E.; Kim, H.-S.; MacLaren, S.; Mason, N.; Petrov, I.; Rogers, J.A. Transfer of graphene layers grown on SiC wafers to other substrates and their integration into field effect transistors. *Appl. Phy. Lett.* **2009**, *95*.

117. Unarunotai, S.; Koepke, J.C.; Tsai, C.-L.; Du, F.; Chialvo, C.E.; Murata, Y.; Haasch, R.; Petrov, I.; Mason, N.;*et al.* Layer-by-layer transfer of multiple, large area sheets of graphene grown in multilayer stacks on a single SiC wafer. *ACS Nano* **2010**, *4*, 5591–5598.

118. Wang, Y.; Zheng, Y.; Xu, X.; Dubuisson, E.; Bao, Q.; Lu, J.; Loh, K.P. Electrochemical delamination of CVD grown graphene film: Toward the recyclable use of copper catalyst. *ACS Nano* **2011**, *5*, 9927–9933.

119. De la Rosa, C.J.L.; Sun, J.; Lindvall, N.; Cole, M.T.; Nam, Y.; Loffler, M.; Olsson, E.; Teo, K.B.K.; Yurgens, A. Frame assisted H_2O electrolysis induced H_2 bubbling transfer of large area graphene grown by chemical vapor deposition on Cu. *Appl. Phy. Lett.* **2013**, *102*.

120. Kobayashi, T.; Bando, M.; Kimura, N.; Shimizu, K.; Kadono, K.; Umezu, N.; Miyahara, K.; Hayazaki, S.; Nagai, S.; Mizuguchi, Y.; *et al.* Production of a 100-m-long high-quality Graphene transparent conductive film by roll-to-roll chemical vapor deposition and transfer process. *Appl. Phy. Lett.* **2013**, *102*.

121. Yan, C.; Kim, K.-S.; Lee, S.-K.; Bae, S.-H.; Hong, B.H.; Kim, J.-H.; Lee, H.-J.; Ahn, J.-H. Mechanical and environmental stability of polymer thin-film-coated graphene. *ACS Nano* **2012**, *6*, 2096–2103.

122. Han, T.-H.; Lee, Y.; Choi, M.-R.; Woo, S.-H.; Bae, S.-H.; Hong, B.H.; Ahn, J.-H.; Lee, T.-W. Extremely efficient flexible organic light-emitting diodes with modified graphene anode. *Nat. Photonics* **2012**, *6*, 105–110.

123. Wehling, T.O.; Novoselov, K.S.; Morozov, S.V.; Vdovin, E.E.; Katsnelson, M.I.; Geim, A.K.; Lichtenstein, A.I. Molecular doping of graphene. *Nano Lett.* **2008**, *8*, 173–177.

124.   Hellstrom, S.L.; Vosgueritchian, M.; Stoltenberg, R.M.; Irfan, I.; Hammock, M.; Wang, Y.B.; Jia, C.; Guo, X.; Gao, Y.; Bao, Z. Strong and Stable doping of carbon nanotubes and graphene by MoO$_x$ for transparent electrodes. *Nano Lett.* **2012**, *12*, 3574–3580.

125.   Lee, B.; Chen, Y.; Duerr, F.; Mastrogiovanni, D.; Garfunkel, E.; Andrei, E.Y.; Podzorov, V. Modification of electronic properties of graphene with self-assembled monolayers. *Nano Lett.* **2010**, *10*, 2427–2432.

126.   Ni, G.-X.; Zheng, Y.; Bae, S.; Tan, C.Y.; Kahya, O.; Wu, J.; Hong, B.H.; Yao, K.; Ozyilmaz, B. Graphene_ferroelectric hybrid structure for flexible transparent electrodes. *ACS Nano* **2012**, *6*, 3935–3942.

127.   Blake, P.; Brimicombe, P.D.; Nair, R.R.; Booth, T.J.; Jiang, D.; Schedin, F.; Ponomarenko, L.A.; Morozov, S.V.; Gleeson, H.F.; Hill, E.W.; *et al.* Graphene-based liquid crystal device. *Nano Lett.* **2008**, *8*, 1704–1708.

128.   Tung, V.C.; Chen, L.-M.; Allen, M.J.; Wassei, J.K.; Nelson, K.; Kaner, R.B.; Yang, Y. Low-temperature solution processing of graphene-carbon nanotube hybrid materials for high-performance transparent conductors. *Nano Lett.* **2009**, *9*, 1949–1955.

129.   Kim, Y.-K.; Min, D.-H. Durable large-area thin films of graphene/carbon nanotube double layers as a transparent electrode. *Langmuir* **2009**, *25*, 11302–11306.

130.   Kholmanov, I.N.; Magnuson, C.W.; Aliev, A.E.; Li, H.; Zhang, B.; Suk, J.W.; Zhang, L.L.; Peng, E.; Mousavi, S.H.; Khanikaev, A.B.; *et al.* Improved electrical conductivity of graphene films integrated with metal nanowires. *Nano Lett.* **2012**, *12*.

131.   Jeong, C.; Nair, P.; Khan, M.; Lundstrom, M.; Alam, M.A. Prospects for nanowire-doped polycrystalline graphene films for ultratransparent, highly conductive electrodes. *Nano Lett.* **2011**, *11*, 5020–5025.

132.   Zhu, Y.; Sun, Z.; Yan, Z.; Jin, Z.; Tour, J.M. Rational design of hybrid graphene films for high-performance transparent electrodes. *ACS Nano* **2011**, *5*, 6472–6479.

133.   Ho, X.; Lu, H.; Liu, W.; Tey, J.N.; Cheng, C.K.; Wei, J. Electrical and optical properties of hybrid transparent electrodes that use metal grids and graphene films. *J. Mater. Res.* **2013**, *28*, 620–626.

134.   De, S.; Lyons, P.E.; Sorel, S.; Doherty, E.M.; King, P.J.; Blau, W.J.; Nirmalraj, P.N.; Boland, J.J.; Scardaci, V.; Joimel, J.; *et al.* Transparent, flexible, and highly conductive thin films based on polymer nanotube composites. *ACS Nano* **2009**, *3*, 714–720.

CHAPTER 11

C$_{59}$N Peapods Sensing the Temperature

Yongfeng Li[1,2,*], Toshiro Kaneko[2] and Rikizo Hatakeyama[2,*]

[1]State Key Laboratory of Heavy Oil Processing, College of Chemical Engineering, China University of Petroleum, Changping District, Beijing 102249, China

[2]Department of Electronic Engineering, Tohoku University, Sendai 980-8579, Japan

ABSTRACT

We report the novel photoresponse of nanodevices made from azafullerene (C$_{59}$N)-encapsulated single-walled carbon nanotubes (C$_{59}$N@SWNTs), so called peapods. The photoconducting properties of a C$_{59}$N@SWNT are measured over a temperature range of 10 to 300 K under a field-effect transistor configuration. It is found that the photosensitivity of C$_{59}$N@SWNTs depends very sensitively on the temperature, making them an attractive candidate as a component of nanothermometers covering a wide temperature range. Our results indicate that it is possible to read the temperature by monitoring the optoelectronics signal of C$_{59}$N@SWNTs. In particular, sensing low temperatures would become more convenient and easy by giving a simple light pulse.

Keywords: azafullerenes; carbon nanotubes; encapsulation; photoresponse

1. INTRODUCTION

Sensing over a wide variety of temperature ranges is a challenging task, and most of the traditional thermometers are usually bulky, too large, and expensive. It has been demonstrated previously that the encapsulation of gallium inside a carbon nanotube is useful for a precise temperature measurement [1], but detecting low temperatures below 50 °C is limited due to the solidification of gallium. Recently, the research on single-walled carbon nanotubes (SWNTs) has uncovered numerous intriguing electronic and optical properties that could be used to develop new "smart" SWNT-based device systems with more degrees of freedom in performance by appropriate modification owing to their unique electrical and structural properties. In particular, compared with the case of multi-walled carbon nanotubes, the conductance change in SWNTs after modification with molecules or atoms makes them a potential material for sensor development [2–6]. In addition, one-dimensional structures and nanometer-range diameters of SWNTs make it possible to develop a high-density nanosensor array within a limited space. For example, SWNTs based on field-effect transistor configurations have shown unique or enhanced sensitivity toward gaseous species attracted much attention [7–10]. However, up to now there are few reports about the sensor performance of SWNTs which is related to the temperature sensor.

In this work, it is the first time for us to develop a temperature sensor based on the configuration of field-effect transistor (FET) with a $C_{59}N@SWNT$ as current channel. Our findings indicate that the photoinduced current of $C_{59}N@SWNTs$ depends sensitively on the temperature, which makes them a promising candidate as a component of nanothermometers.

2. EXPERIMENTAL SECTION

The azafullerene $C_{59}N$ was synthesized by a nitrogen plasma ion-irradiation method, which was confirmed by using a laser-desorption time-of-flight mass spectrometer (LD-TOF-MS, Shimadzu AXIMA-CFR+). The encapsulation of azafullerene $C_{59}N$ inside SWNTs is realized by either a vapor reaction method or a plasma ion-irradiation method [11]. In the case of the vapor reaction method, the purified SWNTs together with $C_{59}N$ azafullerene powders are first sealed in a glass tube under vacuum $\sim 10^{-5}$ Torr. Then the sealed glass tube is heated at 420 °C for 48 h to encapsulate $C_{59}N$ in SWNTs. Raw samples are obtained after the above process, and then purified via a washing process in toluene to remove the excess $C_{59}N$ attached to the surface of SWNTs. The purified peapods are examined in detail by Field

Emission Transmission Electron Microscopy (FE-TEM, Hitachi HF-2000 and quanta 200F) operated at 200 kV.

The electronic transport properties of $C_{59}N@SWNTs$ are investigated by using them as the current channels of field-effect transistor (FET) devices [11,12]. First the $C_{59}N@SWNTs$ sample is ultrasonically dispersed in N,N-dimethylformamide and then spincoated onto FET substrates, each of which consists of Au source-drain electrodes with a channel length of 500 nm on a SiO_2 insulating layer. A heavily doped Si substrate serves as a backgate. Photoinduced transport measurements are performed in the temperature range of 10–300 K under vacuum conditions on a semiconductor parameter analyzer (Agilent 4155C). Light illumination is carried out by a 150 W Xe lamp (LSX-2501) equipped with a monochromator to select the incident excitation wavelength (390~1,100 nm). The light illumination intensity is less than 50 mW/cm^2.

3. RESULTS AND DISCUSSION

A TEM observation result of $C_{59}N@SWNTs$ is shown in Figure 1(a), which indicates that $C_{59}N$ molecules with spherical symmetry are filled into both individual and bundled SWNTs. In comparison with an empty pristine SWNT, $C_{59}N$ molecules with spherical symmetry are clearly observed in the individual SWNTs, forming a one-dimensional chain-like structure inside the SWNTs, as illustrated by arrows in the TEM image. Therefore, the TEM observation provides strong evidence that $C_{59}N$ molecules are encapsulated inside SWNTs. The energy dispersive X-ray analysis is also used to distinguish the elements in the $C_{59}N@SWNTs$ during TEM measurements, as indicated in Figure 1(b). However, only C and Cu originated from the TEM copper grid are detected, and there is no signal of N found in the spectrum due to very small amount of N from $C_{59}N$. A schematic diagram of FET device with a $C_{59}N@SWNT$ as current channel is illustrated in Figure 2(a), and its corresponding AFM image of the FET device is shown in Figure 2(b), in which a $C_{59}N@SWNT$ contacting two Au electrodes is clearly observed, which well confirms that an individual $C_{59}N@SWNT$ indeed plays the role of the current channel with gap width of 500 nm.

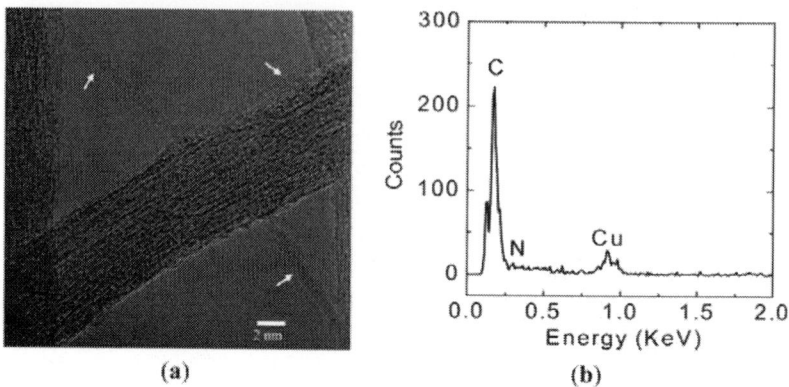

(a)   (b)

Figure 1. (a) A TEM image of $C_{59}N$@SWNTs. (b) EDX of $C_{59}N$@SWNTs.

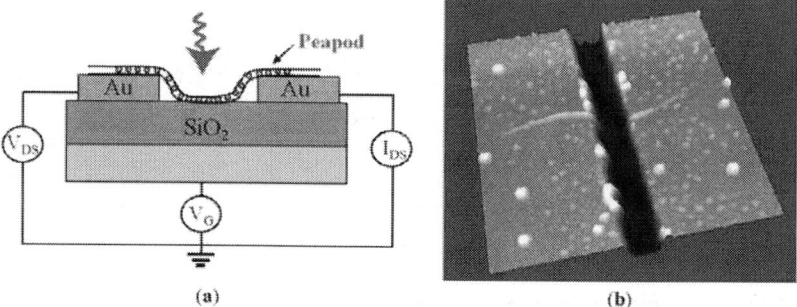

(a)   (b)

Figure 2. (a) Schematic illustration of FET configuration with a SWNT as current channel. (b) An AFM image of $C_{59}N$@SWNT-FET.

The transport properties of pristine semiconducting SWNTs are well known to exhibit the *p*-type behavior, as shown in Figure 3(a) [11], where a characteristic curve of source-drain current I_{DS} *versus* gate voltage V_G is described for source-drain voltage V_{DS} = 1 V. Figure 3(b) presents the transport property of $C_{59}N$@SWNTs where the I_{DS}-V_G curve is measured at V_{DS} = 1 V. In contrast, the transport property of $C_{59}N$@SWNTs drastically changes to an *n*-type semiconductor. This *n*-type characteristic is attributed to the charge transfer between $C_{59}N$ and local parts of SWNTs [11–13]. It is found that such azafullerene-induced characteristics have been observed in many independent SWNTs devices and they have good reproducibility under measurements performed with different source-drain voltages, which has been confirmed in our previous work [11].

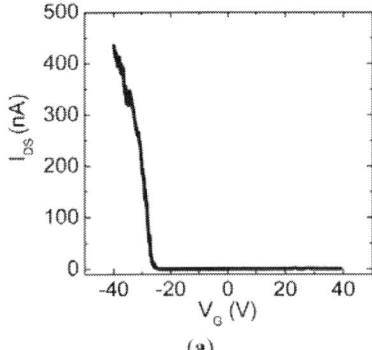
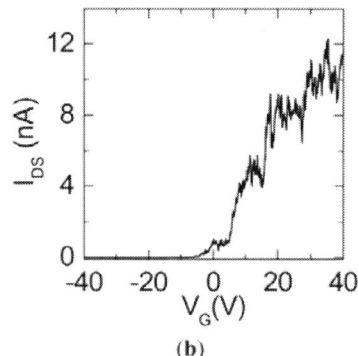

Figure 3. (a) I_{DS}-V_G curve measured with V_{DS} = 1 V for a pristine SWNT-FET device. **(b)** I_{DS}-V_G curve measured with V_{DS} = 1 V for a $C_{59}N@SWNT$-FET device.

Under light illumination, it is noticed that the photoresponse of transfer characteristics for $C_{59}N@SWNT$-FET is strikingly different at different temperatures. Figure 4 shows the transfer curves of a $C_{59}N$ peapod FET device, in which the I_{DS}-V_G curves are recorded for the $C_{59}N@SWNT$-FET in both dark and upon 400 nm light illumination for the two temperatures at 300 K and 10 K, respectively. Obviously, the prominent response of the device at room temperature to light is the decrease of transconductance, and the light irradiation results in ~95% decrease in conductance (Figure 4(a)), which has been mentioned in our previous report [12]. However, at low temperatures such as 10 K, a different photoresponse phenomenon is observed in the transport property of $C_{59}N$ peapod FET device under light illumination, as shown in Figure 4(b), and the source-drain current displays a several times increase under the same light illumination, which is the exactly opposite phenomenon to that observed at 300 K. It is necessary to mention that such a phenomenon has never been observed in pristine SWNT FET devices. This finding indicates that the response of transport properties of $C_{59}N$ peapod FET device significantly depends on the variations of the temperature. On the other hand, the above interesting phenomenon implies a clear photoinduced electron transfer process. To further investigate the photoswitching characteristics at low temperatures, we have further exposed the device to the light pulse (1 s) during sweeping the I_{DS}-V_G curves at 10 K, as seen in Figure 5. As the gate voltage is continually swept (with sweeping speed ~1.4 V/s), the current at V_G = 21 V shows a sudden increase and the current value is about 6 times larger than its original one. After scanning to the high positive gate voltage V_G = 40 V, the measured I_{DS} is two times larger compared with the case of no light illumination. Again, when the 400 nm light pulse exposure is given at V_G = 26 V during I_{DS}-V_G sweeping, the similar sharp increase of current is observed, suggesting that such an effect of current increase is fully reproducible by exposure to the light pulse and disappears without light illumination, demonstrating the complete restoration of photoswitching effect. The results confirm well that the $C_{59}N$@SWNT FET device also exhibits an ultra fast response (on the level of millisecond) to the pulsed light, and the measured current is drastically enhanced under instantaneous UV illumination (400 nm, 1 s), which is entirely consistent with the result observed in Figure 4(b). Moreover, the I_{DS} measured (V_G = 20, V_{DS} = 0.5 V) as a function of time at 10 K under exposure of a light pulse for a $C_{59}N$@SWNT-FET device is shown in Figure 6, suggesting strong evidence for the great increase in current under the light pulse. This finding is well consistent with the measured results in Figure 5, which is different from that observed at room temperature.

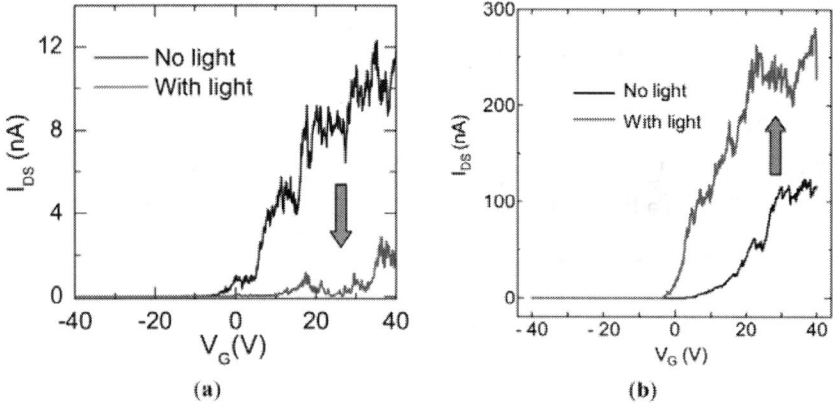

Figure 4. (a) I_{DS}-V_G characteristics (V_{DS} = 1 V) of a $C_{59}N$@SWNT-FET device measured without and with light (400 nm) illumination at room temperature 300 K. **(b)** I_{DS}-V_G characteristics (V_{DS} = 0.5 V) of a $C_{59}N$@SWNT-FET device measured without and with light (400 nm) illumination at low temperature of 10 K.

In order to understand the effect of temperatures, we have further measured photoinduced characteristics of current *vs.* light pulse at different temperatures, as indicated in Figure 7. Interestingly, the current increase is found to depend inversely on the temperature, and becomes gradually negligible when the temperature is increased from 10 to 90 K. As the temperature is further increased to 140 K, a clear negative photocurrent, *i.e.*, a decrease in current is observed, as seen in Figure 7(c), upon pulsed light illumination. Up to 300 K, a significant decrease of current upon the light pulse is observed, just in agreement with the result in Figure 4(a). Therefore, the above results suggest that it is possible to read the temperature by monitoring the optoelectronics signal of $C_{59}N$@SWNT-FET.

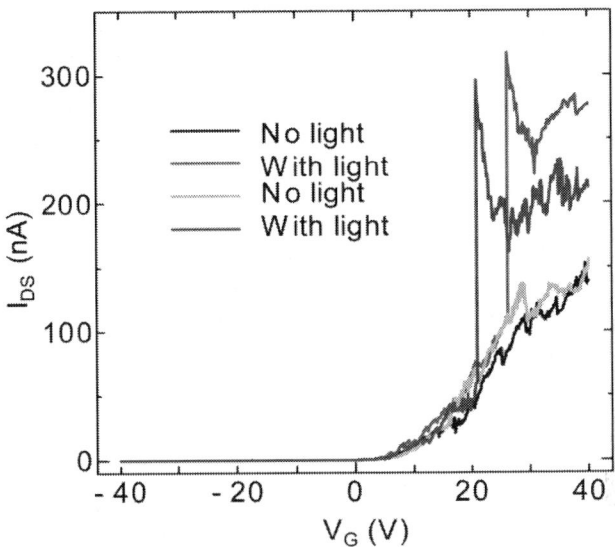

Figure 5. I_{DS}-V_G characteristics (V_{DS} = 0.5 V) measured at 10 K for an n-type $C_{59}N@SWNT$ with a light pulse (400 nm) at V_G = 21 V and 26 V.

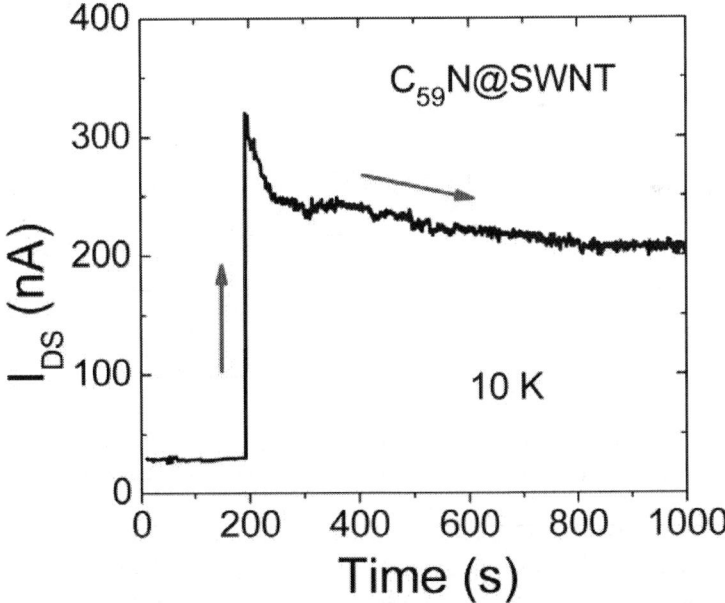

Figure 6. I_{DS} measured as a function of time at 10 K under exposure of a light pulse (400 nm) for a $C_{59}N@SWNT$-FET device with V_{DS} = 0.5 V.

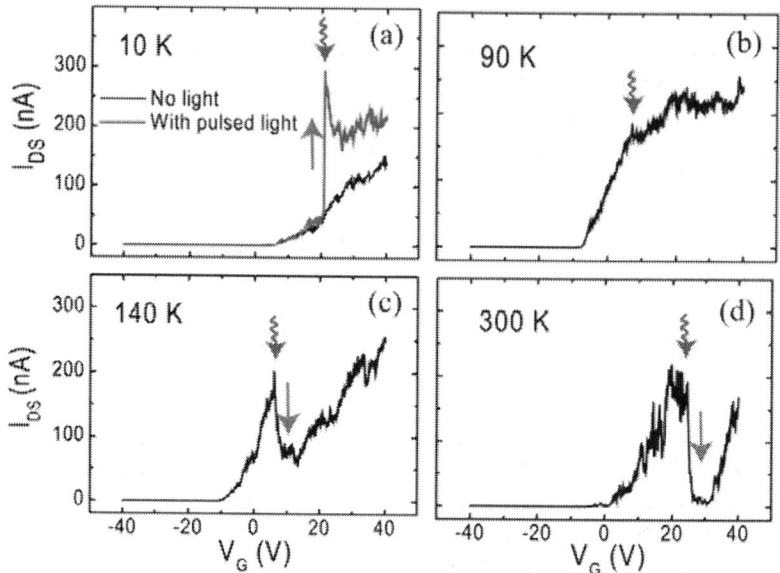

Figure 7. Photoinduced currents observed during tracing the I_{DS}-V_G curves upon a light pulse at (**a**) 10 K, (**b**) 90 K, (**c**) 140 and (**d**) 300 K, respectively.

Figure 8 presents the ratio of the changed current (ΔI_{DS}) caused by instantaneous light illumination to the original current ($\Delta I_{DS}/I_{DS}$) as a function of temperature in the range of 10–300 K. A variation of photoinduced current *vs.* temperature indicates that when the temperature is decreased and increased from 90 K in the range of 10–300 K, the positive and negative photocurrents rise, respectively. In other words, the photocurrent is found to depend inversely on the temperature, and it becomes gradually negligible when the temperature is increased from 10 to 90 K. As the temperature is further increased from 90 K to 300 K, a negative photocurrent is observed upon pulsed light illumination. This finding reveals that it is possible to read the temperature by monitoring the optoelectronics signal of $C_{59}N@SWNTs$. In particular, sensing low temperatures would become more convenient and easy by giving a simple light pulse. In order to understand the photoswitching mechanism, we have further measured the transport properties of C_{60} fullerenes encapsulated SWNT ($C_{60}@SWNT$) under the same experimental conditions of light illumination. There is no big change in the conductance under light illumination when the sample was measured at room and low temperatures, implying that the azafullerene is responsible for the decrease of conduction. According to our previous work [11–14], the n-type transport behavior of $C_{59}N@SWNT$ is considered to be due to the charge transfer from monomer $C_{59}N$ to SWNT by the weak C-C bonding since the azafullerenes $C_{59}N$ can easily lose or gain electrons through regioselective reactions. According to theoretical calculations [15], such bonding can easily undergo homolysis under photolysis or thermolysis conditions, resulting in the formation of azafullerenenyl radical $C_{59}N^{\bullet}$. When the light energy is higher than the bonding energy, the thermolysis of the bond will lead to the stop of charge transfer, leading to a decrease in current at room temperature. To confirm this, we have also measured the transport behavior of $C_{59}N@SWNT$ at high temperatures (supporting information), and find that the current becomes unstable when the temperature reaches 400 K which indicates that the bond between $C_{59}N$ and SWNT will break due to the high system temperature or light absorption. However, the thermal effect is reduced in the low temperature environment, as a result, the weak bonding between $C_{59}N$ and SWNT may remain linked under light illumination. On the other hand, it was reported that the $C_{59}N^{+}$ exhibits distinguishing absorption spectral features in absorption spectra in the range of 1–3 eV [16]. The light energy exerted on the $C_{59}N@SWNT$-FET device is in the range of 1.24–3.1 eV. Therefore, the multiplication effect cannot be ruled out since the quantum efficiency will be greatly enhanced at low temperatures. The above phenomenon might lead to the increase of photoinduced current at low temperatures although the exact mechanism is still unclear at this stage.

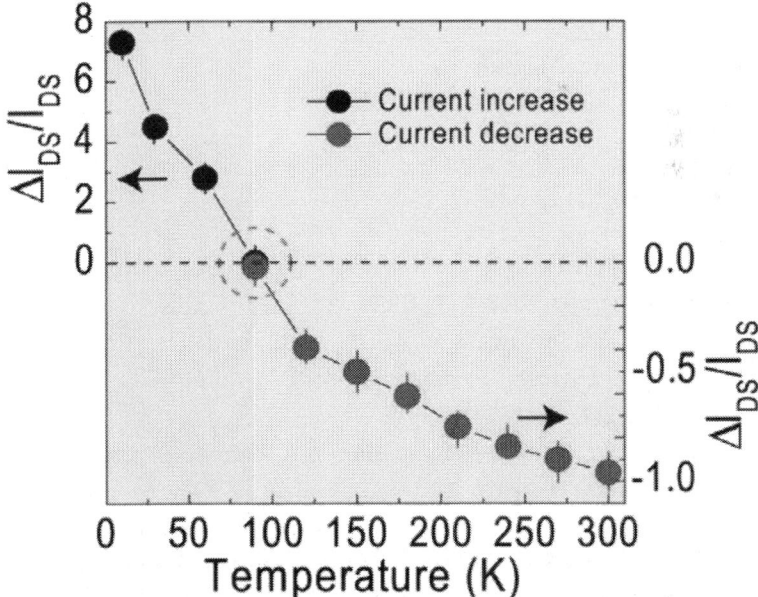

Figure 8. Variation of photoinduced current ($\Delta I_{DS}/I_{DS}$) measured with temperature under exposure of a light pulse (400 nm).

4. CONCLUSIONS

In summary, we have measured the transport properties of azafullerene peapods which act as the channels of FET devices, both in the dark and upon light illumination. It is found that the photosensitivity of $C_{59}N@SWNT$-FET device depends very sensitively on the temperature compared with that of pristine SWNTs. At low temperatures the currents measured with sweeping gate voltages exhibit a remarkable increase, and the variation value of current is significantly dependent on the temperatures. However, the measured current shows a significant decrease when the device is illuminated with light at high temperatures, which is in sharp contrast to the low-temperature photoresponse. Such nanopeapod devices with distinguishing photoinduced properties are expected to have promising applications as a component of nanothermometers covering a wide temperature range.

ACKNOWLEDGMENTS

We gratefully thank for the National Natural Science Foundation of China (No. 21106184), the Science Foundation Research Funds Provided to New Recruitments of China University of Petroleum, Beijing (No. YJRC-2011-18), and Thousand Talents Program.

REFERENCES

1. Gao, Y.; Bando, Y. Carbon nanothermometer containing gallium. *Nature* **2002**, *415*, 599.
2. Pradhan, B.; Setyowati, K.; Liu, H.; Waldeck, D.H.; Chen, J. Carbon nanotube–polymer nanocomposite infrared sensor. *Nano Lett.* **2008**, *8*, 1142–1146.
3. Ulissi, Z.W.; Zhang, J.; Boghossian, A.A.; Reuel, N.F.; Shimizu, S.F.E.; Braatz, R.D.; Strano, M.S. Applicability of birth–death markov modeling for single-molecule counting using single-walled carbon nanotube fluorescent sensor arrays. *J. Phys. Chem. Lett.* **2011**, *2*, 1690–1694.
4. Kauffman, D.R.; Sorescu, D.C.; Schofield, D.P.; Allen, B.L.; Jordan, K.D.; Star, A. Understanding the sensor response of metal-decorated carbon nanotubes. *Nano Lett.* **2010**, *10*, 958–963.

5. Wang, L.; Chen, W.; Xu, D.; Shim, B.S.; Zhu, Y.; Sun, F.; Liu, L.; Peng, C.; Jin, Z.; Xu, C.; Kotov, N.A. Simple, rapid, sensitive, and versatile SWNT–paper sensor for environmental toxin detection competitive with ELISA. *Nano Lett.* **2009**, *9*, 4147–4152.

6. Fu, Q.; Lu, C.; Liu, J. Selective coating of single wall carbon nanotubes with thin SiO_2 layer. *Nano Lett.* **2002**,*2*, 329–332.

7. Mubeen, S.; Zhang, T.; Chartuprayoon, N.; Mulchandani, A.Y.R.; Myung, N.V.; Deshusses, M.A. Sensitive detection of H_2S using gold nanoparticle decorated single-walled carbon nanotubes. *Anal. Chem.* **2010**, *82*, 250–257.

8. An, K.H.; Jeong, S.Y.; Hwang, H.R.; Lee, Y.H. Enhanced sensitivity of a gas sensor incorporationg single-walled carbon nanotub-polypyrrole nanocomposites. *Adv. Mater.* **2004**, *16*, 1005–1009.

9. Star, A.; Joshi, V.; Skarupo, S.; Thomas, D.; Gabriel, J.-C.P. Gas sensor array based on metal-decorated carbon nanotubes. *J. Phys. Chem. B* **2006**, *110*, 21014–21020.

10. Shirsat, M.D.; Sarkar, T.; Kakoullis, J.; Myung, N.V., Jr.; Konnanath, B.; Spanias, A.; Mulchandani, A. Porphyrin-functionalized single-walled carbon nanotube chemiresistive sensor arrays for VOCs. *J. Phys. Chem. C* **2012**, *116*, 3845–3850.

11. Kaneko, T.; Li, Y.F.; Nishigaki, S.; Hatakeyama, R. Azafullerene encapsulated single-walled carbon nanotubes with n-type electrical transport property. *J. Am. Chem. Soc.* **2008**, *130*, 2714–2715.

12. Li, Y.F.; Kaneko, T.; Kong, J.; Hatakeyama, R. Photoswitching in azafullerene encapsulated single-walled carbon nanotube FET devices. *J. Am. Chem. Soc.* **2009**, *131*, 3412–3413.

13. Cuong, N.T.; Otani, M.; Iizumi, Y.; Okazaki, T.; Rotas, G.; Tagmatarchis, N.; Li, Y.F.; Kaneko, T.; Hatakeyama, R.; Okada, S. Orgin of n-type transport behavior of azafullerene encapsualted single-walled carbon nanotubes. *Appl. Phys. Lett.* **2011**, *99*.

14. Li, Y.F.; Kaneko, T.; Miyanaga, S.; Hatakeyama, R. Synthesis and property characterization of $C_{69}N$ azafullerne encapsulated single-walled carbon nanotubes. *ACS Nano* **2010**, *4*, 3522–3526.

15. Andreoni, W.; Curioni, A.; Holczer, K.; Prassides, K.; Keshavarz-K, M.; Hummelen, J.; Wudl, F. Unconventional bonding of azafullerenes: Theory and experiment. *J. Am. Chem. Soc.* **1996**, *118*, 11335–11336.

16.   Xie, R.; Bryant, G.W.; Sun, G.; Nicklaus, M.C.; Heringer, D.; Frauenheim, Th.; Manaa, M.R.; Smith, V.H., Jr.; Araki, Y.; Ito, O. Excitations, optical absorption spectra, and optical excitonic gaps of heterofullerenes. I. C_{60}, $C_{59}N^+$, and $C_{48}N_{12}$: Theory and experiment. *J. Phys. Chem.* **2004**, *120*, 5133–5147.

Molecular Quantum Spintronics: Supramolecular Spin Valves Based on Single-Molecule Magnets and Carbon Nanotubes

Matias Urdampilleta[1], Ngoc-Viet Nguyen[1], Jean-Pierre Cleuziou[1], Svetlana Klyatskaya[2], Mario Ruben[2,3] and Wolfgang Wernsdorfer [1,*]

[1]Institut Néel, CNRS et Université Joseph Fourier, BP 166, F-38042 Grenoble Cedex 9, France
[2]Institute of Nanotechnology (INT), Karlsruhe Institute of Technology (KIT), 76344 Eggenstein-Leopoldshafen, Germany
[3]Institute de Physique et Chimie de Matériaux de Strasbourg (IPCMS), CNRS-Université de Strasbourg, 67034 Strasbourg, France

ABSTRACT

We built new hybrid devices consisting of chemical vapor deposition (CVD) grown carbon nanotube (CNT) transistors, decorated with TbPc$_2$ (Pc = phthalocyanine) rare-earth based single-molecule magnets (SMMs). The drafting was achieved by tailoring supramolecular π-π interactions between CNTs and SMMs. The magnetoresistance hysteresis loop measurements revealed steep steps, which we can relate to the magnetization reversal of individual SMMs. Indeed, we established that the electronic transport properties of these devices depend strongly on the relative magnetization orientations of the grafted SMMs. The SMMs are playing the role of localized spin polarizer and analyzer on the CNT electronic conducting channel. As a result, we measured magneto-resistance ratios up to several hundred percent. We used this spin valve effect to confirm the strong uniaxial anisotropy and the superparamagnetic blocking temperature ($T_B \sim$ 1 K) of isolated TbPc$_2$ SMMs. For the first time, the strength of exchange interaction between the different SMMs of the molecular spin valve geometry could be determined. Our results introduce a new design for operable molecular spintronic devices using the quantum effects of individual SMMs.

Keywords: molecular quantum spintronics; molecular magnets; nanoelectronics devices

1. INTRODUCTION

Single molecule magnets (SMMs) have attracted much interest over the last years because of their unique magnetic properties. These molecular structures combine the classical properties of magnets with the intrinsic quantum nature of nanoscale entities. With a large spin ground state and a magnetic anisotropy well-defined, molecular clusters composed of few magnetic atoms have shown various properties such as the blocking of the spin orientation at low temperatures, quantum tunneling of magnetization (QTM) [1] and interference effects between tunneling paths [2]. Besides, the synthetic chemistry produces controlled molecular structures at high yield and low cost. As a result, a wide range of SMMs systems incorporating transition metal and/or rare earth metal ions with tailored magnetic interactions have been discovered. In addition, the rich variety of quantum systems provided by the molecular magnetism field strongly motivates the use of SMMs for both quantum information storage and processing purposes.

At the same time, studies in spintronics using magnetic materials in electronic devices have made considerable progress from fundamental studies to practical applications. This technology is based on the discovery of magnetoresistive effects, such as the giant magnetoresistance effect, where metallic spin valves are composed of two metallic layers separated by a non-magnetic one. Depending on the relative magnetization orientation of the two magnets (parallel or antiparallel), a drastic change of the electrical resistance is observed. Nowadays, new directions in spintronics aim at transposing the existing concepts and at developing alternative ones with various types of materials, from inorganic to π-conjugated organic semiconductors [3]. Organic semiconductors are promising since they may offer longer spin relaxation times [4] than conventional transition metals as well as new functionalities (e.g., switchability with light, electric field, *etc.*). In this context, SMMs are interesting candidates to study and preserve quantum coherence of the electronic spin in molecular spintronics devices. Such devices lead the way to the electronic detection and coherent manipulation of SMMs spin states, important for quantum computation schemes at the single molecule level.

In this work, proposed recently [5], we realized a device consisting of SMMs anchored by supramolecular interactions on the sidewall of a chemical vapor deposition (CVD) grown carbon nanotube (CNT), itself connected to a three-terminal (transistor) geometry [6,7]. The CNT acts as a path for conducting electrons so that electronic transport does not occur directly through the magnetic orbitals of the SMM. This prevents charge-induced excitations or relaxation of molecular spin states. In particular, we showed that the electronic transport is extremely sensitive to the orientation of local magnetic moments. This property allows the electrical detection of magnetization reversal of individual molecular spins. In addition, a spin valve effect with two molecules leads to very large variations of the conductance with magnetoresistance ratios of up to several hundred percent. Our approach differs from previous realizations of carbon based spin valves [8–10] and does not imply magnetic leads. Indeed, the spin-dependent transport through this supramolecular spin valve is completely determined by the magnetic properties of the molecular species magnetically coupled to the conducting channel of the CNT. Similar results were recently obtained using graphene nanoconstriction decorated with TbPc$_2$ magnetic molecules [11]. In this case, a magnetoconductivity signal as high as 20% was found for the spin reversal. These results show the behavior of multiple-field-effect nanotransistors with sensitivity at the single-molecule level.

The paper is structured as follows. In Section 2, we introduce different methods used to fabricate and measure the supramolecular based junctions. The results section (Section 3) describes the magnetic properties of the pyrene functionalized heteroleptic *bis*-phthalocyaninato-Terbium (III) SMMs (called TbPc$_2$ in the following) used in this study. In Section 3.1, we briefly discuss the magnetization reversal mechanisms of the TbPc$_2$ SMM, namely QTM and the direct relaxation process, using μ-SQUID measurements on diluted crystals of molecules. Section 3.2 exhibits the electronic transport features of the spin valve, being in the closed quantum dot regime. Then, in Section 3.3, we present the spin-valve behavior revealed by the magneto-conductance measurements under magnetic field sweeps. The anisotropic dependence of the hysteretic conductance jumps is also studied and in good agreement with the expected uniaxial anisotropy of TbPc$_2$ SMMs. Finally, Section 4 discusses the spin valve mechanism in further detail and we conclude with a brief outlook.

2. DEVICES FABRICATION

In this project we studied about 150 samples of which 28 showed magnetic signals related to the $TbPc_2$-SMMs and eight of them were studied in detail, manifesting similar behavior concerning their magneto-conductance. In contrast to our first publication [6], the sample presented in this paper was fabricated using CVD nanotubes. Catalyst islands were designed on SiO_2 by creating holes by optical lithography in LOR3A resist, which were filled with Fe/Mo catalyst in nanoporous alumina. After liftoff of the resist, nanotubes were grown in a Firstnano CVD oven at 750 °C, whereby methane was used as a carbon source [12]. SWNTs were located by AFM or SEM with respect to lithographically patterned markers and then contacted with 50 nm Pd by standard electron-beam lithography, defining 200 nm long CNT junctions. A solution containing bis(phthalocyaninato)terbium (III) substituted with pyrene groups [13] with a 10^{-6} molarity was dropped on the device and dried under nitrogen flow. Samples with large resistance (>100 kΩ) at room temperature were micro bounded and measurements were carried out in a dilution fridge with a base temperature of 40 mK. The electronic temperature was estimated to be around 150 mK.

3. RESULTS

3.1. Magnetic Properties of the $TbPc_2$ SMMs

Among the existing SMMs families, single ion lanthanide complexes are among the most simple and robust systems [13]. Here we focus our attention on the $TbPc_2$ SMM based on a single Tb^{3+} ion coordinated to two phthalocyanine (Pc) ligands as depicted in Fdigure 1(a). The SMM behavior originates from the electronic multiplet substructure of the Tb^{3+} ion in its $(4f)^8$ electronic configuration leading to a $J = 6$ magnetic moment. The $TbPc_2$-SMM exhibits a significantly large axial magnetic anisotropy originating from the strong spin-orbit coupling in lanthanide ions and from the ligand field potential made by the two Pc ligands. It leads to a well-defined ground state configuration ($J = 6$, $|J_z| = 6$) separated from the first excited state ($J = 6$, $J_z = 5$) by an energy splitting $\Delta E \sim 600$ K [14]. All our measurements were performed at very low temperatures, so that we only consider the lowest energy substates available with $J_z = \pm 6$, the corresponding Zeeman diagram is plotted inFigure 1(b). Apart from the Terbium(III) ion, the molecule has a spin ½ delocalized over the two phtalocyanines groups [15]. This unpaired electron mediates a magnetic coupling between the Terbium 4f electrons and its environment.

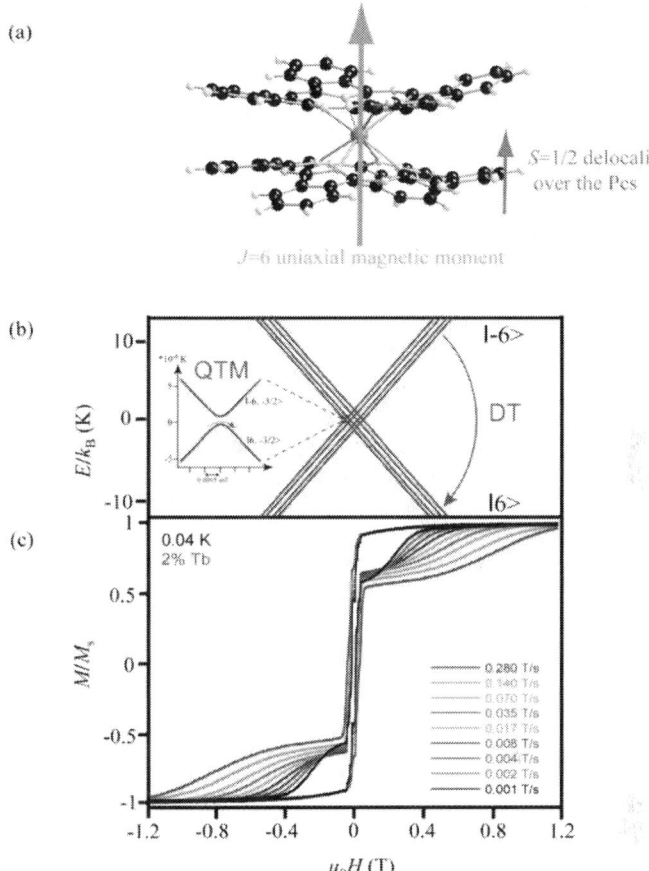

Figure 1. Magnetization reversal mechanisms in the TbPc₂ SMM (alkyl and pyrene substituents are omitted for reasons of clarity). (**a**) Scheme of the TbPc₂-SMM. The terbium ion has a $J = 6$ magnetic moment and an unpaired electron is delocalized over the organic part; (**b**) Zeeman diagram calculated for the TbPc₂ SMM ground state ($J = 6$, $|J_z| = 6$). The interaction with the Tb nucleus spin $I = 3/2$ splits each electronic substate through the hyperfine coupling, providing a path for Quantum Tunneling of Magnetization (QTM) at the anti-crossing of two levels; (**c**) Hysteresis loops of the crystallized TbPc₂-SMM (2% in the YPc₂ matrix) measured at 40 mK for different sweeping rates ranging from 1 to 280 mT.s⁻¹. QTM reflects in staircase-like steps of the hysteresis loops at low magnetic fields, each step corresponding to a level anti-crossing. Molecules, which did not undergo QTM can relax their magnetization to a lower energy state by the direct transition (DT) occurring at larger magnetic fields.

Figure 1(c) presents μ-SQUID magnetization hysteresis loops of the TbPc$_2$ SMMs at a temperature of 40 mK. The SMMs have been diluted in an YPc$_2$ paramagnetic matrix to minimize any magnetic intermolecular interactions. At an applied magnetic field of −1.2 T along the easy-axis, all magnetic moments in the crystal are saturated along the same direction. Around zero-field, a large number of these magnetic moments switch their magnetization orientation through staircase-like steps of the hysteresis loop. The origin of these magnetization steps is the QTM between the $J_z = 6$ and $J_z = -6$ substates. Since the other excited substates are, energy-wise, far above the fundamental ones, transitions involving the excited states cannot occur. As a consequence, the origin of QTM in single ion rare earth based SMMs differs strongly from the 3d-metal-cluster SMM cases. This comes from the interaction of the Tb ion with the ligand field. Besides this, the Tb nucleus owns a nuclear spin of $I = 3/2$ with a natural isotopic abundance of 100%. The strong hyperfine coupling in TbPc$_2$ leads to a splitting of the $J = 6$ electronic multiplet in several $|J_z\rangle|I_z\rangle$ coupled states. These states are visible in the Zeeman diagram in Figure 1(b). QTM can occur through appropriate magnetic field conditions where two such states are brought to resonance at an avoided crossing of two levels (Figure 1(b)). The remaining magnetic moments, which did not undergo magnetization reversal by QTM around the zero magnetic field, start to reverse at about a few hundreds of mT. This transition is visible in the hysteresis loops in Figure 1(c) as a broad field scan rate dependent step. This one phonon mediated mechanism, called the direct transition (DT), is schematized in Figure 1(b).

We highlight that those measurements have been carried out on an assembly of TbPc$_2$-SMMs in a crystallized form. One could thus wonder whether the SMM properties may still be observed on sub-monolayers or isolated molecules deposited on a surface. Indeed, SMMs may lose their magnetic properties when attached to metallic surfaces, as was shown e.g., for Mn$_{12}$-acetate [16]. A slight modification of the ions surroundings can lead to a drastic change of the crystal field potential and thus to an alteration of the magnetic properties [17]. However, in the case of the TbPc$_2$-SMM, it has been recently demonstrated that the structural and magnetic properties are still conserved [18]. The main fingerprints of SMMs, that is, QTM around zero magnetic field, strong axial anisotropy and high superparamagnetic blocking temperature are, indeed, still present. The TbPc$_2$-SMM is one of the most interesting and reliable systems in order to study magnetism at single molecular level. Besides, this molecule is very well suited to be attached to sp^2 carbon nanomaterials, such as carbon nanotubes or grapheme [19] via supramolecular π-π interactions. This strategy has been used to build our supramolecular spin valve and is described in the following.

3.2. Electronic Transport through Carbon Nanotubes Functionalized by TbPc2-SMMs

The original geometry of our devices is presented in Figure 2(a). A CNT, contacted with non-magnetic electrodes, forms a quantum dot (QD), which is laterally coupled to several TbPc2-SMMs through π-π stacking interaction. Indeed, phtalocyanine groups can be functionalized with pyrenes ligands, allowing a supra-molecular anchoring point and a better coupling to π-conjugated systems such as carbon nanotubes [20].

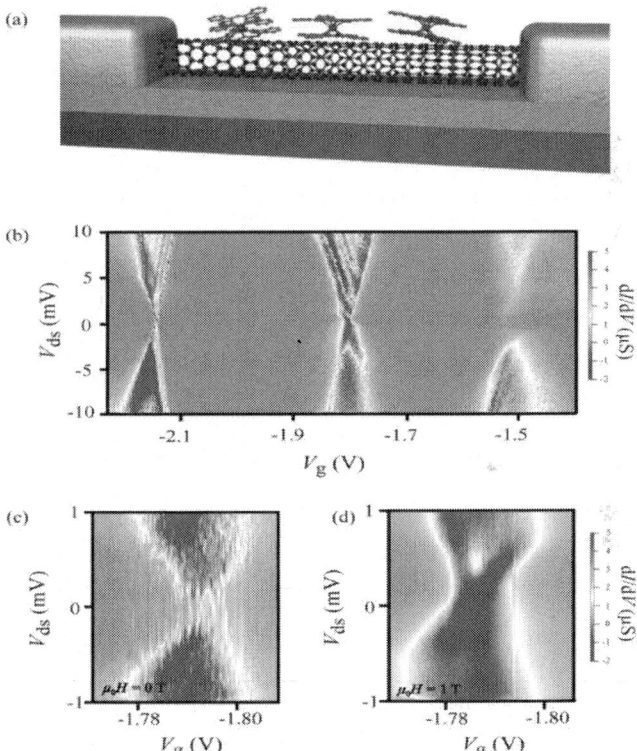

Figure 2. Electronic transport in carbon nanotube quantum dots with grafted TbPc2 molecules. (**a**) Artist view of the device scheme, consisting of an electrically connected carbon nanotube junction, laterally coupled to isolated TbPc2-SMMs; (**b**) Color-scale plots of the differential conductance dI/dV at temperature $T = 40$ mK, as a function of source-drain voltage V_{sd} and back-gate voltage V_g, displaying the charge stability diagram in the Coulomb blockade regime. The typical charging energy is about 20 meV; (**c–d**) Enlarged views of (b), showing the charge degeneracy point around $V_g = 1.79$ V at constant static magnetic fields (c) $\mu_0 H = 0$ T and (d) $\mu_0 H = 1$ T.

When deposited, the TbPc$_2$-SMM anchors on the nanotube in a way that the substituted Pc ligand of the molecule comes directly in contact with the nanotube surface. The energy gain through supramolecular interactions is maximized by the formation of strong aromatic π-π and C-H-π stacking interactions between the substituted Pc and CNT. Because of the efficient hybridization between Pc and CNT orbitals [21], we can assume that the unpaired electron delocalized over the Pc ligands has strong interaction with the conduction electron of the nanotube.

The supra-molecular device structure is cooled down in a ^{3}He/^{4}He dilution refrigerator to 40 mK and a differential conductance measurement is conducted with an Adwin real-time acquisition system, programmed in a lock-in mode. The lock-in amplitude and frequency are set to 50 μV and 33 Hz, respectively. The QD characteristics are measured by bias spectroscopy, that is, the differential conductance dI/dV is plotted in a color code as a function of the back-gate and bias voltages. Figure 2(b) shows the standard Coulomb diamond expected for a CNT QD with a charging energy around 20 meV. Figures 2(c,d) display a zoom on the degeneracy points without and under magnetic field (1 T). At low bias (<1 mV) and without magnetic field (Figure 2(c)), the degeneracy point has a noisy conductance, which fluctuates between two values. Then under 1 T, the conductance becomes stable and closed (Figure 2(d)). These features reveal the presence of an extra tunnel barrier in the QD [22], which can be modulated by the magnetic field.

3.3. Characterization of the Strongly Anisotropic Spin-Valve Effect

The magnetoresistance measurements, at the previously mentioned degeneracy point (see Figure 2), are presented in Figure 3(a). At −1 T, the differential conductance is saturated to its maximum value. Sweeping up the magnetic field at 20 mT/s until zero-field, dI/dV drops down abruptly to its minimum value. When still increasing the field, dI/dV abruptly recovers its original value. The complete measurement from −1 T to +1 T (trace) and back to −1 T (retrace) forms a hysteresis loop, which is characteristic of a spin-valve device.

Another remarkable feature of our device is its anisotropic response: we observed that the field values at which sharp conductance jumps are measured, called switching fields, depend strongly on the direction of the applied field. From a certain angle, our magnetic field is not strong enough to observe switching, and the hysteresis disappears (Figure 3(b)). In this case, the differential conductance does not depend on the field history; dI/dV is minimum for negative magnetic field and maximum for positive magnetic field, demonstrating that the switching near zero-field occurs independently from the one taking place at larger field. Thereby, we have to consider several independent magnetic objects to explain the data.

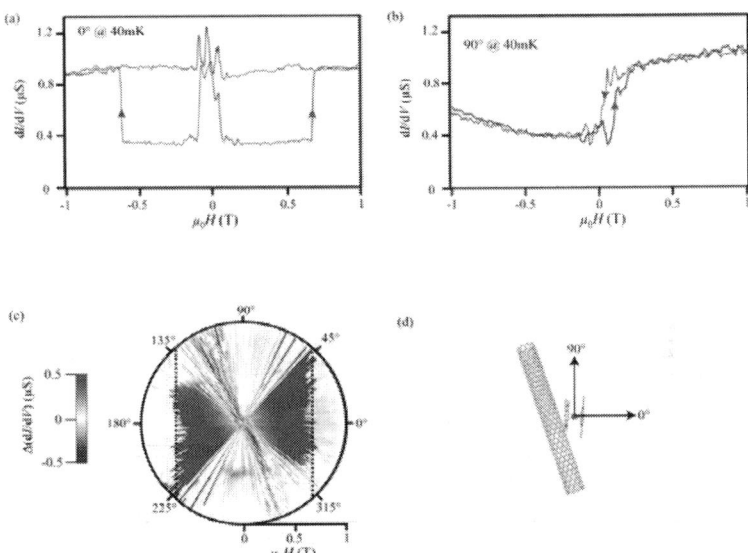

Figure 3. Conductance hysteresis loops of the supramolecular spin valve. (a–b) Differential conductance dI/dV measured at $T = 40$ mK as a function of in-plane magnetic field μ_0H applied respectively along (a) the easy axis direction (0°); and (b) the hard direction (90°) of magnetization. The red and blue arrows indicate the magnetic field sweep directions; (c) Color-scale plot of the dI/dV hysteresis (obtained from the difference between both field sweep directions) as a function of the applied magnetic field angle. The white color code is associated to zero hysteresis (reproducible dI/dV curves); (d) Relative disposition of the molecule with respect to the nanotube. The magnetic hard axis is 30° tilted from the nanotube axis.

In order to go one step further, we have plotted on Figure 3(c) the hysteresis amplitude (difference between trace and retrace) as a function of the applied field direction. It turns out that the switching field occurring at high value (the dashed line on Figure 3(c)) describes a straight line in the field space. The projection of those points along one axis stays constant, which is the fingerprint of the Ising like uniaxial anisotropy of the TbPc$_2$-SMM family.

Figure 4(a) shows the dependence of the hysteresis on the bias voltage applied to the quantum dot. This bias dependence shows that the hysteresis persists until 1 mV, but above this value, any residual hysteresis is smaller than the noise level. Furthermore, this measurement provides additional information: the switching field evolves with the bias voltage. We conclude from this observation that the conduction electrons can excite the molecules for larger bias voltages. Indeed, the energy of the conduction electrons increases with the bias voltage. A part of this energy might be transferred to the molecule via the electron density on the Pc-ligands and therefore to the anisotropic Tb spin system.

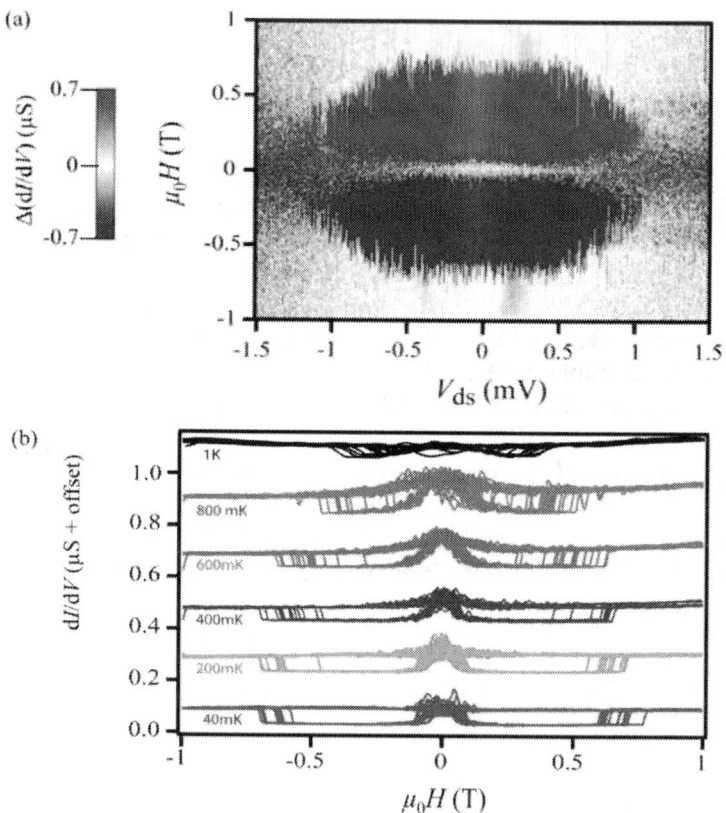

Figure 4. Bias and temperature dependences of the conductance hysteresis loops: (a) Color scale map of the dI/dV hysteresis as a function of in-plane magnetic field and source drain voltage V_{ds}. The magnetic hysteresis are suppressed above $V_{ds} = \pm 1$ mV; (b) 15 records of conductance hysteresis loops for several temperatures ranging from 0.04 to 1 K at a constant sweep rate of 50 mT/s. The curves for $T > 40$ mK are offset by a multiple of 200 nS for clarity.

As the temperature increases, the amplitude of the magnetoresistance decreases (Figure 4(b)). Indeed the spin-phonon interaction increases with the temperature and affect the spin-coherence. The magnetoresistance feature is lost around 1 K, which is below the TbPc$_2$-SMMs blocking temperature but is in agreement with an exchange interaction of a few hundreds of μV. The blocking temperature can be roughly estimated to be around 1.5 K, in agreement with recent XMCD measurements of TbPc$_2$-SMMs monolayers [18].

4. DISCUSSION

The spin-valve features can be explained by a simple model involving two distinct molecules: one with a close to one QTM probability and another one only subjected to DT. Under a high magnetic field, the magnetic moments of both molecules are polarized in parallel. This situation corresponds to the high conductance regime. When the magnetic field is reduced, one of the molecule experiences QTM close to zero field, whereas the other stays in the same state. The device is then in an antiparallel configuration and the conductance is minimum. Finally, when the following switching field is reached, meaning when the second molecule undergoes a DT, the parallel arrangement, and therefore the original conductance value, are recovered. This is, of course, the idealized case. In fact, by looking closely at the QTM region, we can observe several abrupt changes of conductance. It means that several molecules (more than two) are involved in the close to zero field process. However, only one molecule relaxes at larger fields via a DT.

From a microscopic point of view, each of these molecules interacts with the nanotube by creating a localized spin-polarized state in its vicinity (schematized in Figure 5(a)). A dipolar interaction is not sufficient to explain such an effect. Indeed, the dipolar interaction between the $S = 1/2$ radical on the Pcs and the conduction electrons cannot exceed a few tens of mT ($1/2$ μ_B creates a $1/2$ T dipolar field at 1 Å, and ~20 mT at 3 Å). Also, even if we consider the effect of the moment $J = 6$ on the conduction electron, the dipolar field is of the same order of magnitude, which is too small to explain the strength of the effect. Considering an exchange interaction, mediated by the π-electron density in the organic Pc ligand, seems to be more realistic. Indeed, Hu et al. [23] have shown that spin-polarization may occur through the interaction between a π-system delocalized over a carbon chain and latterly coupled spin radical. As shown by Gambardella et al. [24] in a similar system, the electron density on the Pc ligand is able to mediate the magnetic information by a strong exchange interaction. The molecules induce localized states on the nanotube by lifting the spin degeneracy through this interaction (Figure 5(b)). The strength of this interaction can be estimated to be around 200 μV by looking at the gapped degeneracy point in Figure 2(d). The spin level splitting is inhomogeneous along the tube when both molecules are in the antiparallel configuration, as a result of the mismatch between energy levels for a same spin (Figure 5(c)). This barrier can be overcome by applying a bias voltage higher than the exchange interaction between the molecule and the conduction electrons.

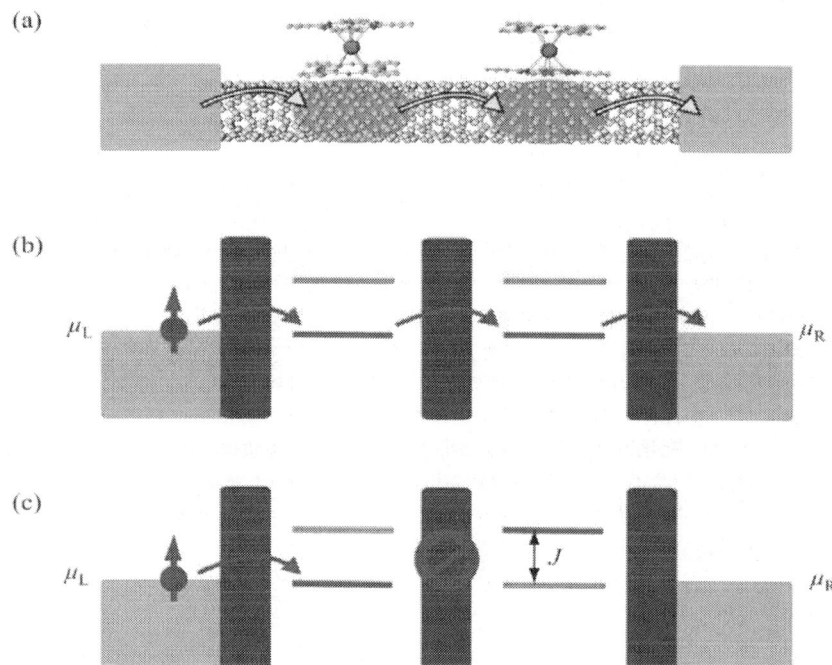

(a)

(b)

(c)

Figure 5. (a) Scheme of the localized dots induced by hybridization between the molecules and the nanotube; (b) Both molecules are polarized in the same manner. It induces a Zeeman splitting identical for both sets of localized states; (c) In the antiparallel configuration the Zeeman splitting is inhomogeneous, preventing spin transport through the device, unless a bias higher than the exchange interaction is applied.

5. CONCLUSIONS

We have reported on the characterization of a fully molecular spin-valve made of a carbon nanotube laterally coupled to a few single-molecule magnets determining for the first time the strength of the exchange energy between the SMMs. The interaction between both objects is strong enough to allow an abrupt modification of the conductance by changing the magnetization direction of only one molecule. As a result, the device exhibits a spin-valve effect and the magneto-conductance ratio can reach a few hundreds of percents. These features are obviously related to the grafted TbPc2-SMMs. Indeed, the anisotropic response clearly corresponds to an Ising-like uniaxial magnetic system. Our results open a pathway toward the fabrication of an all-organic spintronic device by the use of supramolecular self-organization techniques. Moreover, the high sensitivity of the device allows the characterization and the control of a single localized spin. Thus, an entanglement between different spin systems could also be possible using the CNT as tunable bus.

ACKNOWLEDGMENTS

This work is partially supported by the DFG programmes SPP 1459 and TRR 88 "3Met", ANR-PNANO project MolNanoSpin No ANR-08-NANO-002, ERC Advanced Grant MolNanoSpin No 226558, STEP MolSpinQIP and the Nanosciences Foundation of Grenoble. Samples were fabricated in the NANOFAB facility of the Néel Institute and we thank B. Fernandez, G. Julie, T. Crozes and T. Fournier for help in device fabrication. We thank M. Affronte, F. Balestro, N. Bendiab, L. Bogani, E. Bonet, V. Bouchiat, L. Calvet, A. Candini, D. Feinberg, J. Jarvinen, L. Marty, R. Maurand, T. Novotny, R. Piquerel, C. Thirion and R. Vincent for fruitful discussions and software development. We thank E. Eyraud, R. Haettel, C. Hoarau, D. Lepoittevin and V. Reita for technical support.

REFERENCES

1. Christou, G; Gatteschi, D; Hendrickson, DN; Sessoli, R. Single-molecule magnets. *MRS Bull* **2000**, *25*, 66–71.

2. Wernsdorfer, W; Sessoli, R. Quantum phase interference and parity effects in magnetic molecular clusters. *Science* **1999**, *284*, 133–135.

3. Xiong, ZH; Wu, D; Vardeny, ZV; Shi, J. Giant magnetoresistance in organic spin-valves. *Nature* **2004**, *427*, 821–824.

4. Barraud, C; Seneor, P; Mattana, R; Fusil, S; Bouzehouane, K; Deranlot, C; Graziosi, P; Hueso, L; Bergenti, I; Dediu, V; Petroff, F; Fert, A. Unravelling the role of the interface for spin injection into organic semiconductors. *Nat. Phys* **2010**, *6*, 615–620.

5. Bogani, L; Wernsdorfer, W. Molecular spintronics using single-molecule magnets. *Nat. Mater* **2008**, *7*, 179–184.

6. Urdampilleta, M; Klyatskaya, S; Cleuziou, J-P; Ruben, M; Wernsdorfer, W. Supramolecular spin valves. *Nat. Mater* **2011**, *10*, 502–506.

7. Sanvito, S. Filtering spins with molecules. *Nat. Mater* **2011**, *10*, 484–485.

8. Hueso, LE; Pruneda, JM; Ferrari, V; Burnell, G; Valdés-Herrera, JP; Simons, BD; Littlewood, PB; Artacho, E; Fert, A; Mathur, ND. Transformation of spin information into large electrical signals using carbon nanotubes. *Nature* **2007**, *445*, 410–413.

9. Aurich, H; Baumgartner, A; Freitag, F; Eichler, A; Trbovic, J; Schönenberger, C. Permalloy-based carbon nanotube spin-valve. *Appl. Phys. Lett* **2010**, *97*, 153116.

10. Kim, WY; Kim, KS. Prediction of very large values of magnetoresistance in a graphene nanoribbon device.*Nat. Nanotechnol* **2008**, *3*, 408–412.

11. Candini, A; Klyatskaya, S; Ruben, M; Wernsdorfer, W; Affronte, M. Graphene spintronic devices with molecular nanomagnets. *Nano Lett* **2011**, *11*, 2634–2639.

12. Kong, J; Soh, HT; Cassell, AM; Quate, CF; Dai, H. Synthesis of individual single-walled carbon nanotubes on patterned silicon wafers. *Nature* **1998**, *395*, 1–4.

13. Klyatskaya, S; Galan-Mascarós, JRG; Bogani, L; Hennrich, F; Kappes, M; Wernsdorfer, W; Ruben, M. Anchoring of rare-earth-based single-molecule magnets on single-walled carbon nanotubes. *J. Am. Chem. Soc* **2009**, *131*, 15143–15151.

14. Ishikawa, N; Sugita, M; Wernsdorfer, W. Quantum tunnelling of magnetization in lanthanide single-molecule magnets: bis(phthalocyaninato)terbium and bis(phthalocyaninato)dysprosium anions. *Angew. Chem. Int. Ed* **2005**, *44*, 2931–2935.

15. Ishikawa, N; Sugita, M; Okubo, T; Tanaka, N; Iino, T; Kaizu, Y. Determination of ligand-field parameters and f-electronic structures of double-decker bis(phthalocyaninato)lanthanide complexes. *Inorg. Chem* **2003**,*42*, 2440–2446.

16. Ishikawa, N; Sugita, M; Tanaka, N; Ishikawa, T; Koshihara, S-ya; Kaizu, Y. Upward temperature shift of the intrinsic phase lag of the magnetization of Bis(phthalocyaninato)terbium by ligand oxidation creating an S = 1/2 spin. *Inorg. Chem* **2004**, *43*, 5498–5500.

17. Mannini, M; Sainctavit, P; Sessoli, R; Moulin, CCD; Pineider, F; Arrio, MA; Cornia, A; Gatteschi, D. XAS and XMCD investigation of Mn-12 monolayers on gold. *Chem. Eur. J* **2008**, *14*, 7530–7535.

18. Sorace, L; Benelli, C; Gatteschi, D. Lanthanides in molecular magnetism: Old tools in a new field. *Chem. Soc. Rev* **2011**, *40*, 3092–3104.

19. Margheriti, L; Chiappe, D; Mannini, M; Car, P-E; Sainctavit, P; Arrio, M-A; Buatier de Mongeot, F; Cezar, J; Piras, F; Magnani, A; *et al*. X-Ray detected magnetic hysteresis of thermally evaporated terbium double-decker oriented films. *Adv. Mater* **2010**, *22*, 5488–5493.

20. Lopes, M; Candini, A; Urdampilleta, M; Reserbat-Plantey, A; Bellini, V; Klyatskaya, S; Marty, L; Ruben, M; Affronte, M; Wernsdorfer, W; *et al*. Surface-enhanced Raman signal for terbium single-molecule magnets grafted on graphene. *ACS Nano* **2010**, *4*, 7531–7537.

21. Wang, Y; Hu, N; Zhou, Z; Xu, D; Wang, Z; Yang, Z; Wei, H; Kong, ES-W; Zhang, Y. Single-walled carbon nanotube/cobalt phthalocyanine derivative hybrid material: preparation, characterization and its gas sensing properties. *J. Mater. Chem* **2011**, *21*, 3779–3787.

22. Bockrath, M; Liang, W; Bozovic, D; Hafner, JH; Lieber, CM; Tinkham, M; Park, H. Resonant electron scattering by defects in single-walled carbon. *Science* **2001**, *291*, 283–285.

23. Hu, G; Guo, Y; Wei, J; Xie, S. Spin filtering through a metal/organic-ferromagnet/metal structure. *Phys. Rev. B* **2007**, *75*, 1–6.

24. Lodi Rizzini, A; Krull, C; Balashov, T; Kavich, J; Mugarza, A; Miedema, P; Thakur, PK; Sessi, V; Ruben, M; Stepanow, S; Gambardella, P. Coupling single-molecule magnets to ferromagnetic substrates. *Phys Rev Lett***2011**. in press.

Index